THE SCIENCE AND TECHNOLOGY OF *WHISKIES*

THE SCIENCE AND TECHNOLOGY OF *WHISKIES*

EDITED BY

J R Piggott, R Sharp and R E B Duncan

Longman Scientific & Technical

Copublished in the United States with
John Wiley & Sons, Inc., New York

Longman Scientific & Technical,
Longman Group UK Ltd,
Longman House, Burnt Mill, Harlow,
Essex CM20 2JE, England
and Associated Companies throughout the world.

Copublished in the United States with
John Wiley & Sons, Inc., 605 Third Avenue, New York, NY 10158

First published 1989

British Library Cataloguing in Publication Data
The science and technology of whiskies
1. Whiskies. Distilling
I. Piggott, J. R. II. Sharp, R.
III. Duncan, R. E. B.
663'.52

ISBN 0-582-04428-6

Library of Congress Cataloging-in-Publication Data
The Science and technology of whiskies/edited by J. R. Piggott, R. Sharp, and R. E. B. Duncan.
p. cm.
Includes index.
ISBN 0-470-21416-3
1. Whiskey. I. Piggott, J. R. (John Raymond), 1950- . II. Sharp, R. (Russell), 1942- . III. Duncan, R. E. B. (Robert Erskine Burt), 1930- .
TP605.S33 1989
663'.52–dc19 88-36784
CIP

Printed and Bound in Great Britain at the Bath Press, Avon

Printed in 10/12 Palatino

Contents

Preface

We started planning this book at the realisation that the centenaries of the first publications of Nettleton's and Barnard's classic books about Scotch whisky and the whisky industry were approaching, and that no modern equivalents were available. We set out to provide an overview of the whisky industry world-wide; inevitably there are differences as well as similarities between the processes commonly used in the whisky producing countries of the world, originated by custom and often confirmed by legislation. We therefore invited authorities from around the world to write chapters, describing the whisky process in principal and industrial practices where they were sufficiently widespread to be of general rather than purely national interest. If a reader cannot find the process details for a particular nation's industry, we apologise and hope that the generalities we have presented will be of value.

The chapters cover the science and technology of whiskies from materials to the finished bottle ready for delivery to the consumer. The first three chapters deal with the cereals in common use, the process of malting so essential for Scotch and most other whiskies, and the operations required for transforming the malted or unmalted cereal grains into a fermentable extract. The fourth chapter deals with fermentation, and the next two with distillation. Chapter 7 reviews the techniques of process control in the distillery, and is followed by two chapters on maturation, a stage essential for the development of a palatable beverage. Chapter 10 covers the practices of blending and bottling, and it is here, in the blending stage, that the differences between whisky processes around the world become most apparent. The final two chapters deal with issues which are not central to the process, but are of economic importance to the industry; they are energy management and by-product treatment and disposal.

We are pleased to acknowledge the assistance of the contributors, who have done their best to turn our ideas into words; needless to say the successes of the book are theirs, the failures ours; of the publishers, for their help and patience; and of Alan Provan for editorial assistance.

List of Contributors

J Alsaker
Stord Bart za–s, POB 777, 5001 Bergen, Norway (Chapter 12)

G N Bathgate
Distillers Company (Cereals) Ltd, 7 Trinity Road, Elgin, Morayshire IV30 IUF, Scotland (Chapter 2)

Michael Booth
Hiram Walker & Sons Ltd, Walkerville – Box 2518, Windsor, Ontario, N8Y 4S5, Canada (Chapter 10)

Armond R Boucher
Joseph E. Seagram and Sons, Inc., 3 South Corporate Park Drive, White Plains, NY 10604, USA (Chapter 6)

A J Bronsky
The House of Seagram, 800 Third Avenue, New York, New York 10022, USA (Chapter 1)

J H Brown
Scottish Grain Distillers Ltd, Carsebridge Technical Centre, Alloa, Clackmannanshire, Scotland (Chapter 7)

R Cook
Distillers Company (Cereals) Ltd, 7 Trinity Road, Elgin, Morayshire IV30 IUF, Scotland (Chapter 2)

P J Hardy
Scottish Malt Distillers Ltd, 1 Trinity Road, Elgin, Morayshire IV30 IUF, Scotland (Chapter 7)

Kai Harju
Research Laboratories of the Finnish State Alcohol Company, Alko Ltd, POB 350, SF-00101 Helsinki, Finland (Chapter 4)

Matti Korhola
Research Laboratories of the Finnish State Alcohol Company, Alko Ltd, POB 350, SF-00101 Helsinki, Finland (Chapter 4)

Matti Lehtonen
Rajamaki Factories, The Finnish State Alcohol Company, Alko Ltd, SF-05200 Rajamaki, Finland (Chapter 4)

R Matsuyama
Research centre of Suntory Ltd, Wakayama-dai-1, Shimamoto-cho, Mishima-Gun, Osaka, Japan (Chapter 8)

I Morhalo
Hiram Walker & Sons Ltd, Walkerville – Box 2518, Windsor, Ontario, N8Y 4S5, Canada (Chapter 10)

D Nicol
Long John International Ltd, 1780 London Road, Glasgow, Scotland (Chapter 5)

K Nishimura
Research centre of Suntory Ltd, Wakayama-dai-1, Shimamoto-cho, Mishima-Gun, Osaka, Japan (Chapter 8)

Richard J Panek
Joseph E. Seagram and Sons, Inc., 3 South Corporate Park Drive, White Plains, NY 10604, USA (Chapter 6)

J M Philp
United Distillers PLC, Glenochil Research Station, Menstrie, Clackmannanshire FK11 7ES, Scotland (Chapter 9)

R W Schumann
The House of Seagram, 800 Third Avenue, New York, New York 10022, USA (Chapter 1)

Bill Shaw
Hiram Walker & Sons Ltd, Walkerville – Box 2518, Windsor, Ontario, N8Y 4S5, Canada (Chapter 10)

J G Watson
Chivas Brothers Ltd, Strathisla-Glenlivet Distillery, Keith, Banffshire AB5 3BS, Scotland (Chapter 11)

G D Wilkin
William Grant & Sons Ltd, The Girvan Distillery, Girvan, Ayrshire KA26 9PT, Scotland (Chapter 3)

Note on spelling: 'Whisky' is used in this book when the reference is to Scotch or Canadian, while 'whiskey' denotes American or Irish. 'Whisky' has been used where no specific context is applicable.

CHAPTER 1

Cereals for whisky production

A J BRONSKY and R W SCHUMANN

Many fermentable materials which can be used to produce alcoholic beverages are to be found around the world, such as corn (maize), rye, barley, sorghum, wheat, sugar, molasses, honey, fruit, oats and potatoes to name but a few. This chapter will consider only the cereal grains which are currently used in the United States, the United Kingdom and elsewhere to produce distilled alcoholic beverages, that is, corn (maize), rye, wheat and barley. These grains, all from the family of grasses (Gramineae), have the prerequisite characteristic for alcohol production, that is, a high concentration of starch. The composition of these four grains is shown in Table 1.1.

Grains for the distilling industry are obtained either directly from a supplier (which could be a farmer or a grain company) or, in the case of some large distilleries, through their own grain-handling facilities. In either case it is the responsibility of the supplier and receiver to monitor carefully incoming materials, so that defects in the grain do not enter the grain supply system to contaminate the existing grain and affect the outcome of the distillate. Further discussion of quality control procedures will be found later in this chapter.

Corn (maize)

As with most living plants, corn has developed in a rather long, slow evolutionary process. No one really knows where corn came from. Investigations have traced it to Mexico, where a grass called *Zea canina* is found growing wild. Mexico is also the native habitat of Teosinte (*Euchloena mexicana*) and gamma grass (*Tripsacum dactyloides*), two plants closely related to corn (*The*

Table 1.1 Composition of corn, rye, barley and wheat.

*Composition of the four grains**				
% of total:	*Corn*	*Rye*	*Barley*	*Wheat*
Endosperm	82	87	84	85
Germ	12	3	3	3
Bran	6	10	13	12
Chemical composition (dry basis):†				
Nitrogen-free extract	69.2	70.9	66.6	69.9
Starch	72	68	63–65	69
Sugars	2.6	0	2–3	0
Protein	8	12.6	12	13.2
Soluble N % of total	4.7	0	11	0
Crude fibre	2	2.4	5.4	2.6
Fat	3.9	1.7	1.9	1.9
Ash	1.2	1.1	2	1.9

**Beverage Alcohol Production* (1982).
†*Beverage Alcohol Production* (1982); Morrison (1959).

Distiller's Grain Manual, 1942). It is believed that corn evolved from one of these grasses, and that present-day corn is a result of 60 000 years of natural change, cultivation and hybridisation (Besant *et al.*, 1977). Corn was in cultivation when Columbus reached America in 1492 and from that time it was introduced into Europe, Africa, China and Asia Minor.

BOTANY

Corn is a grass and belongs to the same family of plants as timothy, blue grass and wheat. The root system is fibrous and spreads several feet in all directions (*The Distiller's Grain Manual*, 1942; Aldritch and Lang, 1965). The development of the root system depends on the soil type and on the weather. As the plant grows, it extends brace roots to help to keep the stalk upright in the wind.

Female and male flowers are located in different parts of the plant. The silk is the female part and the tassel is the male part. Wind and gravity bring about pollination as the pollen from the tassel is carried to the stigma of the silk. Plants vary in size from 6 to 12 ft (*The Distiller's Grain Manual*, 1942), bearing one or more ears, which are normally 8 to 12 in. long (Fig. 1.1).

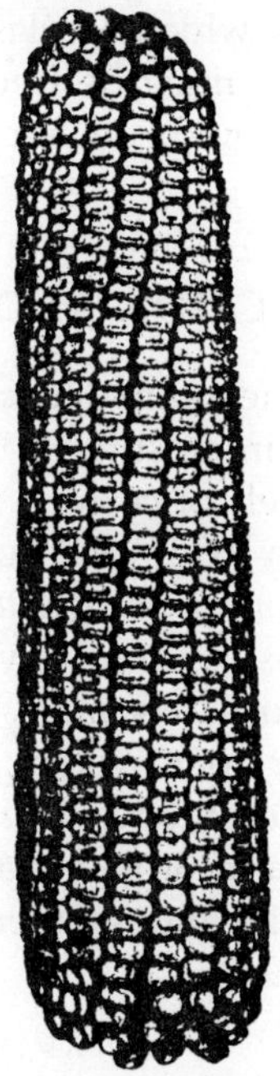

Fig. 1.1 Hybrid yellow dent corn. This ear is representative of crop suitable for distillery purposes.

Corn is classified as:	Family	Gramineae
	Tribe	Maydeae
	Genus	*Zea*
	Species	*mays*

The most common classes are (*The Distiller's Grain Manual*, 1942; *Beverage Alcohol Production*, 1982):

Zea mays indentato	Dent corn
Zea mays indurata	Flint corn
Zea mays amylacea	Soft corn
Zea mays everta	Popcorn

Dent corn is the main type of corn grown in the United States. Dent and flint corn are the only two classes of corn that have recognised US federal grading classifications. Distillers in the US use dent corn primarily, which is classified Grade 2, dried to 14% or less moisture, and must be free from all foreign odours, such as those of must, mould, oil, etc. (Willkie and Prochaska, 1943).

PRODUCTION AND YIELD

On a worldwide basis, the estimated corn production in 1984 is shown in Table 1.2. It is interesting to note that in 1941, the US produced 2.5 billion bushels of corn on 92 727 000 acres compared to 8.3 billion bushels on 69 189 000 acres in 1986 (*Agricultural Statistics 1987, US Department of Agriculture*). This gain in yield was due to more intense use of hybrid seed, fertiliser and mechanisation. Hybrid corn has many advantages: more vigorous growth and stronger root system; greater resistance to ear and stalk diseases, drought and insects; fewer barren stalks, nubbins (imperfect ears) and broken stalks; more uniform maturity, better quality of grain and greater ease in harvesting (*The Distiller's Grain Manual*, 1942).

Table 1.2 World corn production.

	Million 56 lb bushels
1 United States*	7 657
2 EEC	791
3 USSR	476
4 Argentina	433
5 South Africa	256
6 Thailand	177
Rest of world	7 740
World total	17 530

*Of the total of the United States, Iowa produced 19%, followed by Illinois 16%, Nebraska 10%, Indiana 9%, and Minnesota 9%.

Since corn is normally the cheapest cereal grain available in the USA due to the size of crop, availability and high starch content, it is the predominant grain of choice for alcoholic beverages. In addition, it is also mandated by federal regulations that corn be used in bourbon whiskey and in corn whiskey in the United States.

Rye

The earliest cultivation of rye appears to have been in western Asia and southern Russia (Bushuk, 1976). Rye apparently was not cultivated in the fields of the Greeks and Romans and is not mentioned in ancient oriental literature. Rye does well on acid soils and can withstand most weather conditions except heat (*Beverage Alcohol Production*, 1982).

BOTANY

Rye and wheat are closely related in their manner of growth (Fig. 1.2), but unlike wheat, rye cross-fertilises freely. Thus pure strains of rye are hard to develop and maintain.

Rye is classified as:	Family	Gramineae
	Tribe	Hordeae
	Genus	*Secale*
	Species	*montanum*

Fig. 1.2 Rye plant showing flower spike (head) and single floret (grain).

Rye production in the United States is of minor importance, whereas worldwide it is an important crop. Recent estimated production for some of the major producing countries is shown in Table 1.3.

Table 1.3 World rye production.

	Million 56 lb bushels
1 USSR	571
2 Eastern Europe	512
3 EEC	120
4 Other western european countries	45
5 United States*	31
6 Canada	28
7 Argentina	11
Rest of world	12
World total	1330

*34% of the total is grown in South Dakota.

Rye contains a lower percentage of starch than does corn or wheat (Table 1.1), so it is less preferred as far as cost is concerned to corn (*Beverage Alcohol Production*, 1982). The main role of rye in beverage production is its flavour contribution; it contributes a distinctive spicy flavour to whisky. Normally distillers buy No. 2 Plump rye or better. When rye is the predominant grain in the mash, it commonly causes some foaming problems. Rye malt is sometimes used alone or in combination with other grains to impart a special flavour and aroma to whisky. The amount of α-amylase in rye malt is lower than in distiller's barley malt.

Barley

The use of barley by man predates recorded history. Egyptians ate barley bread long before 3000 BC (*The Distiller's Grain Manual*, 1942), and barley was cultivated by Swiss lake-dwellers in the Stone Age (*Beverage Alcohol Production*, 1982). Modern-day barley comes in two types, the 2-row and 6-row varieties. The former was brought to the USA from England, whilst the latter was brought by the Dutch (*Beverage Alcohol Production*, 1982).

BOTANY

The plant is shorter than wheat or rye, but otherwise it is similar (*The Distiller's Grain Manual*, 1942; Cook, 1962). The number of stems which are produced varies with the thickness of the stand, and if the plant has plenty of space it may be as high as 15–30 in.

Fig. 1.3 Barley plant showing flower spikes (heads) of beardless and bearded barley. Enlarged is a spikelet of three florets that grows from a node on one side of the stem which, with its companion on the other side of the stem, forms a 6-row spike. An individual floret (grain) is also shown.

Barley is classified as:	Family	Gramineae
	Tribe	Hordeae
	Genus	*Hordeum*
	Species	*polystichum*

vulgare – common or nodding 6-row barley
hexastichum – club or erect 6-row barley
spontaneum – distichum 2-row barley
intermedium – hybrid barley

Recent varietal distribution of barley grown in the USA is shown in Table 1.4, and world production in Table 1.5. The top producing areas in the USA are: North Dakota 26%, Idaho 15%, Washington 11%, Minnesota 10%, and Montana 10%.

Table 1.4 Varietal acreage distribution of barley in the United States.

Three midwestern states		*Seven western states*	
	%		%
Morex	35.0	Advance	2.0
Glenn	18.0	Klages	8.1
Azure (Blue)	12.8	Piroline	10.7
Robust	12.4	Clark	7.0
Bumper	3.6	Moravian	2.1
Others	18.2	Morex	2.1

Table 1.5 World barley production.

	Million 48 lb bushels
EEC	2030
USSR	1860
USA	597
Canada	473
China	322
Rest of world	2650
World total	7932

The starch content of barley is about the same as rye and less than corn (Table 1.1). The barley bran adds a distinctive flavour to whisky, but the predominant use for this cereal is malting and the resultant important enzyme content of malted barley. Barley malt for the production of whisky is discussed fully in Chapter 2.

Wheat

The origin of wheat is unknown, except that it was cultivated long before recorded history (*The Distiller's Grain Manual*, 1942). Wheat is the major cereal grain, since a great deal goes directly into the human food chain.

BOTANY

Wheat is usually self-fertilised, but may be cross-fertilised. A number of natural hybrids are known and others have been produced (Fig. 1.4).

Wheat is clasified as:

Family .... Gramineae
Tribe .... Hordeae
Genus .... *Triticum*
Species .... *vulgare*

Fig. 1.4 Wheat plant showing flower spikes (heads) of beardless and bearded wheat. Enlargements show a spikelet of florets and an individual floret (grain).

Types of wheat include soft red winter wheat (Stefferud, 1951), hard red winter wheat, hard spring red wheat, durum wheat, and white wheat. Only the soft red wheat and the white wheat have been found to give satisfactory alcohol yields (*Beverage Alcohol Production*, 1982).

Worldwide production is estimated at over 18 billion bushels and the distribution is shown in Table 1.6. The leading producing areas in the USA are as follows: Kansas 17%, North Dakota 11%, Oklahoma 7%, Washington 6%, Texas 6%, South Dakota 5%, Minnesota 5%, Colorado 4%, and Montana 4%.

Table 1.6 World wheat production.

	Million 60 lb bushels
China	3 123
EEC	2 767
USSR	2 756
USA	2 595
India	1 657
Eastern Europe	1 455
Canada	779
Australia	639
Argentina	441
Rest of world	2 446
World total	18 658

Even though wheat contains about the same amount of starch as other grains, the alcohol yield is lower. Like rye, foaming problems are encountered in processing wheat (*Beverage Alcohol Production*, 1982).

US grade designations for cereals

In general, the grade designations for cereals (*Official United States Standards for Grain*, 1978) include, in the following order:

1. the letters 'US';
2. the number of the grade or the words 'sample grade', as the case may be;

3. with the exception of rye, the name of the applicable class or sub-class;
4. the name of each applicable special grade;
5. when applicable, the word 'dockage', together with the percentage thereof.

There are exceptions and additions to this general information. If requested by the applicant, the grade designation for the class barley must include, following the word 'barley', the approximate percentage of each class and of black barley in the mixture in order of predominance. For malting barley, the words 'Plump barley', together with the applicable percentage range, is included.

The grade requirements for mixed wheat are slightly different. Mixed wheat is graded according to the US numerical and US sample grade requirements of the class which predominates in the mixture, except that the factor 'wheat of other classes' is disregarded.

Under 'Remarks' on the inspection certificate there is included:

1. in the case of western white wheat, the name and percentage of the white club wheat and other white wheat in the mixture;
2. in the case of unclassed wheat, the colour or other characteristics which describe the wheat, together with the percentage thereof;
3. in the case of mixed wheat, the name and percentage of the classes that comprise the mixture.

The US grades and grade requirements are summarised in Tables 1.7 to 1.10.

OPTIONAL GRADE DESIGNATIONS

The named cereals may be certified (under certain conditions) when supported by official analysis, as 'US No. 2 or better corn', 'US No. 3 or better barley', etc. The optional grade designations include the name of the applicable class or sub-class (with the exception of rye) immediately preceding the cereal in the grade designation. The special grade designations and dockage, when applicable, are also included (under certain conditions) in the certification.

Table 1.7 Grades and grade requirements for corn*.

Grade	*Maximum limits of*				
	Minimum test weight per bushel (lb)	*Moisture (%)*	*Broken corn and foreign material (%)*	*Damaged kernels (%)*	
				Total	*Heat damaged kernels*
US No. 1	56.0	14.0	2.0	3.0	0.1
US No. 2	54.0	15.5	3.0	5.0	0.2
US No. 3	52.0	17.5	4.0	7.0	0.5
US No. 4	49.0	20.0	5.0	10.0	1.0
US No. 5	46.0	23.0	7.0	15.0	3.0
US Sample grade	US Sample grade shall be corn which does not meet the requirements for any of the grades from US No. 1 to US No. 5, inclusive; or which contains stones; or which is musty, or sour, or heating; or which has any commercially objectionable foreign odour; or which is otherwise of distinctly low quality.				

**Official United States Standards for Grain* (1978).

SPECIAL GRADE DESIGNATIONS

A special grade, when applicable, is supplemental to the grade assigned under the US grade designations for cereals. There are many types of special grades for each cereal.

Corn

Special grades of corn include: flint corn, which is corn of any class that consists of 95% or more of flint corn; and flint and dent corn, which can be corn of any class that consists of more than 5% but less than 95% of flint corn. Corn infested with live weevils or other insects injurious to stored grain is graded as weevily corn while corn of any class consisting of 95% or more waxy corn (as determined by a test approved by the Administrator) is graded as waxy corn.

Barley

Barley which, in whole or in part, has been treated with

Table 1.8 Grades and grade requirements for rye*.

Grade	*Maximum limits of*				
	Minimum test weight per bushel (lb)	*Damaged kernels (rye and other grains)*		*Foreign material*	
		Total (%)	*Heat damaged (%)*	*Total (%)*	*Other than wheat (%)*
US No. 1†	56.0	2.0	0.1	3.0	1.0
US No. 2†	54.0	4.0	0.2	6.0	2.0
US No. 3†	52.0	7.0	0.5	10.0	4.0
US No. 4	49.0	15.0	3.0	10.0	6.0
US Sample grade	US Sample grade includes rye which does not come within the requirements of any of the grades from US No. 1 to US No. 4, inclusive; or which contains more than 16% of moisture; or which contains inseparable stones and/or cinders; or which is musty, or sour, or heating, or hot; or which has any commercially objectionable foreign odour except of smut or garlic; or which contains a quantity of smut so great that any one or more of the grade requirements cannot be applied accurately; or which is otherwise of distinctly low quality.				

*_Official United States Standards for Grain_ (1978).
†The rye in grade US No. 1 may contain not more than 10.0%, in grade US No. 2 not more than 15.0%, and in grade US No. 3 not more than 25.0% of 'thin' rye, which 'thin' rye shall consist of rye and other matter that will pass readily through a sieve 0.032 in. thick with perforations 0.064 in. by 0.375 in.

sulphurous acid or any other bleaching agent, is graded as bleached barley. Bright barley is barley that is of exceptionally good natural colour, i.e. not achieved by bleaching, while any barley badly weathered or stained (except by bleaching) is graded as stained barley. Where more than 4% of the kernels are blight-damaged and/or mould-damaged then that barley is graded as blighted. If the kernels are covered with smut spores to give a smutty appearance in mass or the barley contains 0.2% or more of smut masses in total, then the barley is graded as smutty.

Other special grades of barley are: ergoty (barley containing in excess of 0.1% ergot); garlicky (barley with three or more green garlic bulblets, or equivalent quantity of dry or partly dry

Table 1.9 Grades and grade requirements for the sub-classes 6-row barley, 2-row barley and the class barley*.

Grade	*Minimum limits of*		*Maximum limits of*					
	Test weight per bushel (lb)	*Sound barley (%)*	*Damaged kernels (%)†*	*Heat-damaged kernels (major)(%)*	*Foreign material (%)*	*Broken kernels (%)*	*Thin barley (%)*	*Black barley (%)‡*
US No. 1	47.0	97.0	2.0	0.2	1.0	4.0	10.0	0.5
US No. 2	45.0	94.0	4.0	0.3	2.0	8.0	15.0	1.0
US No. 3	43.0	90.0	6.0	0.5	3.0	12.0	25.0	2.0
US No. 4§	40.0	85.0	8.0	1.0	4.0	18.0	35.0	5.0
US No. 5	36.0	75.0	10.0	3.0	5.0	28.0	75.0	10.0
US Sample grade	US Sample grade shall be barley which: (a) does not meet the requirements for the grades US Nos 1, 2, 3, 4, or 5; (b) contains a quantity of smut so great that one or more of the grade requirements cannot be determined accurately; (c) contains more than 7 stones or more than 2 crotalaria seeds (*Crotolaria* spp.) per 1000 g barley; (d) has a musty, sour, or commercially objectionable foreign odour (except smut or garlic odour); or (e) contains the seeds of wild brome grasses; or (f) is heating or otherwise of distinctly low quality.							

**Official United States Standards for Grain* (1978).
†Includes heat-damaged kernels (major). Frost-damaged kernels (minor) and mould-damaged kernels (minor) shall not be considered as damaged kernels.
‡These limits do not apply to the class barley.
§Barley that is badly stained or materially weathered shall be graded not higher than US No.4.

Table 1.10 Grades and grade requirements for wheat*.

Grade	*Minimum test weight per bushel (lb)*		*Maximum limits of (%)*					*Wheat of other classes§*	
	Hard red spring wheat or White club wheat	*All other classes and sub-classes*	*Heat-damaged kernels*	*Damaged kernels (total)†*	*Foreign material*	*Shrunken and broken kernels*	*Defects (total)‡*	*Contrasting classes*	*Wheat of other classes (total)***
US No. 1	58.0	60.0	0.2	2.0	0.5	3.0	3.0	1.0	3.0
US No. 2	57.0	58.0	0.2	4.0	1.0	5.0	5.0	2.0	5.0
US No. 3	55.0	56.0	0.5	7.0	2.0	8.0	8.0	3.0	10.0
US No. 4	53.0	54.0	1.0	10.0	3.0	12.0	12.0	10.0	10.0
US No. 5	50.0	51.0	3.0	15.0	5.0	20.0	20.0	10.0	10.0
US Sample grade	US Sample grade shall be wheat which: 1. Does not meet the requirements for the grades US Nos 1, 2, 3, 4, or 5; or 2. contains a quantity of smut so great that one or more of the grade requirements cannot be determined accurately; or 3. contains 8 or more stones, 2 or more pieces of glass, 3 or more crotalaria seeds (*Crotalaria* spp.), 3 or more castor beans (*Ricinus communis*), 4 or more particles of an unknown foreign substance(s) or a commonly recognised harmful or toxic substance(s), or 2 or more rodent pellets, bird droppings, or an equivalent quantity of other animal filth per 1000 g wheat; or 4. has a musty, sour, or commercially objectionable foreign odour (except smut or garlic odour); or 5. is heating or otherwise of distinctly low quality.								

**Official United States Standards for Grain* (1978).
†Includes heat-damaged kernels.
‡Defects (total) include damaged kernels (total), foreign material, and shrunken and broken kernels. The sum of these three factors may not exceed the limit for defects.
§Unclassed wheat of any grade may contain not more than 10% of wheat of other classes.
**Includes contrasting classes.

bulblets in 500 g barley); weevily (barley infested with live weevils or other insects injurious to stored grain); tough (barley containing more than 14.5% moisture).

Rye and wheat

Some of the special grades for rye and wheat are similar in definition. When between two and six green garlic bulblets or an equivalent quantity of dry or partly dry bulblets are present in 1 kg rye or wheat then that cereal is graded as light garlicky. If the quantity of bulblets is more than six in 1 kg rye or wheat then the grade used is garlicky.

Rye or wheat which has an unmistakable odour of smut, or that contains a quantity of smut equivalent to 14 smut balls but not in excess of 30 smut balls of average size in a 250 g sample of the cereal, is graded as light smutty. Where the smut exceeds a quantity equal to 30 smut balls in a 250 g sample of the cereal, the grading used is smutty.

Rye or wheat that contains more than 0.3% of ergot is graded as ergoty and rye or wheat, as with corn and barley, that is infested with live weevils or other insects injurious to stored grain is graded as weevily. As applied to wheat, the meaning of the term 'infested' is set forth in the *Grain Inspection Manual*.

Other special grades of rye and wheat include:

(a) Plump rye: rye that does not contain more than 5% of rye and other matter that will pass through a 20 gauge metal sieve with rectangular perforations 0.064 in. wide by $\frac{3}{4}$ in. long.
(b) Tough rye: rye containing more than 14% but not more than 16% moisture.
(c) Treated wheat: wheat which has been scoured, limed, washed, sulphured, or treated in such a manner that the true quality is not reflected by either the US numerical or the US grade designation alone.

Quality control procedures

As the distiller is interested in producing the highest quality product, the starting point is obviously raw material quality, i.e. grains and water. Grain contributes many flavour characteristics to the distilled spirits and only top quality grain should be used for fermentation. The quality control department has the responsibility to evaluate each grain shipment to ensure that quality standards are met.

A typical procedure involves a sample of approximately 120 g being taken from a composite sample of each shipment of grain. After initial information is obtained, i.e. type of grain, date received etc., the sample is allowed to reach at least room temperature. Visual inspection for irregularities such as broken kernels, foreign material, mould growth and insects is followed by the assessment of all odour characteristics. This odour evaluation is repeated after the grain has been warmed to approximately 32°C. Undesirable odour characteristics are recorded, giving 0 to 3 points according to the intensity of the odours. Only when an undesirable odour is detrimental enough to reject a shipment are 3 points given.

Under normal conditions using quality grains processed and distilled without any interference due to contamination etc., distillates produced primarily from corn, such as bourbon or corn whiskey, are described as estery, sweet, mealy, or grainy. Distillates in which rye predominates are described as spicy, estery, sweet or fruity. Characteristics of distillates produced from barley and barley malt are discussed in Chapter 2.

Defects which tend to survive the fermentation and distillation processes and affect the distillate are described as musty, mouldy, stale, binny, rancid, greasy, sour, putrid, spoiled, oily, and those odours associated with chemical contamination. Of those quality defects only binny, greasy and sour, provided they are not strong in intensity, can age out during the maturation period. These defects normally cause the distillates to be rejected by the product quality control department. When this occurs, significant additional costs are incurred due to the loss of original product plus the additional redistillation cost to produce a distillate of lesser value.

References

Aldritch S R, Lang E R 1965 Modern corn production. *The Farm Quarterly* **21**: 61–116

Besant L, Kellerman D, Monroe G 1977 *Grains Processing, Production, Marketing*. Board of Trade of the City of Chicago

Beverage Alcohol Production. 1982. Joseph E Seagram & Sons, Inc. House publication

Bushuk W 1976 *Rye: Production, Chemistry and Technology*. American Association of Cereal Chemists, St. Paul, Minnesota

Cook A H 1962 *Barley and Malt Biology, Biochemistry, Technology*. Academic Press, New York and London

The Distiller's Grain Manual. 1942. Joseph E Seagram & Sons, Inc. House publication

Morrison F B 1959 *Feeds & Feeding* 22nd edn. The Morrison Publishing Company, Clinton, Iowa

Official United States Standards for Grain. 1978. Superintendent of Documents US Government, Washington DC

Stefferud A (ed) *Crops in Peace and War: Yearbook of Agriculture 1950–1951*. US Department of Agriculture, Washington DC

US Department of Agriculture 1987 *Agricultural Statistics 1987*. US Government Printing Office, Washington DC

Willkie H F, Prochaska J A 1943 *Fundamentals of Distillery Practice*. Joseph E Seagram & Sons, Inc. House publication

CHAPTER 2

Malting of barley for Scotch whiskies

G N BATHGATE and R COOK

Introduction

The art and skill of making barley into malt is as old a technology as whisky distilling itself. There is no doubt that malting specifically for the Scotch whisky industry grew in stature in parallel with the craft of distilling, and even today some distilleries retain their own maltings on site, to produce a highly characteristic product for that particular still. Despite the age of the industry the historic documentation of malting in Scotland is sparse. One of the earliest papers devoted to malt was written by Sir Robert Moray in 1678. In this most interesting article, Moray states that: 'Malt is there made of no other Grain but Barley, whereof there are two kinds one which hath four rows of Grains in the Ear, the other two rows. The first is the more commonly used; but the other makes the best malt.' The 4-row barley to which Moray referred is the antecedent of the land race 'Bere', which is still grown in Orkney for use in the manufacture of 'Bere Bannocks'. What we can infer from this is that even 300 years ago the selection of appropriate varieties for malting was important. Moray goes on to discuss the importance of harvesting, barley storage, grading, steeping, germination and kilning, and while his comments lack the support of real scientific evidence, much of what he describes was based on sound empirical knowledge.

In writing about modern practices we can therefore follow this historical precedent, since the same factors mentioned above are as important today as they were in the seventeenth century. The following treatise is divided into four sections, dealing in turn with barley procurement, steeping, germination and kilning. Under each of these general headings, the subject-matter is then divided between discussion of practical malting, plant and machinery, and the science of malting in terms of physiology, biochemistry and related aspects of quality control. This will enable the reader to abstract more easily the information he requires.

Barley procurement

BARLEY HARVESTING AND DRYING

From the scant literature available we learn that malting and distilling in Scotland were historically local affairs with the distiller/maltster purchasing locally grown barley for processing. This was very much the case until the end of the Second World War and the advent of the combine harvester. Up to that point, barley was cut by scythe or reaper, 'stooked' to allow final drying out and maturing, stacked and finally threshed from the stack. While 'stooking' helped to reduce the final moisture at harvest, the grain was probably held in stack at moistures of 15–17% until threshing took place. It was then not uncommon for such grain to grow mouldy in stack and for the threshed grain to have very low germination energy.

The malting of local grain of such dubious quality limited production capacity, particularly in the prolonged germination times required on floor. When compared with high-quality malting barley grown at that time in England for the burgeoning brewing industry, Scottish barley was certainly inferior. In addition, cheap barley with good germination could be readily imported, particularly from Canada, where 6-row, high-diastase barley was plentiful. In the period from approximately the 1950s through to the middle of the next decade, the Scotch whisky industry steadily outgrew the supply of malt from local maltings. The additional malt requirement came either from new pneumatic maltings (Saladin Box type) using imported English or southern Scottish barley, or directly as malt shipped north from English maltsters. The English barleys used were initially Spratt and Plumage Archer followed by Proctor and more latterly the winter barley Maris Otter.

This trend continued until the late 1960s when a radical change in barley handling and malting technology (Griffin, 1982) once again promoted the malting of local barley. This involved using the malting plant as a barley drying facility at harvest to reduce the moisture to around 12%, preserving the germination capacity and ensuring storage safe from infestation.

The first combined barley drying and malting plant was based on a multiple Saladin Box design (Griffin and Pinner, 1965), in which each box acted both as a batch steeping/germination vessel and as a kiln of approximately 50 t capacity for malting or 100 t for barley drying. This initial plant was followed by larger concrete boxes (200 t malting, 400 t barley drying) and subsequently by circular, steel vessels of up to 500 t capacity (Turcan, 1981; Stevens, 1984). At present the largest of these combined drying and malting plants can receive and dry over 2000 t wet barley per day.

At the same time, farming technology was also developing very rapidly. With the introduction of even larger combine harvesters, freshly threshed grain was available at harvest rather than being stored in stack. This also increased the necessity to dry and store grain at harvest.

Maltsters have always felt more comfortable with barley dried in deep-bed batch driers at low temperatures (40–45°C) than with barley dried in continuous driers using the Louisiana State University cross-flow system at higher air temperatures (60–65°C). Although the germination characteristics from both kinds of drier are similar, there is some evidence (Bathgate, 1983) to suggest that barley dried in deep-bed batch driers recovers from dormancy faster than that from continuous driers.

Dormancy does not usually pose a problem to maltsters in more equable climates. However, in Scottish conditions winter barleys, and more recently the spring barley Triumph, are likely to show severe dormancy periods of up to 16 weeks. While dormancy can be reduced by appropriate storage conditions (see later), the constraints of having to dry and store barley on site at harvest are a severe financial limitation to Scottish maltsters. However, the benefits are substantial. By guaranteeing quality and variety and minimising drying and storing costs rather than paying a merchant or farmer to perform this critical task, Scottish maltsters have provided consistently high-quality distilling malts. While temperature and moisture may be a disadvantage, the mild climate and high sunshine hours/day generally produce a bold barley sample with high thousand corn weight and relatively low protein (about 10–10.5%). However, the overall growing period is relatively short, with spring barley, usually sown in early March, being harvested in late August to early September. With the onset of equinoctial gales in autumn, there is a requirement that the barley should be stiff-strawed, be able to hold on to its ear, and be able to reach full maturity during the short growing season. The short-strawed, erectoid variety Golden Promise has filled this role for nearly 20 years and despite being very susceptible to mildew attack, being outyielded by many newer varieties and being outstripped in quality by others, it has admirably demonstrated the agronomic characteristics required for a successful Scottish barley. Nevertheless, the very high achievable yields for other cereal and alternative cash crops has put considerable pressure on Golden Promise. In its heyday, Golden Promise represented over 95% of the Scottish crop but in 1985 it had dropped to 30–40%. No one variety has replaced Golden Promise but several high-yielding, high-quality contenders are vying for supremacy (Table 2.1).

Table 2.1 Current barley varieties grown in Scotland (Scottish Agricultural Colleges Advisory Service – 1986/87).

% Seed sales *Variety*	*1986*	*1987*	*Comments*
Golden Promise	18	13	Declining rapidly (outclassed)
Triumph	20	22	Late maturing (dormant)
Kym	6	3	Becoming outclassed (good on light soil)
Natasha	7	9	Less dormant; Triumph cross
Corgi	2	2	Semi-dwarf; Triumph cross
Doublet	—	8	Very dormant; Triumph cross
Atem	8	6	Medium quality
Camargue	—	1.5	High yield; Triumph type
Other varieties and feed types	39	35.5	Feed varieties declining

As far as barley procurement is concerned, the high cost of transport from England and continental Europe, the establishment of high-capacity, low-cost malting in close proximity to the major distilleries, and the continued availability of high-quality malting barley in the vicinity, should ensure the long-term viability of the industry in Scotland.

A distinction has to be made between barley purchased for malt whisky and that bought for grain whisky. In the latter case, the function of the malt is primarily as a source of enzymes for the conversion of cooked cereal starch to fermentable sugar rather than as the primary source of carbohydrate, which it is in pot-distilling malt.

The enzymic activity of malt is very strongly varietal-dependent and to a lesser extent correlates with the protein content in certain varieties. For example, North American 6-row varieties and certain Scandinavian varieties are unrivalled in respect to diastase compared with current UK cultivars. While the import of Canadian barley has largely disappeared, some 18 000 t malt for grain distilling was imported from Sweden and Finland in 1985.

STORAGE

Once the barley has been dried down to 12% moisture, the grain is transferred to store with only partial cooling to facilitate dormancy breaking. The temperature of the grain may be between 18–25°C at this stage, depending on the expected degree of dormancy. If the grain is severely dormant, e.g. the spring variety Triumph, it has been suggested (Palmer and Taylor, 1980) that the temperature should be increased to around 30°C for a short period. During this interim period of storage samples are taken regularly for determination of germination energy and once dormancy has fully broken the grain bulk is gradually cooled to an ambient temperature of about 5°C over winter. Care must be taken not to blow down too quickly, since crash cooling can sometimes push the barley back into dormancy. Volumes of around $0.15\ m^3\ min^{-1}\ t^{-1}$ are normally utilised for barley cooling, applied either by pressure fan through ducts under the barley or by suction from above the bulk. There are protagonists for both systems but suction has the drawback of being susceptible to air leaks in the adjacent structures of the storage vessel, so air could be sucked through the fans without passing through the grain bed. Blowing under pressure suffers the disadvantage that if the grain is sufficiently warm, the air passing through the grain can absorb additional moisture, which will condense on the roof of the vessel if this is very cold. Condensation droplets falling back on to the grain can cause severe spoilage on the surface layer of barley.

The types of storage vessel ultilised are varied but certainly the cheapest form is flat-floored storage sheds, with under-floor ventilation, followed by circular, steel, flat-bottomed bins with sweep auger clearing and under-floor ventilation. Concrete silos were constructed in the 1960s when costs were relatively cheap but such silos are presently about five times more expensive to construct than flat stores. The latter are roofed with a gable angle equal to that of the natural angle of repose of the heap of barley they contain, so the wall height is kept to a minimum. Circular, steel, hopper-bottomed vessels are almost as expensive as concrete silos but have the obvious advantage of easy discharge. The discharge of barley from flat store and the inability to segregate different grades or varieties of barley are the major disadvantages of flat stores. Normally barley is discharged by tractor shovel, which is slow and labour-intensive. Barley varieties can be segregated to a limited degree by insertion of temporary walls but in doing so the capacity of the store is impaired.

In some instances barley is pre-cleaned prior to drying. This operation is usually a rough dressing to remove rubble, straw residues and awns from the green barley. Since wet barley is notoriously difficult to sieve, most maltsters clean the barley after drying *en route* to storage silo or after storage *en route* to steep. Cleaning prior to storage has the advantage of a dust-free environment in silo, which improves aeration and working conditions generally but raises the problem of disposing of the dressings at harvest. The alternative of cleaning prior to steep allows for sale of the offals over a longer time period and this may have some commercial advantages, but the accumulation of dust and small grains within the bulk makes it more difficult for the aeration fans to give an even airflow through the barley.

When the barley is cleaned, the fractionation of the grain size may vary between maltsters. Generally, barley is initially sieved over 2.2 mm screens and all material less than 2.2 mm in diameter is sold for animal feeding. The main bulk may then be fractionated over a 2.5 mm sieve, depending on the final malt quality required. Barley between 2.2 mm and 2.5 mm amounts to 10–20% of the total, depending on the crop year, and may have a slightly higher nitrogen content than the 'head corn'. When malted separately the small corns yield between 4 and 5% less extract than the head corns, have a faster water uptake and malt faster (approximately one day) than the 2.5 mm fraction. It is not uncommon to malt small corns separately because of the differences in steeping and germination requirements.

QUALITY CONTROL OF BARLEY INTAKE

Barley variety

Selection of barley varieties for malting usually depends on consideration of availability, cost and whether the malt is to be used for the production of malt spirit or grain spirit. The first application requires a barley which is capable of developing a high level of fermentable extract. For grain spirit, where a relatively small proportion of malt is required to convert adjunct starch into fermentable sugars, malts of adequate enzymic content are necessary.

Annually, detailed lists of recommended cereal varieties are produced by the Scottish Agricultural Colleges (SAC, 1984). Varieties which are 'preferred' by the Institute of Brewing (IoB) for malting in Scotland are selected from these lists on the basis of extensive micro-malting and commercial trials. In 1985 the preferred varieties were Kym, Triumph and Golden Promise. Further lists for English-grown barley are published by the National Institute of Agricultural Botany (NIAB, 1984). In practice it is usual for the maltster to micro-malt small batches of new varieties to check that they are capable of producing malts with adequate levels of diastatic power (DP), α-amylase and fermentable extract. (Diastatic power is mainly a measurement of β-amylase activity, although it includes some effects of α-amylase.) Equipment can be purchased for this purpose or it can be custom-designed and manufactured in a well-equipped laboratory workshop.

Over the past few years increasing numbers of barley varieties (Table 2.1) have become available to the maltster. In view of their varying degree of suitability for malting, reliable confirmation of identity is important. Because the grains of many modern varieties are so similar in appearance, even under the microscope, traditional methods of identification based on morphological features now have a limited value for barley and are even less useful when applied to malt. The presence of blue aleurone pigment in some varieties (for example, Kym, Atem and Halcyon), is an aid to their recognition, but by far the most useful methods available are those based on electrophoretic separation of storage proteins extracted from the kernels (Marchylo and Laberge, 1980). Of increasing importance are the relatively more rapid methods which use the technique of reversed-phase high performance liquid chromatography (Marchylo and Kruger, 1984). A useful booklet with detailed descriptions of the cereal varieties is published annually by NIAB (1985).

Drying and storage

Moisture is removed from harvested barley to enhance its keeping properties, approximate maximum permissible moisture contents for 'short' and 'prolonged' storage being quoted as 14.8% and 13.6% respectively. Perhaps with more scientific confidence, the longevity of barley during storage has been described as a function of initial viability, storage temperature and barley moisture content; various prediction equations and nomograms are available based on these parameters (Ellis and Roberts, 1980). By way of example, this method would indicate that barley with an initial viability of 98% and moisture content of 12% can be stored for over a year at 15°C without the germination figure dropping below 95%.

Theoretical and applied advances in cereal grain drying are well documented in the technical literature (Bakker-Arkema *et al.*, 1978; Foster, 1982) and various aspects of grain drying and storage are described in papers published by the Ministry of Agriculture, Fisheries and Food (MAFF) and SAC. Tables of equilibrium humidity are useful in drying calculations and these are available for barley (Pixton and Warburton, 1971). Determination of grain moisture content is carried out before and after drying and as a means of monitoring the drying process. A wide range of portable instruments is available. These are based on the measurement of conductance, capacitance and infra-red reflectance, all of which are proportional to moisture content, whilst other laboratory-based weighing equipment uses microwave or infra-red radiation to increase the speed of drying. These methods have to be calibrated against the standard IoB oven method. As in any other analytical application, due consideration must be given to grain sampling procedure and sample preparation (IoB, 1982a).

Information on preventative measures against insect infestation and the treatment of grain storage problems is contained in the government publications referred to above and lists of approved pesticides are available (MAFF, 1982). It is essential to choose correct formulations, otherwise severe impairment of grain viability can result. In addition it is often specified that barley for the production of distillery malt should be free from pesticide residues.

The effects of various agronomic practices on grain quality have been discussed in some detail, for example by Briggs (1978). Grain damage may not be obvious in barley received by the maltster and various methods have been evolved to demonstrate injuries arising from fungal infection, mechanical effects of harvesting, injudicious drying, etc. Of paramount importance are tests which indicate the ability of grain to germinate fully in a rapid and uniform manner. The most commonly used tests are described in the IoB handbook of recommended methods (IoB, 1982b).

Viability

Germinative capacity (GC) is a measure of the percentage of living corns that are present in a sample of grain and the figure is calculated from the number of corns that germinate after immersion for three days in a dilute solution of hydrogen peroxide. An equivalent figure is determined with greater rapidity by immersing half-corns in a solution of a tetrazolium salt. Visual examination is carried out after several minutes and kernels with embryos stained beyond a minimum requirement are deemed to be viable. Unfortunately this method often indicates viability in heat-damaged corns which are dead. Germinative energy (GE) is a measure of the percentage of grains that can be expected to germinate fully if the barley is malted normally at the time of the test. The test is carried out by germinating 100 corns in filter paper in petri dishes using 4 ml water.

Dormancy

The difference in value between the figures for GC and GE indicates the degree of dormancy that is present in the barley under test. In a variant of the GE test, 8 ml of water are used instead of 4 ml. The figure obtained by subtracting the '8 ml' percentage from the '4 ml' germination percentage is known as water sensitivity (WS) and is generally associated with dormancy.

The extent of dormancy is a varietal characteristic, the expression of which is more common after a long, cool, damp season. It has been reported that high temperatures 12–16 days after ear emergence promote dormancy, whereas if these occur at the later mealy stage, dormancy is reduced (Reiner and Loch, 1976). When harvesting conditions are around 20°C ambient with barley moistures dropping to 10–14% in the field, the dormancy period is quite short, say, four to six weeks. Scottish conditions, with ambient temperatures around 13°C, rarely give barleys with moistures below 15% in the field, and so dormancy is likely to be a problem with susceptible varieties such as Triumph. Dormancy is a physiological condition which prevents barley from germinating when it is presented with an otherwise satisfactory set of environmental conditions. If the barley is required for malting during this period, then dormancy must be broken by artificial means, as mentioned above, by holding the grain at a moisture content of 12–13%, at slightly elevated temperatures. This procedure sometimes follows drying. Water sensitivity is overcome by introducing air-rests into the steeping schedule.

The phenomenon of dormancy lacks a complete explanation and a vast technical literature bears witness to the ongoing effort that is being made to understand it. Some of the major considerations involve the microflora of the barley husks, operation of the pentose phosphate pathway during the early stages of metabolic activity and the presence of inhibitory levels of endogenous phytohormones such as abscisic acid (Khan, 1977).

Besides occurring as a natural condition, dormancy can be induced by certain drying procedures and, on these occasions, the GC (peroxide) test can yield viability figures that are lower than those obtained in the GE tests. In order to avoid induced dormancy it is necessary to dry the barley in such a way that there is always a free transfer of moisture from the moist kernel when it is subjected to drying temperatures above ambient. The germination properties of dried barley have been reviewed recently by Palmer and Taylor (1980). Dormancy is less apparent in germination tests carried out at lower temperatures and the ratio of GE values determined at 16°C and 20°C has been used to distinguish dormant corns from those which have aged during storage, counts on the latter being unaffected by the temperature reduction (Bourne and Wheeler, 1985).

Pre-germination

Another physiological condition of less common occurrence is that in which the kernel starts to grow while the barley is still in the ear. This phenomenon, known as pre-germination, is encountered when a spell of relatively warm, damp weather precedes harvest. Barley containing pre-germinated corns tends to develop unevenly. The affected corns are susceptible to drying damage and may become reservoirs of microbial infection. Barley containing over 5% pre-germinated corns has been described as unsuitable for brewing, although this figure may have been suggested without taking account of the extent of pre-germination (Brookes, 1980). Screening for pre-germination can be carried out by incubating a sample of ground barley with Phadebas reagent (dye-linked amylopectin). The presence of pre-germinated corns as indicated by their α-amylase content can be detected readily by the production of a soluble blue dye (Mathewson and Pomeranz, 1977).

In order to determine pre-germination on a percentage basis, a suitable number of half-corns is incubated in a solution of fluorescein dibutyrate, after which they are examined under ultraviolet light. Pre-germinated corns possess an enzyme activity (possibly lipase) which hydrolyses the non-fluorescent ester to yield fluorescein, which is easily detectable on the cut surface of the kernel. This procedure is simpler than the European Brewery Convention (EBC) method in which agar gels are used (Jensen and Heltved, 1982).

Nitrogen content, corn size and predictive tests

After viability, nitrogen content is probably the next most important barley quality consideration. The starch content of grain, which is in turn the main source of fermentable extract, is inversely related to its protein content, which is indicated by a determination of nitrogen using a standard method such as the

Kjeldahl technique. Equipment for handling large numbers of samples is commercially available and variants based, for example, on alkali digestion have been introduced to increase the speed of analysis. Near-infra-red reflectance techniques (NIR) are popular for 'weighbridge' checks and for these a choice of equipment exists.

Barley for malt distilling is usually selected with a nitrogen content in the range of 1.6% (±0.1%). For grain distilling, barley with a higher content, for example, 1.8–2.0%, is commonly used. Protein is usually calculated as N × 6.25. The relative amounts of some endosperm protein fractions, separated by electrophoresis, have also been used to predict the suitability of certain barley varieties for malting (Wainwright, 1978).

A measure of corn size is normally carried out on barley received for malting. Perforated screens are used for this purpose, following the procedure described in *Analytica* (EBC, 1975). 'Screenings' pass through the 2.2 mm slots and the material which is retained is further classified into categories of greater than 2.2 mm, greater than 2.5 mm and greater than 2.8 mm. In addition, thousand corn weight is determined according to the IoB method, usually with the help of an automatic counting device. Corn size is a varietal characteristic and within a given variety it is affected by season and location. Several malt properties are related to the size of the corn.

Prediction equations, relating potential hot-water extract to the nitrogen content of the barley, have been in existence for several years. Early versions involved additional determinations such as husk content and included a measure of corn size, in terms of thousand corn weight (Pollock, 1962). All included a varietal 'constant', the value of which had to be adjusted from season to season. More recent versions omit the husk content and the thousand corn weight but still contain the varietal constant which is derived from the results of micro-malting trials using barleys of different nitrogen contents. The use of such predictive tests has been reviewed recently by Lloyd (1983).

For grain distillers, requiring a predictive test to screen new varieties for their potential to produce DP, there is a method which relies on the ability of papain to release bound β-amylase from the endosperm of ungerminated barley. Results from this test correlate with values obtained for DP after micro-malting (ASBC, 1977).

Steeping of barley

STEEPING TECHNOLOGY

Basic steeping requirements are the provision of water, a suitable temperature, and a supply of air as a means of removing excess heat, metabolic waste products (such as carbon dioxide and ethanol) and endogenous germination inhibitors.

The old floor maltings still in use demonstrate the limitations imposed on earlier maltsters. The barley was steeped in a single cast-iron cistern with no aeration or carbon dioxide extraction for upwards of 48 hours, and with no more than a single replacement of water. The water itself was drawn straight from the 'burn' or distillery reservoir at 5–10°C and was not adjusted to a required temperature.

From the earlier discussion on Scottish barley and the primitive steeping technology available, germination counts at casting must have been historically very low and very uneven. In this respect, the discovery of water sensitivity (Essery *et al.*, 1954) was probably of greater significance to the distiller-maltster than it was to his English counterpart.

Despite the common image of being a conservative industry, Scottish distillers have never been slow to adopt new concepts in maltings (as we shall see later in discussing malting plant) and newer designs of steep vessels were quickly incorporated into maltings in the early 1960s. These were generally of the conical type with in-steep aerators and carbon dioxide extraction. Since batch sizes in most floor maltings, Saladin Boxes and later drum maltings were fairly similar (i.e. around 20–30 t capacity as barley) this type of steep vessel has proved perfectly acceptable. The problems associated with uneven temperature distribution and airflow during air-rests in larger conical vessels have therefore been avoided.

Steeping techniques in the newer steeping/germinating/kilning vessels (SGKVs) tend to differ from transferred steeps and since the steeping and germination phases tend to overlap in such maltings they will be discussed in the following section on germination. The steeping schedules followed in conical steeps do not differ significantly from those used in the manufacture of brewing malts (Brooks *et al.*, 1976; Axcell *et al.*, 1983). Assuming that an initial water uptake of about 30–35% is required for onset of germination and an initial air-rest of 12–16 hours at 13–15°C is sufficient to promote an even chit, then the subsequent immersions and air-rests are adjusted to give a casting moisture of 45–47% with a uniform chit of over 96%. The casting moisture may vary according to the variety of barley, its nitrogen content and the degree of 'steeliness' or hardness in the endosperm. Since these properties vary from year to year and even between different regions in the same crop year, the maltster will always have to make minor adjustments to his steeping cycles to maintain uniform germination and water uptake.

More recently large flat-bottomed steeps of the Nordon type have been introduced. Such steeps are preferable when the barley tonnage exceeds 50 t, since a more even temperature and aeration profile can be maintained over the grain bed. If sufficient fan power is provided (say to give about 5 m^3 min^{-1} t^{-1} barley steeped) then the initial few hours of germination can be carried out in the steep before transfer to the actual germination vessel. However, care must be exercised in not damaging the grain if rootlets have started to emerge. The 'Giracleur' discharge on this type of vessel is fairly gentle with regard to physical damage to the grain but subsequent conveying to the germination vessel should be carefully managed to avoid excessive rootlet loss.

Transferring from the more conventional conical steep is best done under gravity as a vertical dry discharge to the germination vessel. In older Saladin plants the steeped grain was sometimes pumped over as an aqueous slurry to the germination box. Hydrostatic pressure applied to chitted malt can retard the rate of germination (Yoshida *et al.*, 1979) just as much as the purely mechanical damage mentioned above, and so should be avoided.

SCIENCE OF STEEPING

The purpose of steeping is to initiate uniform germination and to hydrate the endosperm to a level which is suitable for modification. Steeping conditions must subsequently allow satisfactory development of enzymes without encouraging undue malting losses from respiration and growth.

Grain structure and composition

The detailed anatomy of the barley grain and its chemical

composition have been adequately described in the brewing literature (Palmer, 1980). Only an outline is presented here, selected aspects receiving more attention in following sections of the text, so that the reader can envisage the distribution and relative importance of each component of the grain during steeping.

In longitudinal section two distinctive components of the kernel are apparent, namely the small embryo and the large starchy endosperm. Separating the embryonic axis from the endosperm is the scutellum. The aleurone layer surrounds the endosperm; the pericarp-testa and husk enclose the entire kernel. Embryonic axis and scutellum constitute 3% of the grain dry weight. These tissues are rich in lipids (15%), proteins (34%) and soluble carbohydrates (23%). The dry matter balance is made of minerals and insoluble cell wall material. Endosperm tissue comprises 75% of the dry weight, and contains starch (85%), protein (10%) and cell wall material (mainly β-glucan with pentosan). A further 12% of the kernel weight is accounted for by the aleurone layer, over 40% of which is made up of cell wall material (mainly pentosan) and 20% each of protein and triglycerides, while the balance is again made up of mineral constituents. The remaining major components, husk and pericarp, represent 10% of the dry weight. Materials of the husk are derived from dead cell walls and contain high levels of silica.

Water uptake

Before a barley is committed to steep, viability tests are carried out in the laboratory to establish that dormancy has reached a sufficiently low level to allow rapid and even germination to take place, and the degree of water sensitivity is established. Also prior to full-scale steeping, the rate of water uptake is determined by carrying out micro-steeping trials in which moisture content is measured after suitable intervals of time. A guide to the duration of the first steep, in relation to temperature and the pattern of air-rests required to produce uniform chitting, can thus be obtained. The first steep should raise the moisture content of the barley to a value between 32% and 35%. A following air-rest, with ventilation, will enable the kernel to absorb its adhering surface film of moisture under relatively aerobic conditions. Further periods of immersion, with or without air-rests, are carried out to achieve the final moisture content.

Rate of water uptake, particularly during the initial stages varies with barley variety, season and geographical location. The overall nitrogen and glucan content of the barley appears to have little effect on the rate of uptake. Initially, large corns take up moisture faster than smaller ones. At the end of a prolonged steep, however, the smaller corns possess the highest moisture content. It is frequently stated that the most suitable barley varieties for malting are those which take up moisture at the fastest rate. When the moisture content of the grain exceeds 25%, dehydrogenase activity becomes apparent and at higher moistures changes in the carbohydrate, lipid and protein composition of the embryo are produced by the action of other enzyme systems.

Uptake of water during steeping is generally regarded as a three-stage process (Brookes *et al.*, 1976). Initial adsorption, which is a purely physical process, continues rapidly until a moisture level of about 32% is reached. Within the range 5–35°C water uptake is a direct function of temperature, and at 12°C this 'stage of inhibition' can take 10 hours to complete. A lag phase follows, which can last for a further 10 hours, during which time the rate of uptake is reduced almost to zero. A third phase, one of active water uptake, continues at an increased rate to produce a maximum moisture content after a total steeping time of about 50 hours. During steeping the various structural components of the barley kernel take up water at different rates. Soon after wetting, the moisture content of the husk rises to 50% and this value is retained during immersion. The moisture content of endosperm tissue increases steadily during steeping, reaching a maximum value of 38–40% after about 70–80 hours. Embryo moisture level rises rapidly to 60–65% over 24 hours, then, with the advent of germination (chitting), rises to a maximum of about 85%. Moisture content of the scutellum changes in a similar way but always remains lower than that of the embryo. After 24 hours it reaches 55%, then slowly increases to a maximum value of 65%. During steeping, the bulk of moisture is taken up by the husk and endosperm, which are both relatively inactive in metabolic terms. Consequently, the type of steeping employed has no great effect on overall moisture level. However, vigorous aeration or spray-steeping increases the rate of water uptake by the embryo and scutellum and consequently the rate of development of the germ. This significantly affects the overall pattern of moisture distribution in the steeped kernel.

Endosperm tissue must be properly hydrated before it can be modified. The structure of barley endosperm is non-uniform and 'micro-regions' exist where packing densities and ease of solvent penetration are different from those of surrounding areas. With some difficult or 'low-grade' barleys the effect of spray-steeping is to encourage rapid embryo development before sufficient endosperm hydration has taken place. During germination, the effects of this deficiency are exaggerated by transfer of moisture from the endosperm and associated aleurone tissue to the embryo.

Respiration

Shortly after immersion, respiratory activity in the grain rapidly increases both in the embryo and aleurone tissue. Oxygen is quickly removed from the steep water and there is a simultaneous release of carbon dioxide and ethanol until the end of steeping. Ventilation during the immersion and air-resting periods not only supplies oxygen, which is essential for germination, but also dispels carbon dioxide, which has an inhibitory effect on growth and helps to minimise the accumulation of ethanol in the grain. Although ethanol concentrations are rarely lethal they can retard germination rate and produce non-uniform chitting. Until the corns chit, oxygen supply to the centres of respiration is limited. As a consequence the respiratory quotient (RQ) can remain above 1.0 for at least the first 24 hours of steeping. Subsequently, it drops below unity. Heat release is associated with respiratory activity and during the later air-resting periods ventilation has to be increased to control the temperature of the grain.

Germination inhibitors

Abscisic acid (ABA) is present in mature, ungerminated barley, the highest concentration occurring in the embryo (Yamada, 1985). During steeping, ABA passes into the steep water from the husk and its overall concentration in the grain diminishes. A fraction of ABA appears to be actively destroyed and levels of ABA in the grain are reduced faster when the steep is aerated. The onset of germination coincides with a fall in the level of ABA to a relatively low value. It is known that ABA inhibits gibberellin-induced α-amylase synthesis in the aleurone layer but the role of this compound in the early stages of germination is not yet fully understood. Gibberellin synthesis is delayed until germination has commenced and at the end of immersion steeping this hormone is present at just detectable levels.

During steeping, various phenolic compounds, amino acids, sugars and salts are leached from the kernels. These constitute about 0.5–1.5% of the kernel dry matter. While it is unlikely that they would have made a significant contribution to fermentable extract, their continued accumulation in steep water can retard the metabolic rate of the grain.

Effect on malt quality

Dry matter losses arising from rootlet growth and respiration during subsequent stages of malting increase when barley is steeped to a higher moisture content. Enzyme development and endosperm modification are usually favoured by this condition, particularly if the initial uptake of water occurs rapidly. For immersion steeps, early chitting leads to higher hot-water extract values.

Excessive steeping, especially with inadequate aeration, can lead to non-uniform germination since each barley has its own specific moisture requirement. Satisfactory levels of enzyme can be achieved in distillery malts by using a multiple immersion procedure with one or more air-rests to achieve a moisture content of at least 46%. Generally, pneumatic maltings require higher malt moisture levels than floor maltings. Some barley varieties are reported to produce higher levels of α-amylase when steeped to much lower moisture contents, e.g. 38–40%. The restricted modification which follows this practice would obviously limit its possible use for grain distillery application. Spray-steeping, hot-water steeping and re-steeping procedures appear to offer no advantage to the distillery-maltsters, particularly when high-nitrogen barleys are used. Abrasion and related processes which depend for their success on the concomitant use of gibberellic acid are excluded on the basis that additives are forbidden in the production of malt for whisky.

Wet casting procedures (hydraulic transfer), in which steeped grain is pumped from the steep to another location, can damage the embryo and upset the course of malting as mentioned above (Yoshida *et al.*, 1979). Modification is delayed by approximately a day, some enzyme levels do not recover from the process, and fermentability is reduced. After drainage of the final water, a film of moisture is retained by the steeped barley, which restricts gas exchange. The rate at which this film is removed, either by adsorption into the grain or by transfer to the surrounding air, is important for the onset of respiration and growth. Water in this film can account for 2–5% of the kernel moisture content.

Barley germination

GERMINATION TECHNOLOGY

Malting plant in the Scottish distilling industry is as varied as the rest of the UK malting industry. Just about every type of equipment, from floor maltings to completely automatic SGKV and from the original Galland Drum to the continuous Wanderhaufen process, is used. It is not the purpose of this discussion to debate the merits or demerits of each individual malting plant in the industry but it is worth while taking a general view of malting plant in terms of combined steeping, germination and kilning (SGKV), combined germination and kilning (GKV) and completely separate vessels for each process.

Separate steeping, germination and kilning

This is the commonest type of plant in older malting plants in which the germinating vessels were either Saladin Boxes or Drums. Some of the first maltings in the UK to utilise Saladin plants (1957–1962), Galland Drums (1905) and Vickers Boby Drums (1968–1972) were in the north of Scotland. Continuous malting 'streets' of the Wanderhaufen type have been installed in the two distilleries where the malt off-take was originally required as green malt for grain distilling.

The Saladin plants are of the concrete box type, fitted with either spiral screw turners or 'Van Caspel' type bucket turners similar to those used in Wanderhaufen steeps. Humidification and pressure fan ventilation are such that grain temperatures of 14°C + 2°C can be maintained across the bed with no more than 0.5% moisture loss per day. Therefore, with the ability to spray water on to the grain via booms fitted to the turner gantry, moisture levels can easily be maintained in these vessels.

Nevertheless, control of humidity in open vessels can never be as good as in the totally enclosed drum. While the capital installation costs of such vessels are now extremely high, the simple drum turning mechanism and relatively low maintenance costs make the drum a very effective germination vessel. The total inversion of the grain, under very gentle conditions, means even grain temperature, minimum rootlet damage and minimum moisture loss. Furthermore, heat is also dispatched to the ambient surrounding air by conduction through the steel walls; therefore a lower air volume is required to maintain the temperature differential across the bed. The major disadvantages are the comparatively small batch size (30–50 t), a leading factor in the high capital costs, and high energy costs in moving small batches of malt to and from kiln. The drum maltings now in operation are all of the Boby type which have improved aeration/humidification control compared with the original Galland design.

Combined germinating and kilning vessels (GKVs)

Some Scottish maltsters operate germination vessels which can also operate as kilns. Usually such vessels are circular in construction, have batch sizes of 100–300 t capacity as barley steeped and are fitted with two sets of fans, one for germination and the other for kilning.

Compared with straight germination vessels, GKVs have a much larger under-deck plenum to equalise the pressure of kilning air. During the germination phase the duct introducing kiln air is closed off and vice versa during kilning. The vessels used in Scotland have turners which rotate on a boom, like the sweep of the hands of a clock. In other countries, some maltings equipped with GKVs have fixed turners and rotating floors. However, the rotating turners can be lifted out of the malt during kilning. This is an advantage since stationary turners constantly embedded in the malt can result in uneven drying due to faster airflows around the spirals and localised overheating by thermal conduction through the steel shafts.

Airflows during germination are similar to those employed in Saladin Boxes (i.e. around $15\ m^3\ min^{-1}\ t^{-1}$) and during the last few hours of germination the humidification plant may be turned off to allow a partial drying out of malt before starting kilning. Due to the larger amounts of air employed and the large vessels used, humidification is more difficult and moisture losses are generally higher than in small Saladin and Drum plants.

Retention of moisture after steeping is extremely important for proper endosperm modification (Palmer *et al.*, 1985) since additional spray-steeping cannot guarantee that water applied post-steep will penetrate the endosperm. Whilst final moistures can be adjusted by spraying shortly after casting, later application of water, when the rootlets are fully formed, only increases the moisture content of the culms and of the embryo. Spraying late in the germination phase also has the disadvantage of maintaining rootlet development at a stage when a certain amount of withering is beneficial in limiting malting loss.

In malting for pot-distilling malt whisky the main requirement is a high spirit yield. It has been demonstrated (Bathgate *et al.*, 1978) that the fermentable extract in malt is carbohydrate-limited, so if carbon is lost through a high respiration rate, then the potential spirit yield from the malt is depleted. In other words, distilling malt requires long, cool germination with the minimum heat generation. Germination periods of five to seven days at the aforementioned temperatures are therefore common in all germination plants associated with separate steeps.

Combined steeping, germination and kilning vessels (SGKVs)

Faster throughput can be achieved in this type of vessel, essentially because the steeping and germination phases can be overlapped. To gain maximum use of a separate steeping vessel, it should be utilised as often as possible during the process cycle. In other words, one steep vessel should service at least two germination vessels, with a maximum 72-hour turn-round in a six-day germination cycle. However, if the barley is steeped in the actual germination vessel then there is no finite limit to the time at which water can be applied other than the time required to accumulate the necessary volume of water and the filling and emptying rates. When a number of SGKVs are employed and they have large capacities, the daily volume of water required can become the limiting factor in the steeping cycle employed. The very large amount of water required is due to the void volume in the air plenum under-deck. This voidage may amount to 30% of the water required and represents water which is never actually in contact with the grain. Such malting plants are commonly used in Scotland where water is relatively cheap, and disposing of effluent (mainly by pumping out to sea) is not difficult or expensive. Nevertheless, effluent problems can arise with the disposal of suspended solids if the under-deck area is not kept scrupulously clean after kilning. Obviously some culms fall through the wedge wire floors of the SGKV during malt stripping and these must be removed before the plenum is flooded during the subsequent steep. These are minor disadvantages compared with the low capital cost of installation, the flexibility of the plant, both in terms of barley drying and malting, and the high throughput rate.

As long as the barley has good germination energy, only two immersions in water may be necessary in these plants. The target moisture for the first immersion is higher than for separate steeps and moistures of 36–42% are common with an immersion time of 18–24 hours. This allows for a long air-rest during which the full germination airflow can be employed. With flat beds, air distribution and temperature control are excellent and very fast chitting takes place. The 'air-rests' as such cannot therefore be differentiated from germination and may extend to 24 or even 36 hours before the second immersion which brings the moisture to 46–47%.

SCIENCE OF GERMINATION

Aims, theoretical background and control

For malt spirit production, the purpose of germination is to maximise fermentable extract by promoting both endosperm modification and the development of amylolytic enzymes. For grain spirit there is less requirement for full modification but a demand for relatively high enzyme levels. This type of malt must also supply sufficient amino nitrogen for yeast activity during fermentation.

To achieve these objectives, the progress and uniformity of germination are controlled by maintaining optimum moisture levels within the corn, supplying air, removing carbon dioxide and excess respiratory heat and by mechanically turning the grain to prevent matting of rootlets.

The present descriptions outline the main events. Penetration of the pericarp-testa by the coleorhiza marks the start of germination (chitting). Subsequent embryonic development leads to the production of rootlets and growth of the shoot (acrospire). Enzyme synthesis and activation take place during the germination stage and the cellular structure of the endosphere is 'modified' by the action of these various enzymes while the grain respires aerobically.

Barley variety, nitrogen content, corn size and conditions of steeping all affect the rate of malting. Prior to processing micro-malting trials are carried out to assess the performance of the barley during germination. Tests may be conducted at different temperatures or more commonly with malts at different levels of moisture content. Daily analysis of malt samples can then indicate the rate at which modification and enzyme development are proceeding. The uniformity of germination and the progress of growth can also be observed and some information can be gained about malting loss.

The exact site of enzyme synthesis and details of the initial pattern of endosperm modifications have been matters of some controversy for almost a hundred years. The situation still remains incompletely resolved but it is generally agreed that the major site of synthesis and secretion is the aleurone layer, under the influence of gibberellin released from the embryo. Scutellar tissue appears to make some contribution, but it is more likely that in the studies involving isolation of such tissue there has been contamination by aleurone cells. It is now agreed that modification of the endosperm proceeds from the area nearest to the embryo. Gibberellin is synthesised in the embryo, by a process requiring oxygen. Recent studies indicate that the endogenous gibberellin is GA_1 and that gibberellic acid (GA_3) is present in smaller amounts (Yamada, 1982). Following the migration of gibberellin to the aleurone a number of hydrolytic enzymes are synthesized and secreted into the endosperm in an orderly sequence (MacLeod, 1979).

Other enzymes, many of which are already in the endosperm, do not require the intervention of gibberellin for their activity. For example, GA is required for the aleurone synthesis of α-amylase and limit dextrinase, whilst synthesis of barley β-glucanase and α-glucosidase are enhanced by its presence. β-1-3-glucanase and phosphatase, on the other hand, rely on GA only for their secretion. In contrast, β-amylase and carboxypeptidase are present in the endosperm and have no direct GA requirement. During germination latent β-amylase is converted into the active form by the action of proteolytic enzymes.

Fermentable extract is almost exclusively derived from the starchy endosperm of the germinated grain. Release of this material by the process of modification depends on the combined action of several hydrolytic enzymes on cell walls and matrix proteins. In the endosperm, starch occurs as large (20–30 μm) and small (1–5 μm) granules. Although the latter are more numerous they represent a minor weight proportion of the total starch. Both types of granule are embedded in a matrix of reserve protein composed of hordeins and glutelins. This conglomerate of starch and protein is contained within a cellular envelope composed of 70–75% glucan, 20–25% pentosan, and other minor components. Of the cell wall glucan, 70% is β-1,4 linked, the remainder being linked in the β-1,3 position.

Modification of the endosperm follows the release of β-glucanases, endo-proteases and pentosanases from the aleurone layer. A significant amount of β-glucan is degraded during malting, along with 35–40% of the reserve protein, but less than 15% of the starch is broken down. Incomplete modification of the endosperm will result in reduced extraction of starch during mashing and a reduced susceptibility of the starch to enzymic attack.

In terms of modification, considerable research effort has been made to isolate the factors that distinguish a good malting barley. Malting quality is a varietal characteristic and is not related simply to either glucan or protein content. Malting barleys usually modify faster than poor quality barleys even when their nitrogen content is higher. Traditionally, mealiness and steeliness in the barley endosperm have been used to judge quality. Steely grains, which malt less readily, generally contain more nitrogen than corresponding mealy grains. However, the relationship between endosperm nitrogen and steeliness is not consistent betwen varieties. It is possible that in poor grade barleys there are limited areas of high-nitrogen endosperm which display an extreme degree of steeliness and modify at a slower rate. The polypeptide composition of reserve proteins (hordeins) in different varieties appears to be related to their malting potential (Wainwright, 1979). High glucan content does not always signify poor malting quality. Barleys with relatively high levels of cell wall materials modify successfully provided that adequate β-glucanase levels can be achieved. In assessing this provision it is important to distinguish between free enzyme and unsecreted 'non-functional' enzyme (Palmer *et al.*, 1985). The application of electron microscopy has had a limited success in relating endosperm microstructure to malting quality. Within the endosperm the packing density of the starch-containing cells varies, thus affecting ease of hydration, and differences are apparent between varieties in the density of the small starch granule-protein matrix. In areas of well-modified endosperm, cell walls are absent and the amount of protein matrix is greatly reduced.

Increasing the temperature of germination initially improves the rate of modification but final extracts are lower. However, the modification value for kilned malt, as expressed by the differences between values for hot-water extract using coarse and fine mill settings ('coarse-fine' difference), can remain unsatisfactory. This indicates that overall modification has been restricted. High germination moistures, on the other hand, increase endosperm hydration, helping difficult barleys to modify. An adequate supply of oxygen is essential for modification, particularly during the early stages of germination. If carbon dioxide is allowed to accumulate, modification is restricted. Overmodification is less of a problem for the distiller than the brewer, in that the brewer must retain some high-molecular-weight protein in wort so that head retention is preserved in the finished beer.

In malt for distilling, the most important enzymes are those which are involved in the conversion of starch into fermentable sugars. This group comprises α- and β-amylases, limit dextrinase and α-glucosidase.

α-amylase is an endo-enzyme which specifically attacks α-1,4 glucose linkages. α-1,6 linkages are bypassed and are not hydrolysed. This enzyme also reduces the viscosity of starch suspensions and produces dextrins containing up to 12 glucose units, with only a slow production of glucose and maltose. Amylose and amylopectin are converted to the extent of 70% and 55% respectively. α-amylase is absent from ungerminated barley and depends on gibberellin for its synthesis in the aleurone layer during germination. A natural α-amylase inhibitor occurs in the endosperm of ungerminated barley, but the level of this material does not change during germination. Its role may be to retard the digestion of intact starch granules in the endosperm. While it is heat-sensitive, a proportion may survive kilning and mashing (Munck *et al.*, 1985; Weselake *et al.*, 1985).

By contrast, β-amylase is an exo-enzyme, removing maltose units from the non-reducing ends of α-1,4 linked chains. The substrate viscosity is slowly reduced with the formation of linear dextrins and maltose is rapidly produced. β-amylase will eventually convert amylose completely into maltose, whereas amylopectin is 75% converted. It must be noted that β-amylase will not attack raw starch granules. The enzyme is present in ungerminated barley mainly in a latent form, which is converted to free enzyme during germination.

Limit dextrinase specifically attacks the α-1,6-glucosidic linkages in, for example, amylopectin, β-limit dextrins and starch-derived α-limit dextrins. The 'debranching' enzyme effectively increases the fermentability of worts and subsequent alcohol production. This enzyme is absent from barley and it is synthesised during germination in parallel with α-amylase (Lee and Pyler, 1984).

α-glucosidase (maltase) liberates β-D-glucose from the non-reducing end of α-1,4 linked chains. It will also release glucose which is linked terminally in the α-1,6 position. This enzyme occurs in the aleurone layer and embryo of ungerminated barley.

In the range 12–20°C, higher temperatures increase initial rates of enzyme development but suppress maximum attainable values. Procedures in which germination starts off at a relatively high temperature and finishes off at a lower temperature appear to have no advantage for the distiller, but higher enzyme levels are usually produced at higher moisture contents. Barley varieties differ in their ability to develop enzymes and this ability is influenced by seasonal changes and geographical location.

Production of soluble nitrogen (amino-nitrogen) is another important feature of malting, particularly for grain distilling. Malting conditions which promote the development of enzymes are also suitable for soluble nitrogen production. Thus high levels are favoured by low temperature and prolonged germination periods.

Respiration and growth

Respiration supplies the energy for the synthetic processes that take place during germination. At the start of germination, respiratory activity is divided equally between the embryo and the aleurone layer. As germination progresses a higher proportion occurs in the embryo. Depending on the temperature of germination, respiration rate steadily increases, reaches a peak after 3–4 days, then slowly declines. Initially, respiration is faster at higher temperatures. Carbohydrate is the main substrate, which is converted to carbon dioxide, and up to 4% of the initial barley dry matter is lost by respiration during germination. Respiratory activity is generally higher in barleys germinated at higher moisture levels. If oxygen supply is limited or carbon dioxide is allowed to accumulate, respiration is thereby restricted.

During germination, adequate ventilation is required to dissipate the heat produced by respiration. Part of the energy supplied by respiration is used by the barley to produce rootlets and acrospire which can account for 4–5% of the original barley dry weight. Malting conditions affect growth and respiration in a similar manner, although there is not necessarily a strictly parallel development. It can be seen that total malting loss comprises losses which arise from steeping, rootlet growth and respiration and it accounts for a proportion of material that otherwise could have been fermented to alcohol.

Moisture control

In pneumatic maltings, an unavoidable loss of moisture takes place from the germinating grain as a consequence of respiratory heat release. A thermal gradient is produced in the ventilating airstream which increases its moisture-holding capacity. Regardless of the initial air humidity, rapid transfer of moisture takes place from the malt to the surrounding air. Within the kernels water is transported from the endosperm to the embryo at an increased rate, which has an adverse effect on the modification process. Although the former moisture level can be reinstated by spraying, the distribution of moisture will be different because the water is taken up by the rootlets and the outer layers of the corn.

Fermentability

As stated previously, the main object of the malting process is to maximise the production of fermentable extract. During germination, hot-water extract (HWE) as determined by the IoB method rises steadily to a maximum value, then declines as an increasing proportion of soluble material is diverted to the embryo for growth. Before this maximum is achieved, the potentially fermentable fraction of this extract has already started to decline due to a change in the relative amounts of carbohydrate and nitrogenous material, in favour of the latter. Thus fermentable extract reaches a maximum value before the peak value for HWE is achieved.

Concomitant with the increase in HWE there is usually a steady decline in the difference between values for HWE using coarse and fine mill settings. This difference, which is inversely related to the degree of physical modification, continues to decline beyond the point of maximum HWE. If the germination process is continued until minimum difference is achieved, this can result in the loss of extract. Continuing to a lesser degree of modification, for example to the point of maximum HWE, will similarly incur a loss of fermentable extract. At the point of maximum fermentable extract it is likely that an adequate level of modification will have been achieved, as spirit yield predicted from this value correlates with that obtained in distillery practice. Furthermore, by curtailing the length of the germination period, unnecessary malting losses are avoided, thus increasing the yield of spirit based on original grain. This same argument does not necessarily apply to the production of malt for grain spirit, where enzyme levels are more important than fermentable extract. In terms of other analytical parameters malt fermentability is inversely related to cold-water extract and wort soluble nitrogen (Bathgate *et al.*, 1978).

Measurement of green malt quality

Traditionally the extent of acrospire development and the ease by which the kernel endosperm can be 'rubbed' between the finger and the thumb have been used as a guide to the progress of malting. During germination, grain moisture content and temperature must be monitored continuously. In pneumatic maltings it is unlikely that any deficiency of oxygen or local accumulation of carbon dioxide will occur but the latter can be checked by the use of a portable gas analyser. Levels of α-amylase and DP (which mainly represents β-amylase) are measured in green malt by IoB methods. Typical values for green malt DP and amylase are 70–80° IoB and 20–30 DU (dextrinising units) respectively (expressed at 44% moisture). The progress of modification can be followed by treating sectioned corns with a chemical stain such as calcofluor which specifically combines with β-glucan in the endosperm cell wall. Unmodified areas of endosperm fluoresce under UV light after treatment with the dye (Aastrup and Erdal, 1980). Other tests, which are applied to kiln-dried malts, are referred to in a later section.

Malt kilning

KILNING TECHNOLOGY

Traditionally, distilling malts were kilned in natural-draught kilns with distinctive 'pagoda'-shaped chimneys. The malt beds were very shallow (*c*. 30 cm), airflows very low (approximately 48 hours kilning), and temperature control extremely variable. The fuel used was, of course, peat and later a mixture of peat and anthracite was burned in the hand-stoked furnaces. The adsorption of peat smoke on to the malt is perhaps the most distinctive feature of pot-distilling malts, the actual levels of peat smoke varying from area to area. A few natural-draught kilns are still in operation on Orkney, Islay and on the Mainland but with the centralisation of malt production and the introduction of modern malting plant, alternative means of applying peat smoke have been developed.

Apart from the peating aspect, distilling malt is distinctive in being very lightly kilned; that is to say, low air-on temperatures are employed to preserve the diastatic enzymes essential for a highly fermentable extract, and high airflows (40–70 m^3 min^{-1} t^{-1}) are preferred for quick drying down to around 5% moisture.

The equipment used for kilning is dictated by the type of malting plant being operated, i.e. separate dedicated kiln, GKV or SGKV, each of which are discussed in the following sections.

Separate kilns

Most separate kilns are of concrete construction following the 'Winkler' design. In these vessels the green barley is loaded by mechanical spreader on to a wedge wire floor to a depth of approximately 1 m. The square floor is either self-tipping or is stripped by means of a traversing auger.

The kiln air is heated by direct firing into the airstream or indirectly by passing the air through a bank of heat exchangers. Directly fired kilns were the norm and, as mentioned previously, employed peat, anthracite or coke as the primary fuel. When oil became a cheaper option, the kilns built in the late 1950s and early 1960s were fired with light gas oil or heavy fuel oil. Later many of these older kilns were re-equipped to use natural gas, which became plentiful in the wake of the North Sea oil discoveries. By the late 1970s many maltsters were using natural gas. However, the discovery (Spiegelhalder *et al.*, 1979) that direct firing, particularly with gas, led to high levels of nitrosamines in malt, necessitated a rapid change to indirect firing, using either steam or thermal fluids as the heat exchange medium. Light gas oil, heavy fuel oil, natural gas and coal are all used in secondary heating and many maltings now have the ability to change fuels should this be economically expedient.

With the rapid escalation of fuel costs, energy management has become as important as control of quality on the kiln. In consequence the kilning cycles employed are a compromise between the thermally most efficient and that producing the optimum malt quality. Ideally malt should be dried at as low a temperature and as high an airflow as possible. Since the electrical power consumed by the fans varies as the cube of the volume delivered, kilning power consumption can be very high for high-capacity fans. Similarly, the greatest thermal efficiency is achieved at as high an air-on temperature as possible, completely opposite to the quality requirement. The compromise position is to kiln to the 'break' point (i.e. during the free drying period when the air leaving the kiln is saturated) at a temperature of 60–65°C and then to increase the temperature to 70°C to the finish with a reduced fan speed. In this way the maximum drying is done at as low a temperature and as high an airflow as possible, and only when the moisture is below about 10–20% is the volume dropped and the temperature increased.

GKV and SKGV kilns

The general principals described above for independent kilns also apply to GKVs and SKGVs. The major differences lie in the nature of the malt bed. In loading a separate kiln with green malt, great care is taken to have the malt bed evenly and loosely packed so that the passage of drying air is equally distributed over the whole piece.

The problems usually associated with combined germination and kilning vessels are uneveness in bed depth occurring as the barley germinates and expands, areas of compaction caused by mechanical turners, and problems of uneven drying round the turner spirals, as previously mentioned.

If the bed is uneven and has compacted areas, the air will tend to take a 'line of least resistance', leaving wet patches of malt on the surface of the bed. When this occurs, there is very little one can do to remedy the situation, apart from turning the malt when partially dry (e.g. at the break point). Turning at this late stage has minimal effect on malt quality but the high costs incurred in drying an unevenly loaded kiln are the main concern. Thus great care has to be exercised in proper loading and turning right through the process to ensure efficient kilning.

Nitrosamine control

Nitrosamine contamination is a problem particularly associated with the use of natural gas as a kiln fuel. The way in which the gas is combusted can be altered in so-called 'low NO_x' burners. In these burners the ratio of gas to air is carefully controlled to give a very lean mix. This reduces the flame temperature and the amount of atmospheric nitrogen oxidised to nitrogen oxide or nitrogen dioxide (NO_x), necessary precursors for the formation of nitrosamines.

Alternatively, reaction between oxides of nitrogen and hordenine in the malt to produce nitrosamines can be blocked by the application of sulphur dioxide (SO_2) to the malt. If this practice is operated rock sulphur is burnt and the SO_2 produced is fed into the hot airstream during the early stages of kilning. A third control method, which is increasingly favoured, is to use an indirect firing system.

Peating

When malt was manufactured 'on site', the traditional fuel used was local peat together with faggots of heather and other available combustibles, but we again glean from Sir Robert Moray (1678) that 'faggots of broom should be avoided as they impart an unpleasant aroma'. The procurement of fuel was an integral part of the malting scene and was as important a feature of the production year as procurement of barley. In the 'non-malting' season from late spring and summer when the weather was judged to be too warm for floor malting and distilling, the distillery work-force were employed in digging and 'raising' peats for the autumn and winter malting. This still remains as part of the working schedule in Orkney and Islay. Peating is now regarded as an ancillary process and peat is no longer used as a primary fuel. However, it is interesting to note that at current prices for commercially available peat and estimates of calorific value, the cost per equivalent unit of energy would make peat a very competitive fuel if it were available in commercially viable quantities.

All kilns were hand-stoked, so that as each addition of peat was made to the fire there was a short period of intense 'smoking' of the fuel before high temperature combustion. Peat even in a fairly dry state may contain between 25% and 60% moisture. It is therefore a very 'smokey' fuel. When used in this way as the direct fuel for heating kiln air, the peat not only dried the malt but imparted a strong characteristic aroma as some of the smoke was adsorbed by the malt. Later, and especially when oil and gas burners were introduced, separate peat furnaces were installed as a secondary operation to produce smoke rather than primary heat. However, as long as the primary fuel contained elemental sulphur (e.g. anthracite, light gas oil or heavy fuel oil) the nitrogen oxides produced in the secondary burning of peat could not react with the malt to produce nitrosamines.

It was only when gas was used as a primary fuel that the dual effect of nitrogen oxides from both the gas burners and from the peat furnaces was recognised (Ladish, 1981). The peating of malt is a ready source of nitrosamines, possibly arising from residual nitrogenous material in the peat. Great care must be exercised in stoking the fire to prevent flaming and excessive heat development and in applying the appropriate amount of oxides of sulphur to prevent nitrosamine formation. When direct gas burners or indirect firing is employed, then rock sulphur must be burned or gaseous sulphur dioxide applied so that about 10–30 ppm of sulphur dioxide is detectable on the peated malt to prevent the formation of nitrosamines. If an indirect heat recovery system is employed to partially heat the incoming air through a heat exchanger of some sort (either air-to-air or by 'runaround' coil), then less primary fuel and so less sulphur is applied to the malt. Given these safeguards on burning and sulphuring, peated malt can be made with almost undetectable levels of nitrosodimethylamine (NDMA) and can be marketed with the same guarantees as other forms of malt.

The adsorption of peat smoke on to the malt is not a simple operation and is governed by both the intensity of the smoke and the rate of malt drying. Peat smoke will not readily adsorb on to malt when there is a surface water film, i.e. during 'free drying' and adsorbs only slowly when the surface is very dry. Greatest adsorption takes place when the malt is 'hand dry', i.e. between 15% and 30% moisture. Peat smoke must therefore be applied very intensely during the early period of kilning right up to and beyond the 'break point'. The intensity of peating is usually measured in terms of the concentration of phenolic marker compounds in the peat smoke (Bathgate and Taylor, 1977). The relative intensity of peating can be conveniently broken down into three categories: lightly peated (about 1.0–5.0 ppm total phenols, say); medium peated (5.0–15 ppm); and heavily peated (15–50 ppm). The last category varies considerably according to individual requirements, the heaviest being the Islay malt.

Kilning and malt types

As well as the peating grades mentioned above, distilling malts may vary according to their moisture contents. These are as follows:

(a) *Unkilned malt* 'Green malt', with high DP, can be used directly to convert unmalted cereal starch in grain distilling. Such systems are used in grain distilleries with their own maltings on site or within easy transport distance from the distillery. Thus, the malt is used before it either overmalts or is contaminated by spoilage organisms. Green malt produces the same spirit yield, based on original graining, as dried malt. Green malt is equally effective for the conversion of cooked maize, wheat or barley and a slightly lower proportion is required for conversion during mashing.

(b) *'Grain distilling malt'* Alternatively, when grain distilleries do not have a ready supply of green malt, a lightly dried malt is used for cereal conversion. Normally such malts are kilned at as low a temperature as possible (around 50°C) to preserve their enzymic content. The final moisture is normally around a maximum of 6% so that adequate milling on standard malt mills can take place.

Grain distilling barleys were historically imported from Canada since their 6-row varieties were generally higher in DP than those used in the UK. Some malts are still imported from Scandinavia because of the exceptionally high enzyme content (e.g. 100 DU and 200 DP) generated from such varieties as Akka and Pomo. No UK barleys can compete with this level as regards individual varieties, but generally higher nitrogen barleys are employed since the enzymic content is of more importance than extract.

(c) *'Standard' pot distilling malt* The optimum moisture content for distilling malt is 4.5–5.0%, at which there is a balance between fermentable extract and milling ability.

Energy conservation

In the previous paragraphs on kilning, mention has been made of changes in the types of fuel used in kilning and of heat recovery. These changes were usually made for economic reasons (Kidger, 1984). Peat is costly to produce in very large quantities, so with the availability of anthracite and the spread of the railway network in the nineteenth century, a cheap efficient fuel was available right up to the 1960s, when oil became the cheaper option. The dramatic increase in oil prices in the 1970s and the availability of cheaper North Sea gas saw a rapid change to direct firing with this fuel. The nitrosamine controversy, together with the escalating costs of gas, have motivated many capital investments in indirect-fired (i.e. heat-exchanged) systems combined with heat recovery units.

The commonest form of heat recovery is to reclaim some heat from the kiln exhaust air. Since this air leaves the kiln saturated with water in the early stages of kilning, the installation of a tube heat exchanger on the exhaust flue can at least recover the latent heat of evaporation. The heat recovered can be used to pre-heat the incoming fresh air either directly (air-to-air) or by means of a pumped thermal fluid (usually water/glycol) passing through a second heat exchanger (run around coil). These units have now been incorporated into indirect systems in which the primary heating of the kiln air is achieved by steam-heated exchangers or high-temperature thermal fluid exchangers. In maltings where direct gas firing is still used the burners have been changed to 'low NO_x' units (Ladish, 1981).

Some very sophisticated heat recovery engineering has been

successfully installed in some distilleries with adjacent small maltings. In these, surplus heat from still condensers is pumped to heat exchangers at the maltings and in one case nearly all the surplus heat from the distillery is used either to kiln malt or to heat a horticultural greenhouse.

No doubt further changes in fuel usage will occur as the economics change. In some cases there has already been a swing back to the use of coal in combustion with steam heat exchangers.

SCIENCE OF KILNING

Green malt is kilned to produce a dry, storable product and to arrest further development of biological activity when the required enzyme content and degree of modification have been reached. In addition: (a) kilning effects a degree of sterilisation; (b) it is easier to remove rootlets from kilned malt; and (c) the dried product can be milled in a way that suits mashing requirements. Unwanted flavour components are removed by kilning, whilst other desirable ones are introduced, either from existing precursors or from extraneous sources such as peat smoke.

The physics of malt drying is similar to that of barley and mathematical treatments of both are often found together in the brewing literature, along with worked examples (Sfat, 1965; Briggs *et al.*, 1981). Initially, the temperature of the green malt rises above ambient but remains well below that of the drying stream as a result of evaporative cooling. Moisture is transferred from the malt in an unrestricted manner until a malt moisture content of about 20–25% is reached. This phase of kilning, known as the 'free-drying' stage, is terminated by the 'break', which represents the drying front progressing through the malt bed in the direction of the airstream until it 'breaks' at the surface. Water remaining in the malt is released less readily and as a consequence the temperature of the malt starts to rise. This second 'diffusion' stage continues until a moisture content of approximately 10–12% is reached. At this point, most of the water appears to be 'bound' and further moisture is released only by increasing the temperature of the drying stream. By doing this the moisture of the malt may be further reduced to a value between 4% and 6%. In brewing practice, but not for distilling malts, a 'curing' stage at higher temperatures (80–100°C) usually follows.

In biochemical terms, the activity of the embryo may continue until the malt temperature exceeds 50°C. During the free-drying stage, modification and enzyme synthesis continue. After the break, enzyme activity declines and some enzymes are destroyed, the latter process accelerating during the final stages of kilning. The relatively heat-stable enzymes are more susceptible to damage from heat when the moisture content of the malt is high.

Effect of kilning on modification, enzymes and fermentability

Modification continues during the free-drying stage of kilning. Starch, β-glucan and proteins are broken down and their products are transferred to the embryo which is still actively respiring. As malt temperature increases, the synthetic processes of the embryo slow down and low molecular weight compounds such as sugars and amino acids accumulate in the grain. These materials constitute the 'cold-water extract' of the malt. The process of physical modification can be monitored by staining corn sections with calcofluor which is specific for cell wall glucans. Increasing modification is also indicated by a decrease in the numerical difference between values for hot-water extract, using coarse and fine mill settings. Recently the friability meter ('friabilimeter') has gained popularity as a means of assessing modification and for indicating homogeneity in a sample of malt (Baxter and O'Farrell, 1983).

Preservation of amylolytic enzymes during kilning is particularly important in distilling practice. The activity of these enzymes must continue during fermentation, as well as mashing, this being made possible by the absence of a worts boiling stage. Up to 20% of the total alcohol yield can originate from enzymic activity during this period of 'secondary conversion'. Although the effects of kilning temperature on enzyme activity are widely described in the brewing literature (Narziss and Rusitzka, 1977a, b), precise extrapolation of these data may, in practice, be limited. Frequently in these studies the temperature of the drying stream is quoted without reference to flow rate or humidity, actual malt temperature remaining undefined. In addition, malt beds in practice are not uniform but contain thermal and compositional gradients. The range of variation in enzymic activity during kilning is demonstrated by the examples which follow. During the early stages of kilning, the activities of α- and β-amylase, endo-β-1,3 : 1,4 glucanase and the endo- and exo-peptidases continue to increase. In the final malt, because of the relatively mild kilning conditions used in distilling practice, levels of α-amylase and the peptidases may exceed those in the original green malt. Only slight losses occur in the final levels of β-amylase and limit dextrinase, whereas up to 80% of the endo-β-glucanase activity is lost. α-glucosidase (maltase) is very susceptible to temperatures exceeding 50°C and peroxidase and catalase are among the most easily destroyed enzymes. Acid phosphatase, 6-phytase and lipoxidase lose activity even under mild kilning conditions, whereas lipase activity survives.

Concentrations of other compounds, apart from enzymes, change during kilning. Amino acids decrease in quantity, as do some reducing sugars. Formation of melanoidins from these two groups of compounds is enhanced if the malt is overmodified or is allowed to 'stew' by raising its temperature while still relatively wet.

Fermentability appears to be unaffected by mild kilning conditions. The analytically determined figure (IoB, 1982c) is used in conjunction with the value for standard hot-water extract to indicate 'fermentable extract' from which a 'predicted spirit yield' may be calculated. The laboratory method suffers from variations which may be introduced by the yeast, and for sulphured malts results may be artificially depressed because of relatively low pH values which are produced in the lightly buffered hot-water extract. In practice, spirit yields from these malts would exceed those predicted from the test.

Sulphur dioxide

The combustion of fuel oil, which may contain up to 3% w/w sulphur, produces sulphur dioxide. In directly fired kilns the gas is rapidly taken up by green malt, particularly when the malt moisture content is above 20–25%. Levels are retained by the grain throughout subsequent kilning, even when the initial period of exposure is less than two hours. Only a fraction of the total sulphur dioxide present in the malt is indicated by standard distillation methods. Other techniques show that free, combined and non-distillable forms exist in the grain. It is likely that the chemical state of combination of sulphur dioxide determines its effect on worts' properties. Direct kilning may also produce oxides of sulphur which can produce a black flecked appearance known as 'magpie' on the surface of malt near to the kiln floor. Sulphur residues on malt are detectable chemically as sulphate and correlate with the pH of the wort. It is sometimes considered desirable to introduce sulphur dioxide into gas fired kilns where none is produced by combustion, to combat the formation of NDMA. For low-gravity worts, pH is depressed when malts containing sulphur dioxide are used. Soluble nitrogen levels are increased and the activities of proteolytic enzymes and β-amylase may be enhanced by the reducing action of sulphur dioxide, whereas the activity of other enzymes is reduced (Sim, 1983). Sulphur dioxide lightens the colour of malt (which is of no direct interest to the distillers) and significantly reduces the level of surface microbial infection in the malt (Flannigan, 1983).

Nitrosamines

Kilned malts are a potential source of volatile nitrosamines which appear at low levels in the distilled spirit. Nitrosamines are formed when nitrogen oxides, produced by the combination of fuels during kilning, react with hordenine or related compounds such as dimethylamine present in green malt. The formation of *N*-nitrosodimethylamine (NDMA) can be effectively blocked as described above. The detailed chemistry of nitrosamine formation is not yet fully understood but a useful review of this topic has appeared recently in the brewing literature (Wainwright, 1986). In distilling practice the use of peat during kilning may enhance nitrosamine formation.

Flavour

During kilning new compounds are formed and others, which exist in the green malt, are destroyed. These chemical changes almost certainly affect the organoleptic quality of the final distilled spirit. Included in the types of reaction considered to be of importance during kilning are those based on the enzymic and chemical oxidation of unsaturated fatty acids, the combination of free amino acids and reducing sugars, and the thermal degradation of precursors such as *S*-methyl methionine which are synthesised during germination (Tressl *et al.*, 1983).

The chemistry of the formation of flavour compounds is

complex and becomes increasingly so as kilning temperatures are raised. Detailed reviews are readily found in the brewing literature (Bathgate, 1973). Most reported kilning studies relate to the production of bewing malt. In distilling practice omission of the curing stage during malt drying precludes the formation of many of the compounds that accumulate in ale malts. Furthermore, in brewing practice, malt flavour components are modified and supplemented by wort boiling and hopping. These two processes play no part in whisky production, which is, however, distinguished by the distillation stage, which is of paramount importance both as a formative process and as one of separation. Quantitatively, the major congeners are produced during fermentation. In organoleptic terms, however, compounds which make a significant contribution are often present in only trace amounts. Much work remains to be carried out linking whisky character to chemical composition and to tracing the origin of these components.

Linoleic acid (18 : 2) is the most abundant fatty acid present in the barley grain, and malt contains two lipoxygenase enzymes which convert this compound to 9- and 13-hydroperoxides which are then isomerised and modified via ketols to produce trihydroxy fatty acids. These in turn can serve as precursors for the formation of aldehydes such as hexanal and *trans*-2-hexenal (Gardner, 1975).

A range of important flavour compounds is formed by the Maillard reaction, in which free amino acids and sugars combine and undergo chemical transformation by Strecker degradation and other routes leading to the production of unsaturated aldehydes, furans and pyrroles etc. The formation of dimethyl sulphide (DMS) is an example which illustrates how a diversity of factors – ranging from barley variety and nitrogen content, storage time, conditions during steeping and germination and finally kilning conditions – can all affect the level of a compound which will eventually appear in the distilled spirit (Anness and Bamforth, 1982). The mechanisms of DMS formation are better understood than those which produce the chemically related di- and tri-sulphides, which are of more significance organoleptically. Their formation appears to be influenced by residual levels of sulphur dioxide present in the kilned malt (Williams and Gracey, 1982).

Peat-derived phenols are among the compounds which contribute to the peaty flavour of distilled spirits. These compounds are introduced during kilning by exposing the malt to the smoke (reek) from burning peat. Phenols from peat smoke are absorbed efficiently when the malt moisture content exceeds 25%. A minor proportion of the phenol absorbed from peat reek is retained in the peated malt, the balance being neutralised or destroyed by changes (possibly enzymic) which take place in the malt during kilning. Quantitatively the most important phenols are phenol, isomeric cresols and the xylenols. However, other phenols such as guaiacol, although present is smaller amounts, can make a very significant organoleptic contribution. The combustion temperature of the peat affects both yield and relative amounts of phenols produced. Raising this temperature from 400°C to 750°C increases phenol and cresol concentrations several-fold but diminishes the yield of guaiacol. Methods for the determination of phenols in malt have been reviewed in the brewing literature (Thomson, 1983).

Effect of kilning on malt bacteria

Most of the bacteria which infect spirit fermentations originate from malt. Levels of bacteria are too low in kilned malt to achieve this directly but given the correct conditions these bacteria can become established in the distillery plant and ultimately enter the worts. Bacterial infection of malt is greatly reduced by the presence of sulphur dioxide during kilning. In distilling, the absence of wort boiling and hopping makes it more important to observe plant hygiene. This applies particularly when green malt is used to convert cooked adjuncts. Sometimes bacterial infection during fermentation can reduce spirit yield and introduce unwanted flavour components or their precursors.

Yeast nutrients

Malt supplies soluble amino nitrogen which is essential for yeast growth and rapid fermentation. It is also capable of satisfying other requirements, at least in part, for materials such as unsaturated fatty acids, sterols and vitamins, which enable the yeast to function under anaerobic conditions. Nitrogen is the main consideration, particularly in grain spirit fermentations where a relatively small proportion of malt is required to supply part, if not all, of the amino nitrogen. A concentration of

100–150 mg α-amino nitrogen per litre worts is regarded as adequate. The spectrum of malt amino acids which is available to the yeast and which is modified during kilning will influence the content of volatile components present in fermented wash.

Rootlets

Rootlets produced during the germination process represent 3–5% of the original barley dry matter. These are removed after kilning, thus contributing significantly to overall malting loss. They have a spirit producing potential of over 150 litres of alcohol per tonne but are normally sold for cattle feed on the basis of their high protein content (*c*. 25–30%). They are rich in vitamins and have relatively high levels of basic materials including hordenine (Liu and Pomeranz, 1976). Rootlets are capable of acquiring relatively high levels of nitrosamines, phenols and sulphur dioxide during kilning if conditions are conducive to the formation of these compounds. The sulphur dioxide content of rootlets, for example, can be 100–200 times that of the parent malt, thus emphasising the importance of efficient screening for their removal.

References

AASTRUP S, ERDAL K 1980 Quantitative determination of endosperm modification and its relationship to the content of 1,3 : 1,4-β-glucans during malting of barley. *Carlsberg Research Communication* **45**: 369–79

ANNESS B J, BAMFORTH C W 1982 Dimethyl sulphide – A review. *Journal of the Institute of Brewing* **88**: 244–52

ASBC 1977 Barley 8.0, Potential diastatic power. In *Methods of Analysis of the American Society of Brewing Chemists* 7th edn. ASBC, St. Paul, Minnesota

AXCELL B, JANKOVAKY D, MORREL P 1983 Steeping – the crucial factor in determining malt quality. *Brewers Digest* (August): 20–3

BAKKER-ARKEMA F W, BROOK R C, LEREW L E 1978 Cereal grain drying. In Pomeranz Y (ed) *Advances in Cereal Science and Technology*. AACC, St. Paul, Minnesota, vol II, pp 1–90

BATHGATE G N 1973 The biochemistry of malt kilning. *The Brewers Digest* **48**: 60–5

BATHGATE G N 1983 Current malting technology and its consequences for the brewing and distilling industries. In Priest F G, Campbell I (eds) *Current Developments in Malting, Brewing and Distilling, Aviemore Conference 1982*. Institute of Brewing, London, pp 25–34

Bathgate G N, Martinez-Frias J, Stark J R 1978 Factors controlling the fermentable extract in distillers malt. *Journal of the Institute of Brewing* **84**: 22–9

Bathgate G N, Taylor A G 1977 The qualitative and quantitative measurement of peat smoke on distillers malt. *Journal of the Institute of Brewing* **83**: 163–8

Baxter E D, O'Farrell D D 1983 Use of the friabilimeter to assess homogeneity of malt. *Journal of the Institute of Brewing* **89**: 210–14

Briggs D E 1978 The reception and storage of whole plants and grain. In *Barley*. Chapman and Hall, London, pp 369–418

Briggs D E, Hough J S, Stevens R, Young T W 1981 The technology of malting and kilning (including barley drying). In *Malting and Brewing Science*. Academic Press, London, vol 1. pp 178–80

Brookes P A 1980 The significance of pre-harvest sprouting of barley in malting and brewing. *Cereal Research Communications* **8**: 29–38

Brookes P A, Lovett D A, MacWilliam I C 1976 The steeping of barley. A review of the metabolic consequences of water uptake, and their practical implications. *Journal of the Institute of Brewing* **82**: 14–26

Bourne D T, Wheeler R E 1985 Physiological and biochemical indication of vigour in barley and their relationship to malt modification. *Proceedings of the 20th European Brewing Convention Congress, Helsinki*, pp 627–34

EBC 1975 Barley 2.7, Sieving test. In *Analytica – EBC* 3rd edn. Schweizer Brauerei-Rundschau, Zurich

Ellis R H, Roberts E H 1980 Improved equations for the prediction of seed longevity. *Annals of Botany* **45**: 13–30

Essery R E, Kirsop B H, Pollock J R A 1954 Studies in barley and malt effect of water on germination tests. *Journal of the Institute of Brewing* **60**: 473–81

Flannigan B 1983 Quantification of mould contamination in sulphured and unsulphured malt. *Journal of the Institute of Brewing* **89**: 364–5

Foster G H 1982 Drying cereal grains. In Christensen C M (ed) *Storage of Cereal Grains and their Products*. AACC, St. Paul, Minnesota, pp 79–116

Gardner H W 1975 Decomposition of linoleic acid hydroperoxides, enzymic reactions compared with nonenzymic. *Journal of Agricultural and Food Chemistry* **23**: 129–36

Griffin O T 1982 Clova Maltings – a design for maritime climate. *Technical Quarterly, Master Brewers Association of the Americas* **19**: 41–4

Griffin O T, Pinner B C 1965 The development of a static malting. *Journal of the Institute of Brewing* **71**: 324–9

IoB 1982a Barley 1.1, Sampling. In *Recommended Methods of Analysis*. Institute of Brewing, London

IoB 1982b *Recommended Methods of Analysis*. Institute of Brewing, London

IoB 1982c Pale malt 2.13, Determination of the fermentability of unboiled malt worts. In *Recommended Methods of Analysis*. Institute of Brewing, London

Jensen S V A, Heltved F 1982 Visualization of enzyme activity in germination cereal seeds using a lipase-sensitive fluorochrome. *Carlsberg Research Communication* **47**: 297–303

Khan A A (ed) 1977 *The Physiology and Bio-chemistry of Seed Dormancy and Germination*. Elsevier, Amsterdam

Kidger P A 1984 Proven energy savings in the maltings. *The Brewer* (December): 507–11

Ladish W J 1981 Technical report of the USBA ad-hoc committee on nitrosamines – studies on grains and the malting process. *Technical Quarterly, Master Brewers Association of the Americas* **17**: 96–107

Lee W J, Pyler R E 1984 Barley malt limit dextrinase: varietal environmental and malting effects. *ASBC Journal* **42**: 11–17

Liu D J, Pomeranz Y 1976 Structural and mineral composition of malt sprouts. *Brewers Digest* **51**: 52–7

Lloyd W J W 1983 Some predictive tests in barley and malt analysis. In Priest F G, Campbell I (eds) *Current Developments in Malting, Brewing and Distilling, Aviemore Conference 1982*. Institute of Brewing, London, pp 157–69

MacLeod A M 1979 The physiology of malting. In Pollock J R A (ed) *Brewing Science*. Academic Press, London, vol 1, pp 145–232

MAFF 1982 *List of Approved Products and their Uses for Farmers and Growers*. Reference Book 380, HMSO, London

Marchylo B A, Kruger J E 1984 Identification of Canadian barley cultivars by reversed-phase high performance liquid chromatography. *Cereal Chemistry* **61**: 295–301

Marchylo B A, Laberge D E 1980 Barley cultivar identification by electrophoretic analysis of hordein proteins. *Canadian Journal of Plant Science* **60**: 1343–50

Mathewson P R, Pomeranz Y 1977 Detection of sprouted wheat by a rapid colorimetric determination of α-amylase. *Journal of the Association of Official Analytical Chemists* **60**: 16–20

Moray Sir Robert 1678 An account of the manner of making malt in Scotland. *Philosophical Transactions* **12**(142): 1069

Munck L, Mundy J, Vaag P 1985 Characterization of enzyme inhibitors in barley and their tentative role in malting and brewing. *ASBC Journal* **43**: 35–8

Narziss L, Rusitzka P 1977a Variation of enzymic activities during drying and curing of malt. *Brauwissenschaft* **30**: 1–12

Narziss L, Rusitzka P 1977b Effect of kilning temperature on the behaviour of malt enzymes. *Brauwissenschaft* **30**: 105–11

NIAB 1984 *Farmers Leaflet No. 8 Recommended Varieties of Cereals 1985*. National Institute of Agricultural Botany, Cambridge

NIAB 1985 *Detailed Descriptions of Varieties of Wheat, Barley, Oats, Rye and Triticale*. National Institute of Agricultural Botany, Cambridge

PALMER G H 1980 The morphology and physiology of malting barleys. In Inglett G E, Munck L (eds) *Cereals for Food and Beverages*. Academic Press, New York, pp 301–38

PALMER G H, GERNAH D I, MCKERNAN G, NIMMO D H 1985 Influence of enzyme distribution on endosperm breakdown (modification) during malting. *ASBC Journal* **43**: 17–28

PALMER G H, TAYLOR J A 1980 Dormancy, drying and pre-germination of malting barleys. *The Brewer* **68**: 105–10

PIXTON S W, WARBURTON S 1971 Moisture content/relative humidity equilibrium of some cereal grains at different temperatures. *Journal of Stored Products Research* **6**: 283–93

POLLOCK J R A 1962 The analytical examination of barley and malt. In Cook A H (ed) *Barley and Malt Biology, Biochemistry, Technology*. Academic Press, London, pp 407–9

REINER L, LOCH V 1976 A method for predicting the duration of dormancy in malting barley. *Brauwissenschaft* **29**: 164–8

SAC 1984 *Recommended Varieties of Cereals 1985*. The Scottish Agricultural Colleges, Edinburgh

SFAT M R 1965 Bio-engineering principles of malt kilning. *Brewers Digest* **40**: 50–4

SIM G B 1983 The effect of sulphuring on malt enzyme activities. In Priest F G, Campbell I (eds) *Current Developments in Malting, Brewing and Distilling, Aviemore Conference 1982*. Institute of Brewing, London, pp 265–72

SPIEGELHALDER B, EISENBRAND G, PREUSSMANN R 1979 Contamination of beer with trace quantities of *N*-nitrosodimethylamine. *Food and Cosmetic Toxicology* **17**: 29–30

STEVENS B 1984 Roseisle Maltings. *Brewing and Distilling International* **14**: 34–6

THOMSON F J 1983 The estimation of total phenols on malt as a guide to degree of peating of distilling malts. In Priest F G, Campbell I (eds) *Current Developments in Malting, Brewing and Distilling, Aviemore Conference 1982*. Institute of Brewing, London, pp 273–8

TRESSL R, BAHRI D, HELAK B 1983 Flavours of malt and other cereals. In Piggott J R (ed) *Flavour of Distilled Beverages: Origin and Development*. Ellis Horwood Ltd, Chichester, pp 9–32

TURCAN W 1981 Choice of barley and malting plant for northern Scotland. *The Brewer* (February): 40–2

WAINWRIGHT T 1978 Barley analysis and malt quality. *Technical Quarterly, Master Brewers Association of the Americas* **15**: 184–9

WAINWRIGHT T 1979 The selection of barley varieties for malting and brewing. *The Brewer* **65**: 314–17

WAINWRIGHT T 1986 The chemistry of nitrosamine formation and nitrosamines in malt and beer. *Journal of the Institute of Brewing* **92**: 49–62, 73–80

WESELAKE R J, MACGREGOR W, HILL R D 1985 Effect of endogenous barley α-amylase inhibitor on hydrolysis of starch under various conditions. *Journal of Cereal Science* **3**: 249–59

WILLIAMS R S, GRACEY D E F 1982 Factors influencing the levels of poly-sulphides in beer. *ASBC Journal* **40**: 71–4

YAMADA K 1982 Determination of endogenous gibberellins in germinating barley by combining gas chromatography-mass spectrometry. *ASBC Journal* **40**: 18–25

YAMADA K 1985 Endogenous abscisic acid in barley and use of abscisic acid in malting. *Agricultural and Biological Chemistry* **49**: 429–34

YOSHIDA T, YAMADA K, FUJINO S, KOUMEGAWA J 1979 Effect of pressure on physiological aspects of germination barleys and quality of malts. *ASBC Journal* **37**: 77–84

CHAPTER 3

Milling, cooking and mashing

G D WILKIN

Introduction

Within the field of alcoholic drinks, potable spirits form a significant if diverse group, of which 'whiskies' constitute an important subdivision. For example, in Britain around 50% of the total spirits sector is held by whisky. With time the various groups of potable spirits have become defined, partly by custom and partly by legislation. Without such protection and definition spirits could have become an amorphous group with little distinctive character. Whiskies, by definition, must be made from a cereal base. In Scotland, that cereal base is prepared for fermentation in one of two ways. Either it is malted or it is cooked. If all of the cereal is malted – in which case it is invariably barley – then malt whisky will be produced. If only a proportion is malted and the remainder is cooked – in which case this portion may be any cereal – then grain whisky of one kind or another is the product. Effectively, however, the substrates of interest in this chapter are barley and malted barley, corn (maize) and wheat. A subsidiary difference between malt and grain whisky concerns whether the worts, ready for fermentation, will be free of suspended solids – malt whisky – or will retain solids from the mashing process – grain whisky (Wilkin, 1977).

The preparation of worts for the production of beers and similar drinks is well described in the literature (e.g. Briggs *et al.*, 1981). The production of worts for ultimate conversion into whisky requires similar equipment and, to an extent, similar processing. A common factor associated with beer and malt whisky production is the requirement for clear wort. The brewed product, being consumed directly, necessitates the requirement for clarity and, with malt whisky, the fermented wash which goes through batch pot stills faces a condition where 'burning' occurs when suspended solids are present in excess of the controlling ability of the small mixer within the pot still.

Whilst in the early days of grain spirit mashing it was believed necessary to simulate brewing and malt whisky methodology, as technology improved, the advantages of inclusion of total grain solids in wort led to increasing 'underletting', effectively bypassing the settlement stage in the mashtun. Once this stage was reached the grain mashtun merely became a large and somewhat inefficient mixing vessel. This in turn led to the development of continuous techniques where the mashtun became superfluous.

On the other hand, with the development of more sophisticated methods, the conventional mashtun in brewing and malt whisky production came under attack. These procedures of mash filtration will be dealt with in more detail in later sections of this chapter. The increasing requirements for energy economy are better served by continuous controlled systems than by the traditional batch methods, and these aspects will also be dealt with below.

Malt whisky process

BATCH MASHING

The first process to be considered in the preparation of a mash is that of milling. In malt whisky production, particularly in the Scottish tradition, it is important that the barley has been well modified during malting, as no additives can be used in the mashing procedure to help out a poor quality malt. Under these circumstances the objective of milling is to obtain efficient recovery of the fermentable extract contained in the malt. The manner of milling is dependent to some extent on the equipment used subsequently to separate suspended solids whilst effectively draining out the fermentable sugars. Wilkin (1983) discussed milling and the subsequent separation of solids in the context of extract recovery economics.

In the malt whisky process the great majority of distilleries use batch rather than continuous processes. Thus conventional mashtuns and conventional and not so conventional lauter tuns are used in batch infusion mashing, and the Strainmaster and

Mashfilter in decoction mashing. Wilkin (1983) has discussed the advantages and disadvantages of such deep or shallow draff bed mechanisms as batch mashing systems, and Barnes (1977) has commented that the technology of wort separation would benefit from a greater collaboration between the brewer and the filtration specialist. Many problems would indeed be eased by more careful consideration of the fundamentally different factors associated with deep- and shallow-bed filtrations. The rate of filtration is directly related to the square of the particle size and inversely proportional to the bed depth, whilst leaching efficiency increases with depth but is inversely related to particle size (Dixon, 1977). In operating shallow-bed (mashfilter) and deep-bed (mashtun) systems the rate of extract recovery becomes the practical controlling factor. Poor control of sparging flows, through the finer particles of a mashfilter bed, results in loss of extract recovery from the draff. Given careful control of sparging, high extract recoveries result.

A programme for mashing in a conventional mashtun would be as follows. A batch of malt, typically 5–11 t, is milled and loaded into the mashtun. A first water of 4–4.5 t water per tonne of malt is added to give a mash at 64–65°C. This is drained and a second water at 72–74°C is added at the rate of 1.5–2 t per tonne of malt. This is followed by a third and possibly fourth water at 80–90°C, the combined volume of the two exceeding that of the first water and being adequate to give the intended starting gravity of the fermentation. The total time required for such a mashing cycle, including filling, mixing, draining, and final discharge of the draff (solid residues), is 8 to 11 hours.

Several Scotch whisky distilleries have installed lauter tuns, following the lead of breweries, with the objective of reducing the mashing cycle time. The necessary faster drainage rates are achieved by having a shallower bed and by the design of the knife blades attached to the rotating rakes which cut and loosen the bed. A different approach is adopted in the Strainmaster design of tun, in which the accelerated run-off is achieved through a greatly increased drainage area (Scott, 1967; Whitear and Crabb, 1977).

As an alternative to these processes, Mashfilter systems have been described by McElevey (1974), Dixon (1977) and Bayerlein *et al.* (1980). An extreme variant of these is perhaps the Nordon 'Hypress' (Waesberghe, 1978) for which it is claimed that a bright wort can be obtained in 60–80 min with 2–3% better extract recovery than more conventional filters. It possesses an ability to press spent grains to 45–50% moisture content, a useful economy in by-product processing. Poor control of sparging flows through any mashing system will result in loss of extract from the malt. Mashfilter units operate at high rates and the temptation exists to push them past their optimum sparging conditions, when quite alarming losses of extract will occur.

The degree of milling required for efficient operation of these systems increases from mashtun to filter press. A considerable amount of careful monitoring of performance is required to ensure that the optimum grind is achieved and maintained. Sieve analyses of milled malt for malt whisky and grain whisky are shown in Table 3.1.

Many publications have appeared over the past decade about brewery mashing techniques and may be relevant to many aspects of malt distillery mashing. The control of volatile ester synthesis when wort of high specific gravity is produced was discussed by Anderson and Kirsop (1974), and Crabb and Hudson (1975) discussed the influence of mashing conditions on wort and beer quality. Barrett *et al.* (1975), Whitmore (1976) and Barnes (1977) discussed filtration and separating techniques, and Webster (1981) the prediction of the lautering performance of malt.

CONTINUOUS MASHING

A continuous mashing system in malt whisky production used by William Grant & Sons Ltd in their Ladyburn distillery, in collaboration with Alfa Laval, produced theoretical extract recoveries from malt over a period of one month's operation. Figure 3.1 shows the final arrangement of the mashing system as set up for the final plant trial. Prediction of expected plant recovery was made by yield testing samples of a single malt supply sufficient for the duration of the trial. These tests showed the expected yield from the malt to be 396 to 398 litres of alcohol per tonne.

Malt at 5% moisture was hammer-milled on a Christy Norris mill with a 2.5 mm round-hole screen at 1089 kg (2400 lb) per hour. The sieve analysis specification was:

British Mesh	14	16	30	Tray
% on sieve	1.0	4.0	20.0	75

Table 3.1 Typical sieve analyses for roller-milled and hammer-milled cereals.

Method and material	*Retained on sieve (%) British mesh* 8	12	16	30	60	100	*Tray*
4-roller mill, setting 1 malted barley, 5% moisture	n.s.	22	14	32	17	8	7
4-roller mill, setting 2 malted barley, 5% moisture	n.s.	9	4	33	29	7	18
6-roller mill malted barley, 5% moisture	n.s.	13	7	26	25	13	16
hammer mill, 2.5 mm screen malted barley, 7% moisture	n.s.	2	4	21	37	17	20
hammer mill, 3.5 mm screen 8 t per hour maize grits, 15% moisture	n.s.	5	8	31	44	10	2
hammer mill, 2.5 mm screen 6.6 t per hour maize grits, 15% moisture	0.2	0.7	4.4	30.7	n.s.	n.s.	64
hammer mill, 2.0 mm screen 6.5 t per hour maize grits, 15% moisture	0.4	0.5	2.1	31.0	n.s.	n.s.	66
hammer mill, 2.5 mm screen 7 t per hour whole maize, 15% moisture		5.9	17.5	32.4	n.s.	n.s.	44.2

n.s. = not specified

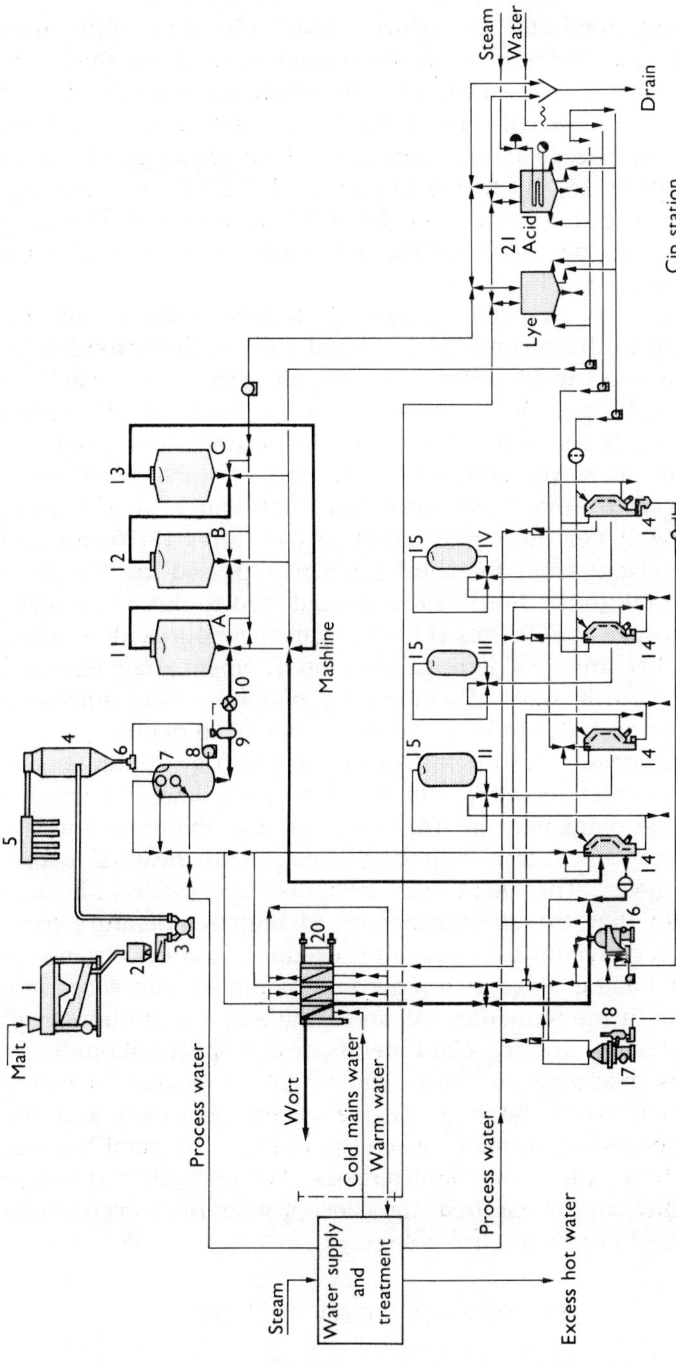

Fig. 3.1 Alfa-Laval Centribrew plant for Scotch whisky.

The liquid feed to the slurry tank (7) was 5819 litres (1280 imp. gal) per hour from the second sieve. From the slurry tank the mix was passed through three holding tanks (11, 12, 13) in series and down the mashline to the first sieve (14). Wort from the first sieve passed through a heat exchanger (2) after centrifugation (16, 17) to the fermenter. Liquid feed to the first sieve was returned wort from the third sieve (14) at 5046 litres (1110 imp. gal) per hour with a malt slurry feed of 5910 litres (1300 imp. gal) per hour.

Draff from the first sieve passed to the first re-slurry tank and from there to the second sieve. Liquid feed to the second sieve was returned wort from the fourth sieve at 1546 litres (340 imp. gal) per hour with a slurry feed of 6546 litres (1440 imp. gal) per hour. Draff from the second sieve passed to the second re-slurry tank and from there to the third sieve. Liquid feed to the third sieve was hot water at 1046 litres (230 imp. gal) per hour plus slurry at 6455 litres (1420 imp. gal) per hour. Draff from the third sieve was passed via the third re-slurry tank to the fourth sieve. Liquid feed to the fourth sieve was hot water at 500 litres (110 imp. gal) per hour with a slurry feed of 2409 litres (530 imp. gal) per hour. Spent draff emerged from the fourth sieve. Backwashing of sieves was automatic every 7 min in order to keep the sieves free from debris.

The equipment is of interest in that it was fully continuous, it gave full extract recovery into the fermenter, and it formed a total recycle plant with internal sparging and therefore with no accumulation of 'sparge' requiring storage with heating to preserve hygiene. The plant was designed specifically for malt whisky, in that the short exposure to high temperature compared with conventional mashtun procedures and the absence of any wort boiling stage ensured continuation of any secondary conversion in the fermenter. As an example of a continuous malt wort production unit the plant met its design specification.

The disadvantages of such equipment in a malt mashing environment were the high degree of sophistication and the expertise necessary to maintain and to operate the plant because of its entirely different characteristics. While further development of the plant is required, the concept warrants consideration in this era of improving technology.

SECONDARY CONVERSION

It is characteristic of whisky production that a degree of conversion of carbohydrate to fermentable sugars occurs after mashing. The importance of this secondary conversion was emphasised by Ramsay and Berry (1983); when producing small volumes of fresh wort for laboratory experiments they pointed out that boiled or sterilised wort cannot be used for this purpose. Secondary conversion in the fermenter is inhibited by destruction of residual wort enzymes and particularly α-amylase at low pH (Gallagher *et al.*, 1942). Similar phenomena were discussed, along with the beneficial effects of the use of maize and malt infusions to increase buffering in poorly buffered media. The addition of one unit per millilitre of an unspecified penicillin to wort produced a pH suitable for secondary conversion (Tsuji and Taketsuru, 1980).

Grain whisky process

Producing spirit from a mash of mainly unmalted cereal requires a substantially different worts preparation procedure from that described above. The starch in the unmalted cereal has to be released from its matrix so that the enzymes, introduced by the addition of malt, can function effectively. This is traditionally achieved by a high-temperature cooking process. In the Scottish industry distillers have adopted different viewpoints on whether milling of the grain is necessary prior to cooking. Milling facilitates the uptake of water and heat and is favoured by some producers. However, those distillers who do not mill the grain claim that, at the high pressures and temperatures used, they achieve equally effective cooking. The grain is disrupted to a sufficiently finely divided state during discharge from the cooker to give efficient extraction during mashing.

BATCH COOKING

The normal form of cereal cooking in a Scotch whisky distillery is shown in Fig. 3.2. The sequence of the operation is simple and fairly lengthy, the reason for the time taken being due mainly to residence in the cooking vessels and in the mashtun itself, and at no point is the operation other than batch.

Process

In the following description the cereal is assumed to be maize. It

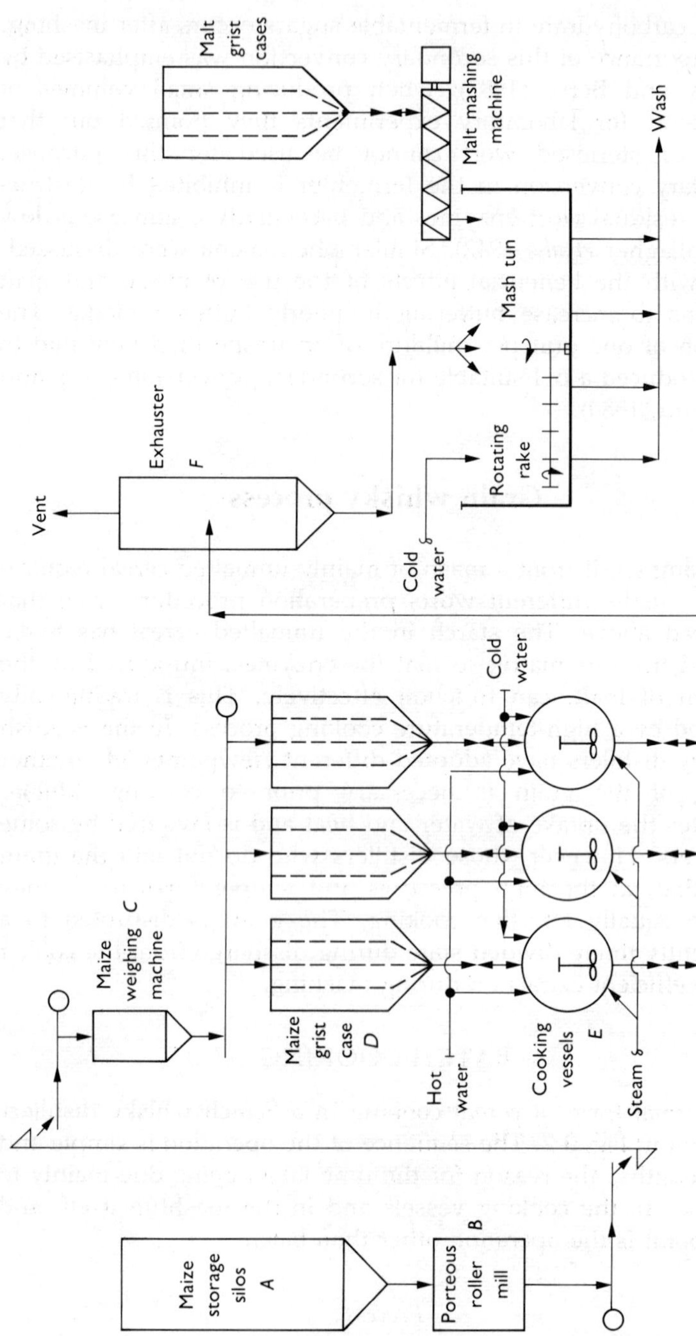

Fig. 3.2 Cereal batch cooking equipment for Scotch whisky. (*Courtesy of William Grant & Sons Ltd.*)

could equally well be wheat or barley. Whole maize is stored in silos (*A*) after initial cleaning to remove cobs and dust. From the storage silo the whole maize corns are dropped into standard Porteous roller mills where the grain is crushed (*B*). The crushed grain is conveyed via a tip weighing machine (*C*) to the intermediate maize grist cases (*D*). The weighed grist is then dropped into the batch pressure cookers (*E*), which are each equipped with mixers to prevent sticking and burning of the grain. Water is added, about 2.5 t per tonne of grain, and steam is injected to bring the pressure in the batch cookers to 207 kPa (30 psig), this pressure being held for about two hours. The cooked maize is then blown by pressure via the steam exhauster (*F*) to the mashtun for mashing. In the mashtun, with the temperature adjusted to 62.5°C, a slurry of ground malt is added. Here again different distillers adopt different practices. Most use dried malt but some have developed procedures for using green malt. This saves the cost of kilning the malt, which is substantial, but requires careful scheduling of malting and mashing programmes to ensure that malt is available at its optimum quality. In either case the distiller wants to use the minimum amount of malt, because of its cost, and constantly monitors the conversion efficiencies achieved. On a dry weight basis some 15% of malt is commonly used.

CONTINUOUS COOKING

A number of years ago William Grant & Sons Ltd took the decision to introduce continuous cooking and mashing into their Girvan grain distillery. This system replaced the cooker system shown in Fig. 3.2.

Process

In Fig. 3.3 are shown the systems for feeding hammer-milled malt to the continuous system. System A is the malt feed and system B is the maize feed as used in the operational plant. It will be noted that some part of the malt slurry from the malt slurry tank is recycled to the maize slurry tanks as pre-malt. The remaining malt slurry meets the cooked, cooled and pre-liquefied maize prior to the conversion tubes.

Figure 3.4 shows the present system of continuous cooking, with energy recovery, pre-cooker tube malting and liquefaction

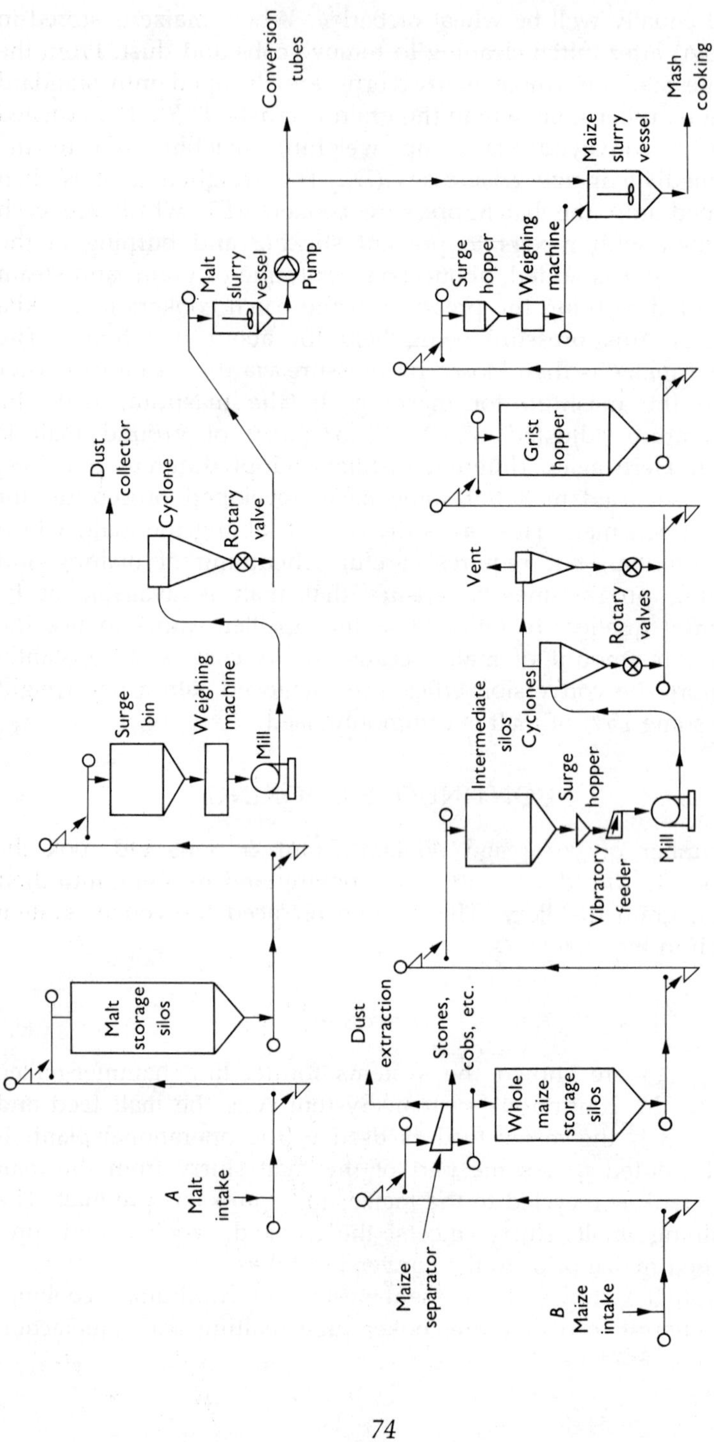

Fig. 3.3 Cereal processing for continuous cooking. (*Courtesy of William Grant & Sons Ltd.*)

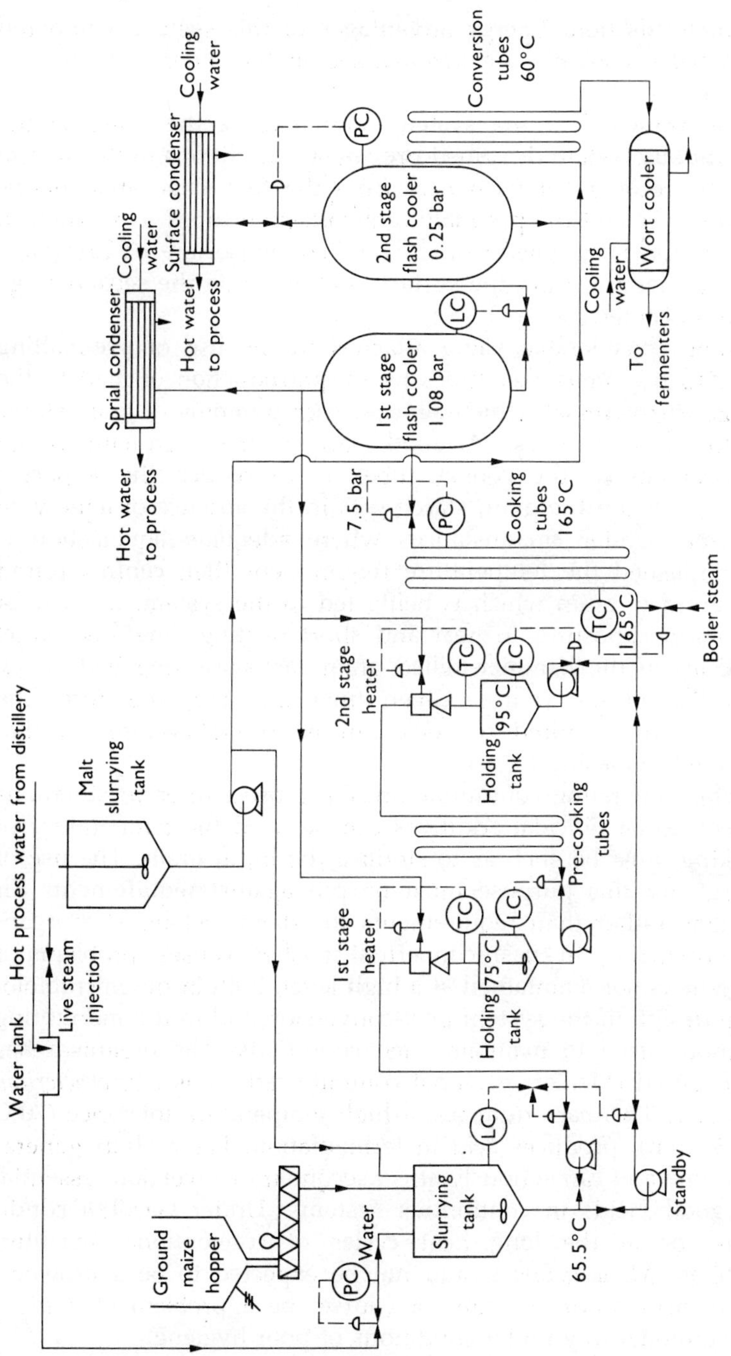

Fig. 3.4 Continuous cooking plant with heat recovery. (*Courtesy of William Grant & Sons Ltd.*)

by malt addition. Energy advantages of this system, compared with batch cooking, are summarised in the 'Energy' section of this chapter.

The principle of the system now used is the same as the continuous system described previously. Additions to the system are the recovery of flash heat from the first stage atmospheric flash cooler and the pre-malting liquefaction tubes. Hot water to the process is supplied from a spiral condenser on the first stage flash cooler and from the surface condenser on the second stage vacuum cooler.

Some considerable value attaches to the use of pre-malting prior to the pre-cooker tubes by the introduction of malt to the maize slurry vessel, which allows easier pumping of grain slurry to the cooker tubes. A good part of the high-temperature requirement in the cooker tubes is to reduce the apparent viscosity when the grain starch gels in the absence of liquefying enzyme. Under circumstances where adequate liquefaction is accomplished the temperature requirement then centres round the grind of grain which is being fed to the system. Because of the short exposures to heat and short wetting time any 'large' particles in the hammer-milled grain can show ungelled starch miscellae in the centre. Given finer grinding, therefore, the temperature required to cook can be brought closer to that required to mash.

When operating continuous mashing procedures in situations where no bactericidal additives can be used the temperature of cooking must be such as to sterilise the input grain. The use of energy for this purpose must be put against requirements for hygiene rather than requirements for starch gelling. An organism occurring in flash-converted mashes causes problems if hygiene is not maintained at a high level, both in the microbiological design of the system post-conversion and in the monitoring methods used to maintain such conditions. The organism has been typed (Mercer, personal communication) as *Microbacterium lacticium*. This bacterium has a high temperature tolerance (70°C for 30 min), produces acid in fermentation, has a short generation time and can seriously affect secondary conversion, essential for good yields in continuous systems. Under Gasohol conditions, or in the long heat cycles of conventional mashtun systems, *M. lacticium* would not be expected to be a problem. Other acid producers can, of course, be a problem in a more conventional way under conditions of poor hygiene.

Energy consumption

Much has been written in the field of energy saving and many arguments lost and won over the question of energy conserving equipment and over the routes to be followed to maximise the potential and actual savings contemplated or attained. Payne (1978) covered the field of energy conservation in a readable manner and specifically heat distribution and utilisation. This section will be confined to energy comparisons of batch and continuous cooking equipment installed at the Girvan distillery of William Grant & Sons Ltd.

MAIZE COOKING

Batch process

The data refer to Fig. 3.2.

(a) Maize feed = 24.38 t (24 tons)
(b) Specific heat of maize
= 1.465 kJ kg^{-1} $°C^{-1}$(0.35 BTU lb^{-1} $°F^{-1}$)
(c) Specific heat of water
= 4.187 kJ kg^{-1} $°C^{-1}$(1.00 BTU lb^{-1} $°F^{-1}$)
(d) Feed water temperature = 15.6°C (60°F)
(e) Cooking temperature = 134.4°C (274°F) at 207 kPa (30 psig)
(f) Water feed = 61 236 kg (135 000 lb)
(g) Specific enthalpy of evaporation of steam at 134.4°C (274°F)
= 2161.3 kJ kg^{-1} (929 BTU lb^{-1})

Heat required by cooker therefore
$= [(24\,380 \times 1.465) + (61\,236 \times 4.187)] \times (134.4 - 15.6)$ kJ
$= (35\,716.7 + 256\,395.132) \times 118.8$ kJ
$= 3.47 \times 10^7$ kJ
Steam required at 207 kPa (30 psig) = $(3.47 \times 10^7)/2161.3$
= 16 055 kg, neglecting radiation losses.
Allowing for radiation loss steam required = 16 330 kg steam.

Continuous process

The data refer to Fig. 3.4, cooking 24.38 t (24 tons) maize through continuous cooking plant at a design feed rate of 17.07 t (16.8 tons) per hour.

(a) *With no heat recovery*
Steam required to cook at 160°C (320°F) at a feed rate of 17.07 t (16.8 tons) per hour = 7257.6 kg per hour.
Therefore to cook 24.38 t (24 tons), steam required
$= (24.38 \times 7257.6)/17.07 = 10\,366$ kg steam

(b) *With heat recovery*
Steam required to cook at 140°C at a feed rate of 17.07 t (16.8 tons) per hour = 5670 kg per hour
Therefore to cook 24.38 t (24 tons), steam required
$= (24.38 \times 5670)/17.07 = 8100$ kg steam
Thus the steam usage at current process conditions is reduced to approximately 50% of that used in batch cooking.

RESTRICTIONS ON ENERGY CONSERVATION

In the production of potable spirits the main criterion is – or should be – quality. It is therefore fundamental to energy conservation in whisky production that no stage should be able to pass foul liquors or vapours to a product still, or to any other leading into it. The result of this would be the inclusion of 'off' flavours or aromas at a detectable level in the product. The stabilisation of product types by custom and legislation also restricts severely the degree of technical advance possible in established products. Many energy-saving routes, practicable for example when making Gasohol, are not possible in established products. Thus a totally integrated plant can be designed, for example the Alfa Laval Biostill system, utilising heat sources foreign to potable spirit systems, such as the diversion of vapours from beer columns to product columns. Such advances can be considered in the production of potable alcohol only after proof that there is no risk to product quality. If the only relevance is the recovery of alcohol, possible routes for energy conservation are dealt with in detail by Lyons (1981) and the Solar Energy Research Institute (1981).

Use of wheat

The use of barley and wheat adjuncts as sources of starch during mashing has been examined by Palmer (1985). It was found that larger quantities of wheat flour could be used if pentosans, large numbers of small starch granules and proteins were first removed. It was also suggested that the economics of such a process are controlled by the profitable disposal of the extracted gluten.

The effects of glucans, pentosans and gums on mash viscosity and processing, the effects of high levels of unmalted grain, the limited efficiency of commercial preparations for the reduction of viscosity of cereal mashes, together with the possible use of low glucan wheats and the application of brewery methods and research to distillery practice have been studied (Macher, 1974). Forrest *et al.* (1985) discussed the problems encountered in using wheat in mashing, and showed that brewhouse performance could be improved by increasing the mean particle size of the wheat products by hammer milling or roll milling whole wheat. This demonstrated the value of preventing components such as cell wall pentosans from affecting filtration. They also showed that use of a combination of commercially available pentosanases and cellulases improved brewhouse performance, presumably by solubilising those components which escaped from the increased size particles.

A comparison of the commercial milling of English wheat with that possible in laboratory mills showed the difficulty of separation of the various fractions (Butcher and Maris, 1975). It is difficult to simulate commercially milled fractions with those produced by laboratory milling.

When whole mash processes using 100% wheat with no other cereal source are used, the problem of pentosans becomes very much more difficult (Forrest *et al.*, 1985). This is particularly true in Scotch whisky production where these components interfere wherever a filtration stage occurs during the process.

In a review of malt quality and wort filtration the values of single-pass filtration and the influence of non-yeast particles, such as β-glucans and various proteins, were discussed. The point was made that viscosity does not have an obvious relationship to filterability. The influence of protein binding on the accessibility of the smaller starch granules to amylolysis, and its consequent effects on filterability, was also discussed (Anon., 1985).

A simple viscometric technique for estimation of the soluble β-glucan content of raw barley has been reported (Morgan and Gothard, 1977). A useful device for the determination of filterability is provided by the 'capillary suction time' method, also used to determine filterability of sewage sludges. This method has been used to determine filterability of floated 'creams' from maize and wheat processing. It can serve as a rapid means to determine the likely results of any proposed process change on filterability. It has been found that this method showed differences in filtrate behaviour when viscosity determinations showed little or no differences. Some engineering aspects of filtration have been reported by Knowles (1974).

The development of continuous processing

Herman *et al.* (1942) described work carried out 'to bring a grain distillery up to date from an engineering standpoint'. The key finding with respect to the cooking of maize was that suspensions of ground grain in water could be cooked in a continuous process in 30–60 sec at 185°C. Gallagher *et al.* (1942) studied percentage conversion when 90% corn and 10% malt mixtures were held at 62.8°C for different periods of time. Their experiments showed that very good fermentation efficiencies and yields were obtained with conversion times as short as 60 sec. These yields were equal or superior to those obtained when the conversion time was increased to one hour. The equipment used in these experiments was described in a separate publication (Anon., 1941).

Kolachov (1943) carried out a programme to develop a continuous process for the efficient production of a readily fermentable mash. The major factors, in his opinion, were cooking temperature, cooking time and fineness of grind. This emphasis on the importance of fine grinding in continuous cooking operations recurs from these early experiences right through to the most recent reports on cold cook procedures.

Unger *et al.* (1944) summarised the main objections to batch cooking. These included large size of plant, long processing times, destruction of enzymes, manual operations, multiple purpose of single items of plant, and intermittent demand for steam. They designed a laboratory pressure cooker capable of rapidly cooking a batch of grain at temperatures up to 190°C. Maize could be cooked most efficiently at 153–154°C in 10–15 min, while at 176–177°C the time required was less than 5 min. Figure 3.5 shows the layout of the pilot plant which was developed. These pilot studies gave yields of up to 413 litres alcohol per tonne (5.54 US proof gallons per 56 lb bushel), with

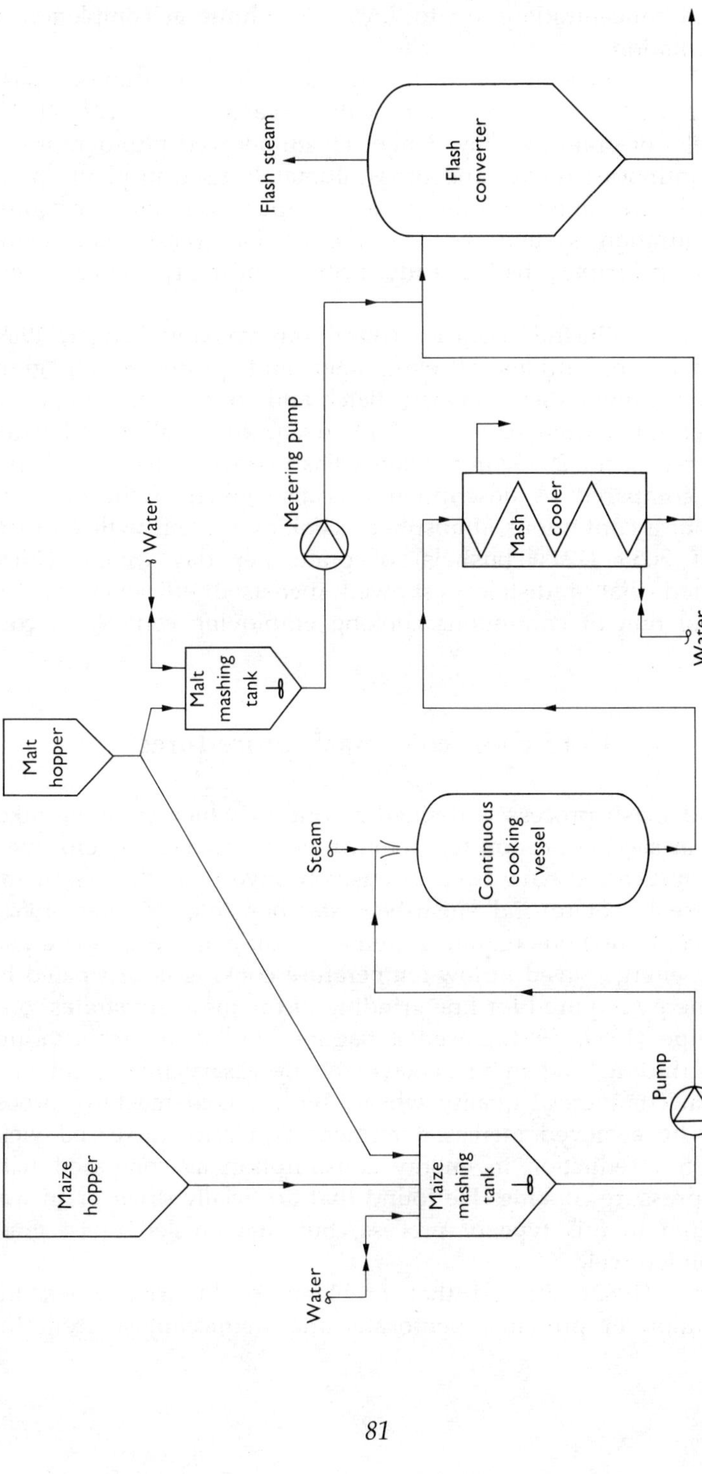

Fig. 3.5 Pilot plant for continuous cooking. (*Courtesy of William Grant & Sons Ltd.*)

alcohol concentrations up to 7.6% by volume at completion of fermentation.

The authors went on to design and build a continuous flash cooker and conversion unit with a capacity of 127 t (5000 bushels) of grain per day. Unger (1948) showed photographs of the continuous cooker and tubes ultimately used in plant operations. Interestingly, the design details for a continuous mashing and filtration system for a sterile aerobic yeast plant using similar principles had already been published (Unger *et al.*, 1942).

Further published data on continuous cooking (Unger, 1969) included descriptions of equipment and processes for grain handling, milling and cooking. Batch and continuous cooking, at temperatures from 62.8°C (145°F) to 182.2°C (360°F) and batch conversions for 20–30 min against flash conversions for 2–5 min were compared. A description was also given of the development of a continuous atmospheric cooking system with a capacity of 305 t (12 000 bushels) of grain per day. Smith (1969) reported that a distillery showed increased efficiency by the introduction of continuous cooking employing centralised control.

Cold cook/cold mash procedures

A cold mash process is defined as one in which mashing takes place at saccharification temperature with endogenous enzymes. These processes have been extensively investigated in Germany for cereal, potato and sugar-beet starches (e.g. Macher, 1982). Cold cook methods require extreme grinding and thus some part of the energy saved in low-temperature cooking is dissipated by the energy required for fine grinding of the grain substrates.

Kreipe (1980, 1981a) used a hammer mill and carborundum disc grinding system to achieve the necessary fine grind of a normal commercial quality wheat. With a cold mashing procedure, he achieved increased fermentation rate, increased yield and 75% reduction in energy consumption as compared with high-pressure cooking. He found that artificially dried grain was unsuited to this type of process, but that ensiled moist grain responded well.

Gams (1982) showed that problems could arise when the separation of proteins, pentosans and hemicellulose from the starch granules was poor. Under these circumstances saccharification of the starch granules was incomplete. The advantages under these circumstances of using an enzyme preparation with protease and hemicellulase, as well as amylase, activity were demonstrated using a preparation known as 'Oloclast'. This material could give good starch solubilisation at a temperature of 45–50°C.

The use of kilned wheat malt as a saccharifying agent for grain distilleries has been studied, and the greatest saccharifying effect was at 55°C and pH 5.5. Highest alcohol yields were obtained with the wheat malt at 10% at a saccharification time of 30 or 60 min. Yields of 380 litres of alcohol per tonne of wheat were claimed, together with a comment that such wheat could be used effectively in a simple cold mashing procedure (Pieper, 1984).

The importance of particle size in cold mashing distillery procedures is critical (Wolf *et al.*, 1981). It was shown that grinding to flour particle size is preferred but that particles up to 1.12 mm diameter give acceptable results at normal saccharification temperatures.

A cold mashing procedure for cereals containing low endogenous enzyme activity by the substitution of triticale for 5% to 20% of the low-enzyme cereal has been examined (Kirste *et al.*, 1984). It was claimed that this procedure was applicable to the manufacture of cereal-derived spirits.

No-cook procedures

No-cook procedures can be defined as those involving a premashing stage in which all the grain substrate to be utilised has been subjected to some form of enzymic modification, resulting in the endogenous activation of enzyme systems capable of producing yeast-fermentable substrates. With advances of technology, particular in enzymology, cold cook and no-cook procedures have become more similar. In terms of Scotch whisky the borderline is formed by the exclusive use of malted grain as the sole converting enzyme source. In other types of whisky no such borderline exists, as a result of the permitted use of natural enzymes at the mashing stage. All such technology is, of course, directly relevant to the production of neutral spirit.

Rutherford (1981) compared practical aspects of malt whisky

(no-cook) mashing with similar processes in brewing, and included discussions of mashing liquor, equipment and procedures. Wilkin (1983) emphasised the distinctions between brewing and Scotch whisky mashing process requirements. For example, the normal brewer's requirement to leave residual carbohydrates in beer and the presence of the wort boiling stage change the approach to mashing, particularly in relationship to mashing temperatures and considerations of hygiene. From the latter aspect, where wort is not boiled in order to permit secondary conversion in the fermenter, extract losses may occur through bacterial proliferation. However, because of the increasing sales of 'Pils' beers with low carbohydrate, the whisky type of process has become much more relevant because of the minimal occurrence of residual carbohydrates.

When use of natural enzymes is allowed considerable energy cost savings are possible. A process development has been reported (Kreipe, 1982), in which the grist is hammer-milled as a slurry, α-amylase is added and the mash pre-heated using as heat source the hot mash from the saccharification unit. Steam is used to heat the saccharification tank, after which amyloglucosidase is added.

Further examples of the use of natural enzyme technology were described by Kreipe (1977), and Pieper (1970) discussed the use of microbial enzymes for processing wheat. Kreipe (1981b) described in detail the distillation of cereal and potato mashes. Such data, whilst not directly relevant to the processing of wheat to Scotch whisky, nevertheless add measurably to our knowledge of those factors which make the use of wheat in some respects more difficult than the use of maize, and thus allow improvements in processing while remaining within the current definition of Scotch whisky.

A process based on fine comminution in a carborundum disc colloid mill, gelatinisation by steaming followed by enzymic liquefaction and saccharification, resulted in an energy saving of 70% to 75% compared with a high-pressure process (Klisch and Drawert, 1982). Two quoted disadvantages were heavy wear on the carborundum discs, and the requirement for relatively high-capacity electricity supplies for driving the colloid mills.

A further publication compares calculated energy consumption for starch saccharification in distilleries using conventional high-pressure steam processing and those using cold mash proecedures (Anon., 1981). It was also stated that the saving in energy by use of the cold mash methods is independent of the relative fuel costs and of electricity; however, the increased alcohol yield achieved by the cold mash procedure may be of greater importance than the actual energy savings.

Another area of cost savings lies in the use of specific organisms to convert otherwise unfermentable substrates to fermentable material, in those countries where this is legislatively possible. An example of such saving is described in a patent (Hirte, 1981) where use is made of a cellulolytic and amylolytic complex in ethanol production. The process described allows conversion of non-fermentable compounds such as cellulose, hemicellulose, pectins and other polysaccharides with glucosidic bonds to fermentable sugars by enzymic hydrolysis. The relevant enzyme complex is produced by an organism quoted as *Penicillium citroviridum*.

Acknowledgement

For the data concerning energy consumption I am indebted to the Chief Engineer, Mr W B Stowers, and his draughtsman Mr A Hunter of William Grant & Sons Ltd's Girvan Distillery.

References

ANDERSON R G, KIRSOP B H 1974 The control of volatile ester synthesis during the fermentation of wort of high specific gravity. *Journal of the Institute of Brewing* **80**: 48–55

ANON. 1941 Continuous cooking of cereal grains. *Chemical and Metallurgical Engineering* **48**(10): 142–5

ANON. 1981 Energiebilanz beim Kaltmaischverfahren. *Branntweinwirtschaft* **121**: 275–80

ANON. 1985 Poor malt and low process times clog up filtration. *Brewing and Distilling International* **15**(7): 18–19, 25

BARNES P J 1977 Brewery wort separation. *Filtration & Separation* **14**: 321–4

BARRETT J, BATHGATE G N, CLAPPERTON J F 1975 The composition of fine particles which affect mash filtration. *Journal of the Institute of Brewing* **81**: 31–6

BAYERLEIN W U, HERTRICH J D, LLOYD M 1980 Modern mash filter practice. *Technical Quarterly, Master Brewers Association of the Americas* **17**: 130–48

BRIGGS D E, HOUGH J S, STEVENS R, YOUNG T W 1981 *Malting and Brewing Science*. Chapman Hall, London, vol 1 Malt and Sweet Wort

BUTCHER J, MARIS P I 1975 A comparison of commercial milling of English wheat with laboratory test milling. *Journal of Flour and Animal Feed Milling* **157**(9): 32–4

CRABB D, HUDSON J R 1975 The influence of mashing conditions on wort and beer quality. *Journal of the Institute of Brewing* **81**: 96–104

DIXON I 1977 Recent trends in mash filtration. *Brewers' Guardian* **106**(10): 43–7

FORREST I S, DICKSON J E, SEATON J C 1985 Novel techniques for improvement of wheat flour performance in the brewhouse. *Proceedings of the 20th European Brewing Convention Congress, Helsinki*, pp 363–70

GALLAGHER F H, BILFORD H R, STARK W H, KOLACHOV P J 1942 Fast conversion of distillery mash for use in continuous process. *Industrial and Engineering Chemistry* **34**: 1395–7

GAMS T C 1982 Ein Verzuckerungsenzym mit niedrigem Temperaturoptimum und Begleitenzymen im drucklosen Stärkeaufschluss. *Branntweinwirtschaft* **122**: 238–9

HERMAN A, STALLINGS E M, WILLKIE H J 1942 Chemical engineering developments in grain distillery. *Transactions of the American Institute of Chemical Engineers* **38**: 791–812

HIRTE W 1981 *Anwendung eines cellulolytischen und hemicellulolytischen enzymkomplexes zur aethanolgewinning*. GDR Patent DD 151472

KIRSTE K H, LAATSCH H U, ZIMARE U, FLAMME W 1984 *Breakdown of starch-containing raw materials in cereal distilleries*. GDR Patent DD 205928

KLISCH W, DRAWERT F 1982 Alcohol manufacture from a current viewpoint. Energy-economic process for mash preparation. *Branntweinwirtschaft* **122**: 102–5

KNOWLES P A 1974 Engineer's guide to filtration. In *Proceedings of Europort '73 Conference (Amsterdam). Transactions of the Institute of Marine Engineers* Series B: 7–15

KOLACHOV P 1943 Recent advances in the distillery industry. *Record of Chemical Progress* **4**: 17–25

KREIPE H 1977 Ten years use of microbial enzymes in distilleries. Retrospect of an important development. *Branntweinwirtschaft* **117**: 330–2

KREIPE H 1980 Is the cold mash process once again of interest? *Branntweinwirtschaft* **120**: 354–6

KREIPE H 1981a Practical experience with wet milling and the modified cold mashing process in grain distilleries. *Branntweinwirtschaft* **121**: 182–4

KREIPE H 1981b *Cereal and potato distilleries*. Verlag Eugen Ulmer, Stuttgart

KREIPE H 1982 Heat recovery in starch hydrolysis by a non-pressurized process. *Branntweinwirtschaft* **122**: 166–7

LYONS T P 1981 Raw materials. In Lyons T P (ed) *Gasohol, a step to energy independence*. Alltech Technical Publications, Lexington, Kentucky

MCELEVEY C A 1974 A modern mash filter. *Technical Quarterly, Master Brewers Association of the Americas* **11**(4): 233–9

MACHER L 1974 The glucan problem in cereal distilleries. *Branntweinwirtschaft* **114**: 49–55

MACHER L 1982 Technology of cold mash methods. *Branntweinwirtschaft* **122**: 120–1

MORGAN A H, GOTHARD P G 1977 A rapid, simple viscometric technique for indirect estimation of soluble β-glucan content of raw barley. *Journal of the Institute of Brewing* **83**: 37–8

PALMER G 1985 Making more use of wheat in spirits production. *Brewing and Distilling International* **15**(1): 26–8

PAYNE G A 1978 *The energy managers' handbook*. IPC Science and Technology Press, Southampton

PIEPER H J 1970 *Microbial amylases in the production of alcohol*. Verlag Eugen Ulmer, Stuttgart

PIEPER H J 1984 Kilned wheat malt: an interesting saccharifying agent for cereal distilleries. *Branntweinwirtschaft* **124**: 256–61

RAMSAY C M, BERRY D R 1983 Development of a small scale mashing and fermentation system for studies on malt whisky production. *European Journal of Applied Microbiology and Biotechnology* **18**: 207–13

RUTHERFORD A G 1981 Mashing for malt whisky production. *Brewer* **67**(797): 82–6

SCOTT P M 1967 The valley-bottom lauter-tun. *Brewers' Guild Journal* **53**: 339–54

SMITH R 1969 Continuous cooking brings increased efficiency. *Food in Canada* **29**(2): 23–7

SOLAR ENERGY RESEARCH INSTITUTE 1981 *A guide to commercial-scale ethanol production and financing*. Solar Energy Research Institute, Golden, Colorado

TSUJI K, TAKETSURU T 1980 Studies on grain whisky. Part 3 Buffering action of maize and malt infusions. *Journal of the Society of Brewing, Japan* **75**(1): 81–3

UNGER E D 1948 Continuous mashing and flash conversion in the distillery. *American Brewer* **81**(9): 21–3, 56, 58

UNGER E D 1969 Raw materials and mashing operations in the grain distilling industry. Abstracts of Papers. *American Chemical Society* **158**: MICR 24

UNGER E D, STARK W H, SCALF R E, KOLACHOV P J 1942 Continuous aerobic process for distiller's yeast. *Industrial and Engineering Chemistry* **34**: 1402–5

UNGER E D, WILLKIE H F, BLANKMEYER H C 1944 The development and design of a continuous cooking and mashing system for cereal grains. *Transactions of the American Institute of Chemical Engineers* **40**: 421–43

WAESBERGHE I 1978 Hypress shows promise as a mash separator. *Brewing and Distilling International* **8**(2): 41

WEBSTER R D J 1981 Prediction of the lautering performance of malt. *Journal of the Institute of Brewing* **87**: 52–6

WHITEAR A L, CRABB D 1977 High gravity brewing – concepts and economics. *Brewer* **63**(748): 60–3

WHITMORE D J 1976 Advances in filtration and separation techniques in the brewing industry. *Process Biochemistry* **11**(7): 14–15, 20

WILKIN G D 1977 Products of Scotch whisky synthesis. *Process Biochemistry* **12**(4): 22–4, 35

WILKIN G D 1983 Raw materials – milling, mashing and extract recovery. In Priest F G, Campbell I (eds) *Current Developments in Malting, Brewing and Distilling, Aviemore Conference 1982*. Institute of Brewing, London

WOLF H, MANGER H, HOFFMAN H 1981 Degree of comminution of raw materials in cold mashing. *Lebensmittelindustrie* **28**: 556–8

CHAPTER 4

Fermentation

— M KORHOLA, K HARJU and M LEHTONEN —

Introduction

Fermentation in whisky manufacture is basically similar to any other alcohol fermentation where sugars hydrolysed from grain starch are metabolised by yeast into ethanol, carbon dioxide, by-products and cell material. The specialities of various whisky fermentations arise from the use of specific raw materials, yeast strains, fermentation conditions and local bacterial contaminations. Another important feature is that the wort (wash) is not boiled as in brewing; therefore even though the hydrolysis of starch takes place to a rather high degree at the mashing stage, the malt enzymes are not destroyed but continue their work during subsequent fermentation (Pyke, 1965). Similar fermentations (simultaneous saccharification and fermentation – SSF), but with added microbial enzymes are used in the production of other grain spirits, e.g. vodka.

Whisky fermentation can be divided into three phases: (1) active yeast growth and fermentation (acceleration) phase; (2) linear fermentation phase; and (3) a stationary or even decline phase (Watson, 1981; Ramsay and Berry, 1983a). Fermentation is an important source of flavour substances; for whiskies these include higher alcohols, fatty acids and esters (Suomalainen and Nykänen, 1966; Lehtonen and Suomalainen, 1979; Nykänen and Suomalainen, 1983).

Most of the available information on whisky fermentation concerns the Scotch whisky industry, thus this review is heavily weighted in that direction. The Scotch whisky industry, in particular, has been very conservative, which has been required and strengthened by local legislation (Berry and Ramsay, 1983). However, the efficiency of the process, both in terms of the yield of spirit obtained and the speed at which fermentation is completed, has increased over the years. Inevitably, one of the most significant improvements that has happened is the current use of known yeast cultures, be it with brewer's yeast, a mixture of brewer's and distiller's yeast or a specific strain of distiller's yeast alone.

Some reviews of whisky manufacture also contain useful information on the general performance of whisky fermentation. These reviews include those of Scotch grain whisky manufacture by Pyke (1965), Scotch malt whisky manufacture by Simpson (1968) and by Berry and Ramsay (1983). On a laboratory scale the fermentation for malt whisky has been described by Ramsay and Berry (1983a). American and Canadian whisky manufacture has been discussed by Brandt (1982). The properties and roles of distiller's yeast in whisky fermentation has also recently been reviewed by Watson (1984). The harmful role of bacteria in whisky fermentation is outlined by Dolan (1976, 1979).

Yeasts and their properties

Distiller's yeasts are tremendously efficient, small biochemical factories capable of operating at temperatures up to 46°C, in the pH range 3–10, and in the presence of 0–15% alcohol by volume or a sugar content of 0.1–25% (de Becze, 1964). One gram (dry weight) of distiller's yeast consists of $3.2–4.6 \times 10^{10}$ cells, thus one cell weighs about 2.5×10^{-11} g or 25 pg (picograms) (calculated from de Becze, 1964).

The main requirements for yeasts used in whisky fermentation include efficient substrate utilisation, fast rate of fermentation, high ethanol yield and tolerance, good flavour production and resistance to infections.

WILD YEASTS

Malted barley is known to contain a variety of micro-organisms, including yeasts (Haikara *et al.*, 1977). The number of yeast cells in Finnish malts was 10^4 per gram. However, no information is available about the taxonomy of those yeasts, nor for that matter of any other wild yeasts in connection with whisky manufacture. However, a wide variety of species is probably present. Examples from molasses and cane juice (Lehtonen and Suomalainen,

1977) and from grapes, musts and wines (Kunkee and Goswell, 1977) show this. The role of wild yeasts in distillery fermentations has recently been discussed by Watson (1984).

At least in the later stages of the fermentation, when the alcohol content becomes high, the natural yeasts are not thought to survive as well as distiller's yeast. However, their effect on the aroma might be noticeable. Some wild yeasts like *Hansenula anomala* and *Candida* spp. can produce 5–10 times the level of ethyl acetate, compared with *Saccharomyces* strains (Sponholz and Dittrich, 1974; Yoshioka and Hashimoto, 1981).

WHISKY YEASTS

The yeast strains in whisky fermentations are traditionally divided into primary strains and secondary strains (Sharp and Watson, 1979). Primary yeasts are those which are used as the main fermenting or ethanol producing yeast. The most famous of these is the DCL Yeast Division 'M' yeast, which in publications is often coded as D1 yeast strain (Watson, 1981).

Taxonomically the primary strains are *Saccharomyces cerevisiae*. The secondary strains are either *S. cerevisiae*, which is used in an ale or stout (top yeast) fermentation, or *S. uvarum* or *S. carlsbergensis*, which are used in a lager (bottom yeast) fermentation. This division is now, however, outdated and in the new treatments of Barnett *et al.* (1983) and Yarrow (1984) all of these have been put under the name *S. cerevisiae*. The same has happened to *S. diastaticus*, which with its amylolytic ability, is interesting for strain construction purposes (Erratt and Stewart, 1978; Watson, 1981; Tubb, 1983). The unification of all the above mentioned (and many other) species to *S. cerevisiae* does not, however, mean that the practical differences of sugar utilisation, fermentation rate, alcohol tolerance and flavour production have disappeared. The practical technological nominations, i.e. distiller's, brewer's etc., thrive strongly, even though a sophistication in the sense of a more complete characterisation of genetic and physiological properties of each strain is required in the future. First reports on the molecular biology of distiller's yeast strains have recently appeared (Bussey and Meaden, 1985; Keiding, 1985).

Distiller's yeasts used in whisky fermentation are difficult to compare due to the fact that most producers keep their strains proprietory. Most of the published work (Sharp and Watson,

1979; Ramsay and Berry, 1981; Watson, 1981; Berry and Ramsay, 1983; Lavery and Berry, 1983; Ramsay and Berry, 1983b; Watson, 1984) on strain characteristics will be covered under the section on 'Metabolism'.

Some of our unpublished work indicates that at 30°C in anaerobic fermentation of a liquid medium containing 1% yeast extract, 2% peptone and 30% glucose, whisky yeast strains typically grow from an initial cell density of 1×10^6 cells per ml to a density of $2–5 \times 10^7$ cells per ml, produce 10.8–11.8 wt% ethanol and some glycerol. The amount of glycerol found in these fermentations (8–11 g per litre) is considerably more than that obtained in malt wort by Merritt (1966): 2.7–3.4 g per litre, or in grain wort by Pyke (1965): 2.5 g per litre. However, the final ethanol contents in these mashes were also quite low (about 5%). As explained by Oura (1977), the amount of glycerol produced indicates both the extent of cell growth and fermentation. Overall, the tested strains seem on these criteria to be of very high-quality distiller's yeast strains.

Secondary strains are mainly used in conjunction with the primary strain in malt whisky distilleries. They are poorer in yield, slower fermenting and less ethanol-tolerant brewer's (surplus) yeasts. Their main function is to provide flavour to the final product (Sharp and Watson, 1979). Baker's yeast is also used in conjunction with distiller's yeast, but with a slightly different rationale. In fermentations at Alko, distiller's yeast was proven to speed up the initial phase of fermentation and by being more ethanol-tolerant than baker's yeast to complete the fermentation in the presence of a high ethanol concentration. Currently we use a specially developed distiller's yeast strain (Korhola, 1983). However, the economics dictate – baker's yeast is cheaper than distiller's yeast – that distillery plant managers normally use some baker's yeast in addition to distiller's yeast. In whisky fermentations mixed populations of distiller's and brewer's yeasts are used.

Production of secondary yeast, i.e. spent brewer's yeast, is not of any great concern to the whisky manufacturer as long as the microbiological quality is good enough. For instance, the viability has to be over 60% (Sharp and Watson, 1979). After the initial aeration or oxygenation of the wort, beer fermentation is an anaerobic process, therefore the yeast bought by the distiller is anaerobically propagated. The secondary yeast is often used to ensure a constant inoculations level. Most primary strains are propagated aerobically. However, as pointed out by Watson (1984), they are often not really aerobic but aerated fermentations on a pasteurised substrate. In a baker's yeast, as well as in modern distiller's yeast propagation, the final commercial stage is highly aerated with a limiting carbon source, thus resulting in almost fully oxidative metabolism.

Metabolism

GROWTH AND FERMENTATION

During whisky manufacture the pitched yeast, after a lag period, first grows and ferments rapidly, then slows down to less tumultuous metabolic activity and finally ceases growth but may continue some fermentation and may even die and autolyse (Watson, 1981; Ramsay and Berry, 1983a). The extent of growth is directly proportional to the amount of available α-amino nitrogen in the wort (Berry and Ramsay, 1983).

Yeast growth and alcoholic fermentation require the availability of nutrients. The mash in whisky fermentation is a rich enough source of sugars, nitrogenous compounds, vitamins, salts like sulphate and phosphate, and minerals like potassium, magnesium, calcium, zinc etc. (Pyke, 1965). Excellent reviews on the metabolic effects of nutrients have been written by Jones *et al.* (1981) and by Jones and Greenfield (1984). Another review covering the whole field of yeast nutrition and solute uptake is by Suomalainen and Oura (1971).

The metabolism of yeast during alcohol production is partly growth associated, partly fermentation associated. In biochemical terms, this is manifested by synthesis of glycerol and succinate in order to keep the reduction/oxidation (redox) state of the cell in equilibrium. According to an analysis made by Oura (1977) glycerol formed in consequence of growth makes 1.5% of the total end-products (minus CO_2) and glycerol formed to balance the excretion of succinate makes 4.3% of the total end-products (minus CO_2). Thus glycerol constitutes 5.8% of the total end-products (minus CO_2), succinate 0.9%, yeast cell material (dry matter) 1.2% and ethanol 92.1%. The biochemical pathways leading to the main end-products of alcoholic fermentation are shown in Fig. 4.1 with emphasis on the redox balances (Oura, 1977).

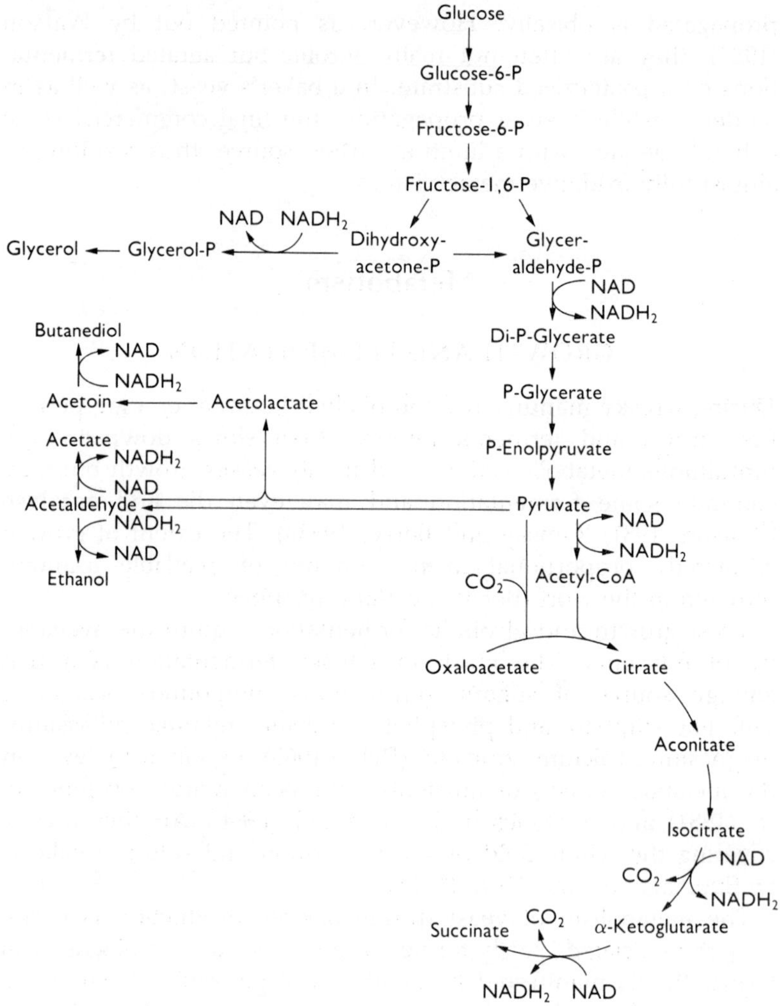

Fig. 4.1 Reactions of glucose to the products that are quantitatively significant during alcohol fermentation (modified from Oura, 1977).

One cofactor in alcoholic fermentation is adenosine diphosphate (ADP), which, together with inorganic phosphate (P_i), is essential for alcoholic fermentation (Oura *et al.*, 1982). Thus, the rate of fermentation can be increased by encouraging yeast growth, glycogen synthesis or some active uptake process. Each of these processes requires ATP and thus all result in regeneration of ADP for fermentation (Oura *et al.*, 1982).

SUGAR UPTAKE

Sugars are taken up by the yeast cell by facilitated diffusion (simple monosaccharides) or by an energy-requiring (active) proton symport mechanism (maltose and sucrose) and some by unknown mechanisms (Cooper, 1982). The sugars encountered by yeast in a whisky mash are mainly maltose, maltotriose, maltotetraose, maltopentaose and higher oligosaccharides plus some glucose, fructose and sucrose. Utilisation of sugars by yeasts has been extensively reviewed by Barnett (1976, 1981). Whisky yeast can take up the mono- and disaccharides and evidently also the trisaccharide maltotriose. Uptake of maltotetraose or maltopentaose is not widely found but it is claimed to occur (Erratt and Stewart, 1978). A similar 'super permease' system is thought to operate in the DCL 'M' yeast (Watson, 1981), but could not be convincingly demonstrated on a laboratory scale and was challenged by Lavery and Berry (1983). Because enzymes other than natural malt enzymes are not allowed to be added to Scotch whisky manufacture there has been strain development work on dextrinase- or glucoamylase-containing strains. Such work has been successful (Watson, 1981), but no details were given. Another aspect of strain development is that glucoamylase with its α-1,4 activity is probably not alone a significant improvement, but that a useful strain should in addition contain a debranching (α-1,6) activity (Tubb, 1983).

The order of sugar utilisation in fermentation is sucrose, fructose, glucose, maltose and higher oligosaccharides (Erratt and Stewart, 1978; Watson, 1981). However, sucrose is most likely first hydrolysed to fructose and glucose by the invertase (β-fructofuranosidase) located outside the plasma membrane (Lampen, 1968).

The active sucrose uptake system mentioned above is probably of minor importance; at least, when the amount of sucrose diminishes the amounts of fructose and glucose temporarily increase, indicating the action of invertase (Watson, 1981).

FLAVOUR METABOLISM

Minor metabolic products of major importance in whisky fermentation are the higher alcohols, fatty acids and their esters (Suomalainen and Lehtonen, 1979).

The metabolic relationships prevailing in the whisky yeast cell producing flavour are excellently shown in Fig. 4.2 (Ramsay, 1982; Watson, 1984).

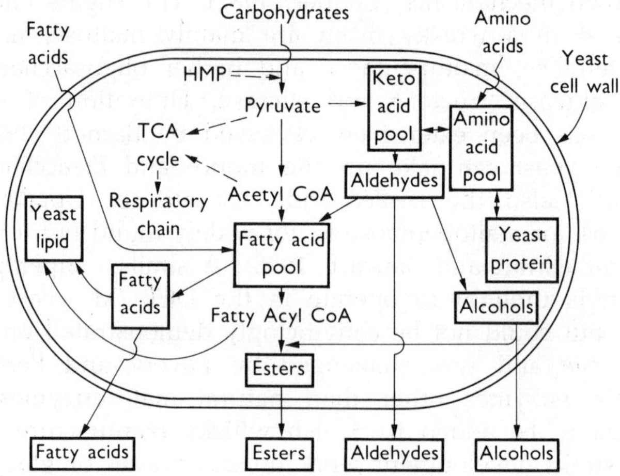

Fig. 4.2 Basic routes by which yeast forms the major flavour groups during fermentation (Ramsay, 1982; Watson, 1984).

Higher alcohols are formed from carbohydrates and external amino acids during amino acid biosynthesis and catabolism, respectively. *n*-Propanol is formed in connection with threonine, isobutanol with valine, 2-methyl-1-butanol (optically active amyl-alcohol) with isoleucine and 3-methyl-1-butanol (isoamylalcohol) with leucine anabolism and catabolism, as shown in Fig. 4.3 (Webb and Ingraham, 1963).

The effect of yeast strains and physiological conditions on production of higher (fusel) alcohols have been compared in several publications (Merritt, 1966; Berry and Ramsay, 1983; Ramsay and Berry, 1981, 1983a, b; Watson 1984). Maximum quantities were produced at 25–30°C in agitated cultures with a high pitching rate in a mash containing high concentrations of amino acids and insolubles. Also, yeast strains and species differ in their ability to produce higher alcohols. Parfait and Jouret (1975) have shown that *Schizosaccharomyces pombe* yeast produces relatively small amounts of fusel alcohols. Different mutations can change the biosynthesis of fusel alcohols. Guymon and his

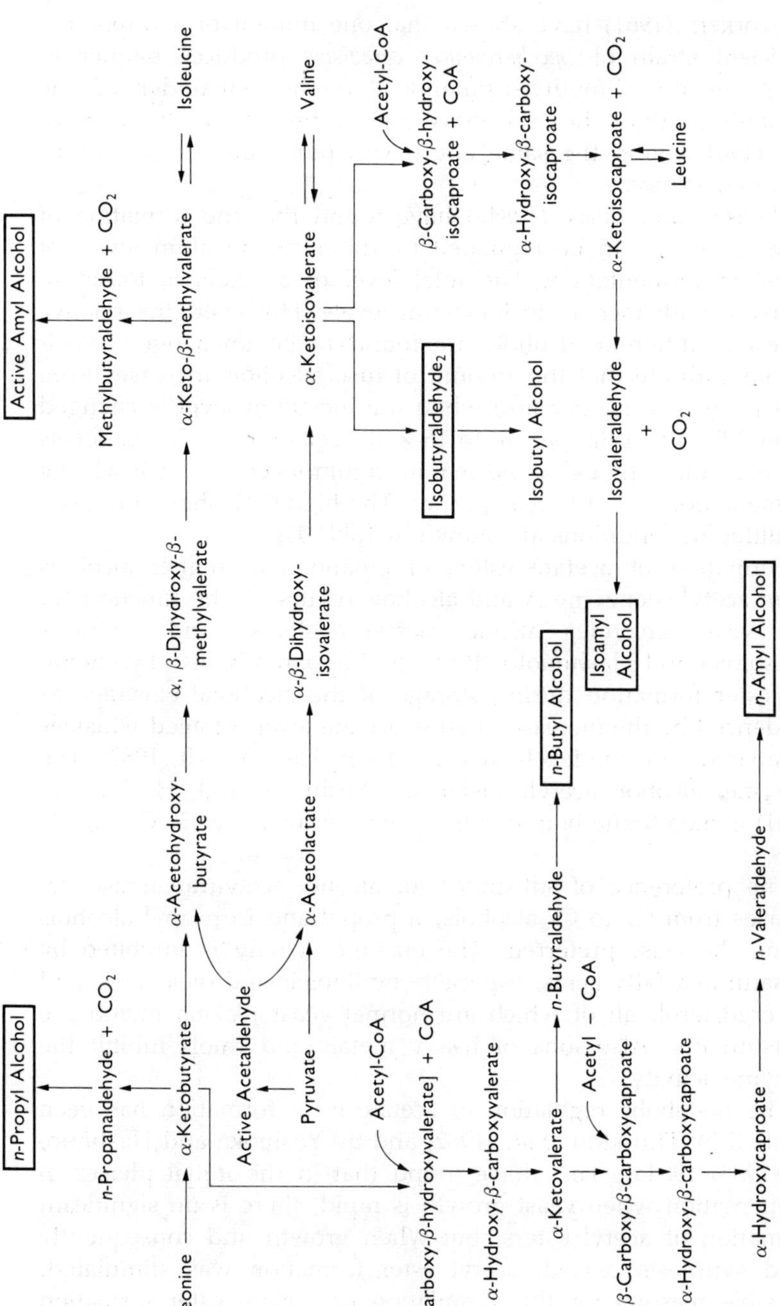

Fig. 4.3 Pathways of formation of major higher alcohols (fusel oils) and *n*-butyl and *n*-amyl alcohols (Webb and Ingraham, 1963). (Optically) active amyl alcohol is 2-methyl-1-butanol, iso-amyl alcohol is 3-methyl-1-butanol and *n*-amyl alcohol is 1-pentanol.

co-workers (1961) have shown that one mutant of a threonine-deficient strain of *Saccharomyces cerevisiae* produces neither *n*-propanol nor 2-methyl-1-butanol, but the remainder of the mutants produce these alcohols quite normally. Webb and Kepner (1961) found the same effect with pure cultures of *Saccharomyces cerevisiae*.

Ramsay and Berry (1984a) have found that the formation of fusel alcohols can be regulated by different inoculum levels of yeast in fermentation. The total level of alcohols is found to increase with increasing inoculum levels. However, the relative levels of different alcohols are found to be unchanged. Their results indicate that the amount of fusel alcohols increases from 288 mg to 504 mg per litre when the inoculum level is changed from 0.5×10^7 cells per ml to 16×10^7 cells per ml. These levels are 0.25 and 8 times the normal inoculum level in Scotch whisky fermentation, 2×10^7 cells per ml. The higher alcohols produced in different conditions are shown in Table 4.1.

Formation of acetate esters of ethanol and higher alcohols from acetyl-coenzyme A and alcohols occurs by the functions of the yeast enzymes alcohol acetyltransferase and esterases (Yoshioka and Hashimoto, 1981), and apparently also by chemical ester formation during storage of the alcoholic beverage as evidenced by the increase in ethyl acetate levels of aged whiskies compared with the levels in unaged whiskies (Brandt, 1982). The enzyme alcohol acetyltransferase (Yoshioka and Hashimoto, 1981) is membrane bound, functioning optimally at 30°C and pH 7–8.

The preference of substrates for alcohol acetyltransferase decreases from C_6 to C_1 alcohols, *n*-propyl and isopropyl alcohols being the least preferred. The enzyme activity is inhibited by unsaturated fatty acids, especially by linoleic and oleic acids and by ergosterol, all of which are normal yeast plasma membrane constituents. Also ions of heavy metals and thiols inhibit the enzyme activity.

The metabolic regulation of acetate ester formation has been studied by Thurston *et al*. (1982) and by Yoshioka and Hashimoto (1983, 1984a). They have found that in the initial phases of fermentation when yeast growth is rapid, there is no significant formation of acetyl esters, but when growth and consequently lipid synthesis ceased, acetyl ester formation was stimulated. Possible reasons for the stimulation of acetate ester formation were regulation of the activity of the alcohol acetyltransferase

Table 4.1 Amounts of higher alcohols in fermented mash produced by whisky yeasts.

Reference	*Higher alcohols, mg per litre (range in different conditions)*						
	n-Propanol	*iso-butanol*	*2-Methyl-1-butanol (optically active amyl alcohol)*	*3-Methyl-1-butanol (isoamyl alcohol)*	*Amyl alcohols*	*Total fusel alcohols*	*β-Phenyl-ethanol*
Pyke, 1965	34	50		135			
Merritt, 1966						40–159	
Ramsay and Berry, 1981	13.0–28.8	31.3–110.1			97–268	141–407	61.6–81.3
Berry and Ramsay, 1983		37–67			117–200	185–500	
Ramsay and Berry, 1983b		35–66			133–190		62–97
amino acids added		40–330			120–500		50–650
Watson, 1984	13.3–20.7	35.9–67.0			104–142	153–230	
Ramsay and Berry, 1984a	16.0–30.2	45.6–111			119–267	180–408	47.9–95.8
Ramsay and Berry, 1984b	17.5–21.9	46.6–84.5			112–174	176–280	48.7–72.4
Korhola, unpublished*	30–60	45–115	20–70	80–165	100–235	175–410	

*YPD 30 medium, 30°C, anaerobic shaking, four different strains.

(Yoshioka and Hashimoto, 1983) or an increase in the ratio of acetyl-coenzyme A to coenzyme ASH (Thurston *et al.*, 1982). The amount of acetyl-coenzyme A in the cell does not seem to affect directly the amount of acetate esters formed during fermentation (Yoshioka and Hashimoto, 1984a).

Besides acetates of fusel alcohols, yeast also contributes greatly to the amount of fatty acids and fatty acid esters in the whisky (Suomalainen and Lehtonen, 1979). The proposed scheme of biosynthesis of fatty acids and their esters is shown in Fig. 4.4 (Lynen, 1967; Nordström, 1964).

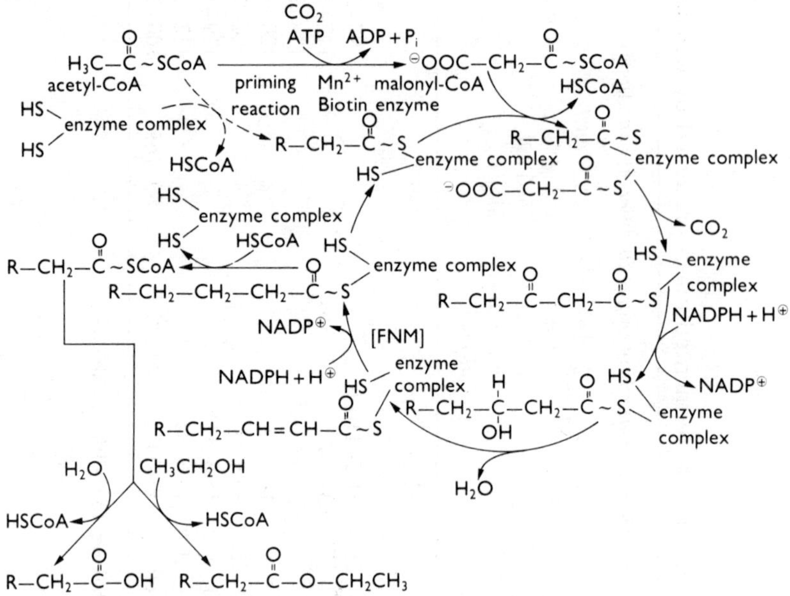

Fig. 4.4 Proposed scheme for formation of fatty acids and their ethyl esters (after Lynen, 1967 and Nordström, 1964).

Factors influencing the formation of esters are the type of yeast used, the quality of wort, pH, fermentation temperature, the size of inoculum and the aeration and mixing of wort (Ramsay and Berry, 1983b, 1984a, b). The same factors also affect the formation of fatty acids. Ramsay and Berry (1984b) have found that the amount of fatty acids formed in fermentation decreases with increasing temperature. Krauss and Forch (1975) have shown for beer fermentation that all conditions which accelerate the fermentation cause a decrease in the free fatty acid content. The distillation technique used also affects the amounts and ratios of aroma compounds in the final product. There will be slightly less esters when filtered worts are fermented, compared with unfiltered worts (Ramsay and Berry, 1983b). The optimum temperature for the formation of ethyl caproate and ethyl caprylate is between 20–25°C, whilst for acetates it is between 25–30°C. Maximum concentration of acetates will be between pH 5.0–6.0 (Ramsay and Berry, 1984b). The levels of ethyl caprylate, ethyl caprate and ethyl laurate decreased at inoculum levels above 2×10^7 cells per ml, which is normal yeast inoculum level in Scotch whisky production (Ramsay and Berry, 1984a).

The formation of esters during sugar fermentation has been studied in some articles (Nykänen and Nykänen, 1977; Nykänen *et al.*, 1977; Yoshioka and Hashimoto, 1984b). The proportion of fatty acid ethyl esters transferred to the medium decreased with increasing chain length; ethyl caproate was all in the medium, whilst ethyl laurate was all in the yeast cell (Nykänen *et al.*, 1977).

Aeration is a very important factor in the formation of unsaturated fatty acids and obviously also their ethyl esters. Aerobically grown pitching yeast is very rich in unsaturated fatty acids, palmitoleic acid and oleic acid, which are scarcely found in anaerobically grown yeast (Ahvenainen, 1982; Ohno and Takahashi, 1983).

Below are some examples of fatty acid ethyl ester concentrations obtained and the effect of different yeasts on their formation. Some commercial whiskies were analysed for their fatty acid ethyl ester contents, and the results are shown in Table 4.2. The Scotch whiskies had much more ethyl palmitoleate than the Irish and Finnish whiskies and bourbons. Figures 4.5A–D present the gas chromatograms of two Scotch whiskies and two Finnish whisky distillates. Compounds in the chromatograms are neutral compounds, mainly esters and alcohols. The figures look very much alike, except for the peaks of ethyl palmitoleate, which are virtually absent in the Finnish examples.

Some experiments on a laboratory scale were made in order to test how ethyl palmitoleate concentrations could be increased in distillates. Worts were fermented with different yeasts and with

Table 4.2 Fatty acid ethyl ester concentrations of some whiskies (mg per kg EtOH).

	Scotch whiskies			*Irish*	*Finnish whiskies*		*Bourbons*	
	Blended 1	*Blended 2*	*Malt*	*whiskey*	*Brand 1*	*Brand 2*	*Brand 1*	*Brand 2*
Ethyl caproate	3.1	3.7	6.7	1.1	2.8	1.7	4.2	6.7
Ethyl caprylate	13.1	23.3	46.8	3.4	11.2	17.8	10.7	19.6
Ethyl caprate	37.2	56.5	99.9	2.7	18.5	14.5	20.6	29.1
Ethyl laurate	25.2	34.7	51.8	0.6	8.3	17.8	13.0	11.2
Ethyl myristate	3.5	3.7	4.2	1.3	1.6	1.3	4.5	2.6
Ethyl palmitoleate	7.1	10.4	11.2	n.d.	0.8	0.7	0.5	n.d.
Ethyl palmitate	7.2	5.5	4.8	2.2	3.4	2.0	17.9	3.0
Ethyl stearate	0.2	n.d.	n.d.	n.d.	n.d.	n.d.	0.4	0.7

n.d. = not detected

and without aeration. Yeasts used were baker's yeast, distiller's yeast, bottom fermenting brewer's yeast and top fermenting brewer's yeast. At the same time these yeasts were tested in industrial fermentations. Results from labaratory fermentations are presented in Table 4.3. Bottom fermenting brewer's yeast was the best yeast to produce ethyl palmitoleate, especially when aerated. Top fermenting brewer's yeast and distiller's yeast did not produce ethyl palmitoleate at all, even when aerated; possibly the ester is formed but escapes with the air. It was noticed that the age of the yeast had a strong effect on the production of ethyl palmitoleate and also on the other esters. Fresh bottom fermenting brewer's yeast produced about 20 times more ethyl palmitoleate than three-week-old yeast.

Industrial fermentation experiments were not aerated. Distiller's yeast was used in all cases together with the tested yeast in a 1 : 1 ratio. Brewer's yeasts and baker's yeast produced about equal amounts of the ethyl esters of lower fatty acids from caproic acid to lauric acid, but the ethyl ester concentrations of myristic and palmitic acids were about five times higher when brewer's yeasts were used, compared with baker's yeast. Ethyl palmitoleate concentrations were only a fraction of those of ethyl palmitate. Differences between bottom and top fermenting brewer's yeasts were much smaller than those found in laboratory experiments. Perhaps the use of distiller's yeast smoothed out these differences.

Current practices

YEAST PRODUCTION

Spent brewer's top or bottom yeast is purchased usually from a local brewery. Distiller's yeast is manufactured in a yeast factory with modern control instrumentation, or on site in aerated vats with little control over the production. The production scheme of distiller's yeast manufacture in the baker's yeast plant at Alko Ltd is shown in Fig. 4.6.

This streamlined system utilises the ethanol formed during P2 stage for cell mass growth at P3 stage (ethanol drops during P3 stage from the initial 0.7% to a final 0.01%). The yeast grows at P3 stage with an overall growth rate (μ) of 0.14 per hour. The

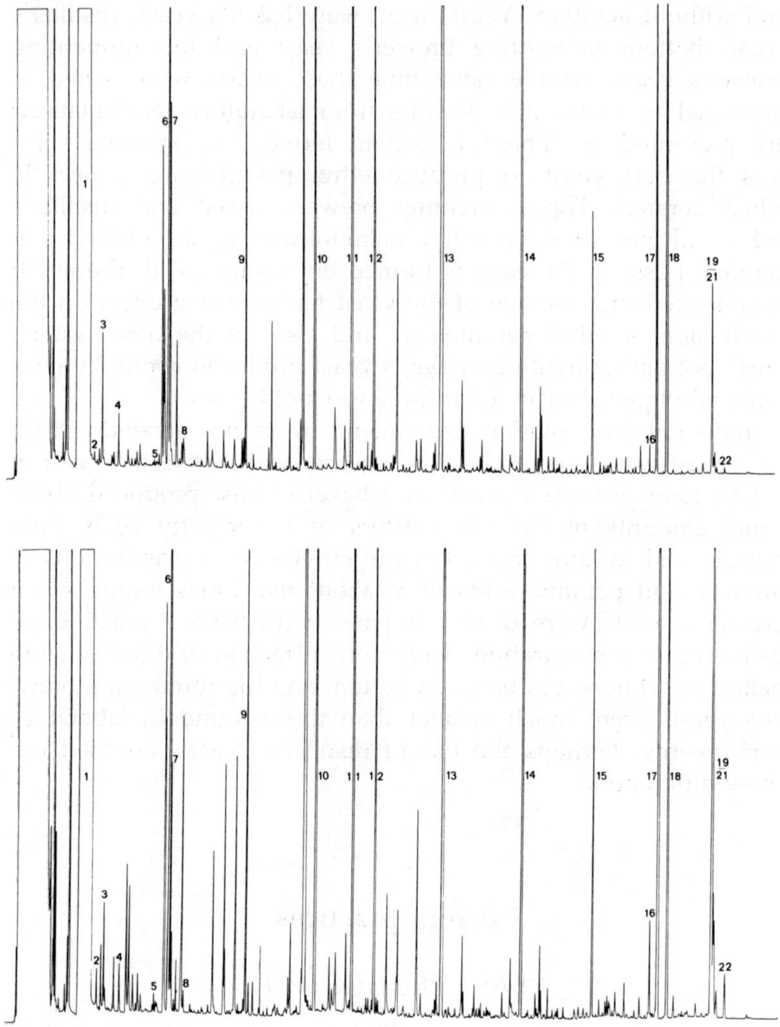

Fig. 4.5 Chromatograms of neutral compounds of various whiskies. A: Blended Scotch whisky. B: Malt whisky. C: Finnish whisky distillate fermented by baker's yeast and by distiller's yeast. D: Finnish whisky distillate fermented by brewer's bottom fermenting yeast and by distiller's yeast.

1 iso-amyl alcohol + optically active amyl alcohol (3-methyl-1-butanol + 2-methyl-1-butanol)
2 ethyl iso-butyrate
3 iso-butyl acetate
4 ethyl butyrate
5 ethyl iso-valerate
6 iso-amyl acetate
7 optically active amyl acetate
8 ethyl valerate
9 ethyl caproate
10 methyl caprylate, internal standard

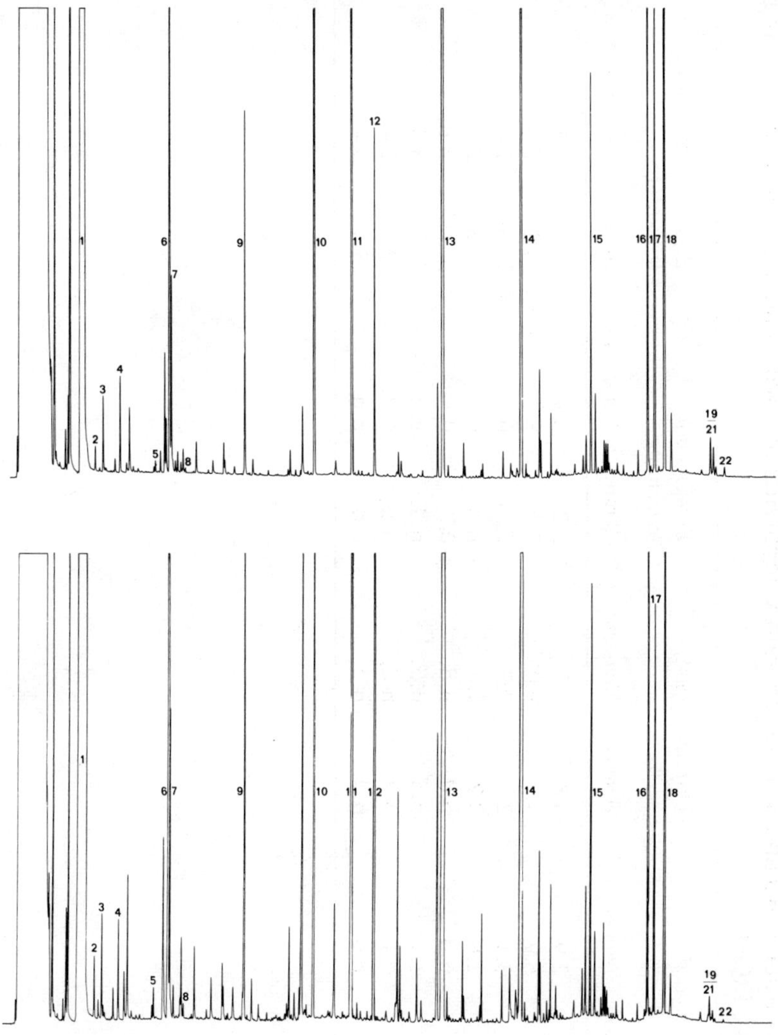

11 ethyl caprylate
12 *β*-phenethyl acetate
13 ethyl caprate
14 ethyl laurate
15 ethyl myristate
16 ethyl palmitoleate
17 ethyl palmitate
18 methyl heptadecanoate, internal standard
19–21 ethyl oleate, ethyl linoleate and ethyl linolenate
22 ethyl stearate

Chromatographic conditions: cross-linked methyl silicone capillary column, length 50 m, internal diameter 0.31 mm, 0.52 μm, FID, carrier gas He, splitless mode. Temperature programme: isothermal at 50 °C for 8 min, heating 4 °C per min to 240 °C. Samples were extracted with iso-pentane.

Table 4.3 Fatty acid ethyl ester concentrations in laboratory fermentations with and without aeration using different yeasts (mg per kg EtOH).

	Baker's yeast		*Distiller's yeast*		*Brewer's yeast, top fermenting*		*Brewer's yeast, bottom fermenting*	
	Aerated	*Non-aerated*	*Aerated*	*Non-aerated*	*Aerated*	*Non-aerated*	*Aerated*	*Non-aerated*
Ethyl caproate	2.2	2.9	3.8	5.3	2.2	3.9	1.1	3.4
Ethyl caprylate	1.8	6.9	4.1	9.0	3.1	8.0	4.2	6.8
Ethyl caprate	0.4	3.1	2.0	7.7	0.3	0.7	8.6	7.6
Ethyl laurate	0.4	2.8	0.2	1.6	n.d.	0.3	1.8	2.7
Ethyl myristate	n.d.	n.d.	n.d.	n.d.	0.7	0.5	n.d.	n.d.
Ethyl palmitoleate	1.3	1.1	n.d.	n.d.	n.d.	n.d.	20.3	8.5
Ethyl palmitate	0.4	0.4	n.d.	n.d.	n.d.	n.d.	2.3	2.4
Ethyl stearate	n.d.	n.d.	n.d.	n.d.	n.d.	n.d.	1.9	1.4

n.d. = not detected

nitrogen content of final yeast is typically 7.5 g per 100 g yeast dry matter. An average yield for the large scale (P2 + P3) is 65 kg Y_{27} per 100 kg beet molasses (50% sucrose). The final yeast is transported in 20 kg blocks to a distillery by cooled lorries or used in an adjacent distillery.

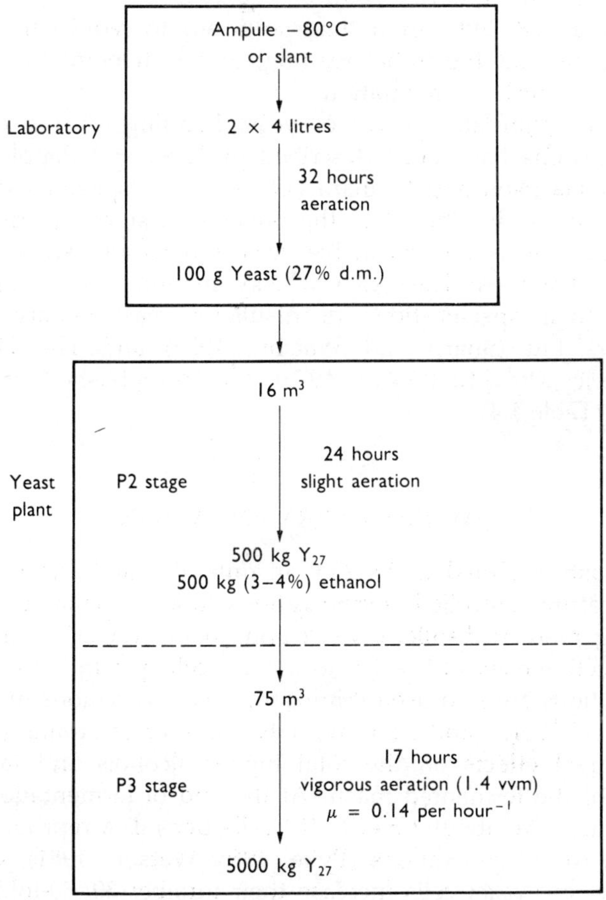

Fig. 4.6 Production scheme of distiller's yeast at the baker's yeast plant of Alko Ltd (K. Edelmann, unpublished).

An interesting account of yeast propagation for an American whiskey manufacturer has been given by Brandt (1982). First a regular fermenter mash is fortified with approximately 15% malt.

The yeast propagation mash is then inoculated with a lactic acid producing bacterial culture, and maintained at 53°C until a pH of 3.9 is reached. The soured mash is then heated to 63°C and held until needed or pasteurized at 82°C for 30 min and cooled to 23–24°C. The non-pasteurized mash has also to be cooled. The soured yeast propagation mash is then inoculated with a suitable volume of liquid starter culture. The yeast is propagated at not more than 28–29°C for a time sufficient to reduce the Balling reading by half the initial reading, at which point the yeast is ready for whisky fermentation.

Various alternative ways of yeast handling during distillery fermentations have been described by Reed and Peppler (1973) and by Harrison and Graham (1971). The propagated distiller's yeast may be distributed to the buyer as a slurry, pressed cake or crumbs or as a dried active yeast (Sharp and Watson, 1979; Brandt, 1982). At least two whisky manufacturers have published their specifications of distiller's yeast quality: Chivas Brothers Ltd (Sharp and Watson, 1979) and The Highland Distilleries Co. Ltd (Dolan, 1979). Our unpublished practice is added (Table 4.4).

WHISKY FERMENTATION

The mash is cooled to 20–25°C in untemperated and to 30°C in temperature-controlled fermentations and inoculated with a varying ratio of distiller's yeast and spent brewer's yeast to an initial cell density of 5×10^6 to 2×10^7 cells per ml. Single strain inoculations are also used (Simpson, 1968). Variations in inoculation level (Berry and Ramsay, 1983) and yeast strains (Fig. 4.5) have great effects on the total higher alcohols and fatty acid esters of the fermented mash. At the end of fermentation (30 to 72 h) there are about $1.5–2 \times 10^8$ cells per ml, a reproduction of from 3 to 3.3 generations (Pyke, 1965; Watson, 1981). Conventionally, the yeast cells increase their number 30–50-fold or 5–6 generations during fermentation (de Becze, 1964).

In malt whisky fermentation the initial free amino nitrogen level is about 0.3 g per litre (Watson, 1981). The amount of α-amino nitrogen controls the extent of yeast growth (Berry and Ramsay, 1983). Another key factor in controlling the extent of yeast growth in fermentations is the amount of available oxygen (Kirsop, 1981). This, however, is not of great concern in whisky fermentations – contrary to beer fermentations – because the distiller's yeast is propagated aerobically and thus is rich enough in sterols and unsaturated fatty acids.

The temperature of fermentation may vary in older plants from 20°C at the pitching to as high as 37°C, with 33°C being normal at the end of fermentation (Dolan 1979; Watson, 1981; Ramsay and Berry, 1983b). In temperated fermentations the temperature is kept constant at 30°C. The pH of the wort in malt whisky fermentation is initially around 5.0–5.5 (Dolan, 1976; Ramsay and Berry, 1983b) and passes through a minimum of *c*. 4.0 close to halfway through the fermentation, before increasing to around 4.2 to 4.5 close to the end of fermentation and then possibly dropping again due to acid production by contaminating lactic acid bacteria (Dolan, 1976, 1979). The specific gravity of the mash is lowered from the initial *c*. 1.050–1.055 to 0.998–1.000 at the end of fermentation (Simpson, 1968; Dolan, 1976; Watson, 1981). The yield of alcohol is 7–10% by volume (Simpson, 1968; Dolan, 1976, whose values were converted to volume per cent by the equation: 1° proof = 0.5714 vol%).

After fermentation distillation is carried out without separation of yeast from the mash.

Infections

Contamination of the fermentation is detected by unusually high acid production with a consequent drop in pH, unusually low specific gravity (0.998 or lower), unusually low alcohol yield, foul smell or simply by the presence (visual inspection or cultivation methods) of undesirable microbes in the mash (Dolan, 1976, 1979; Sharp and Watson, 1979; Watson, 1981). Yeast death and autolysis resulting in a drop in specific gravity can also result from the combination of high alcohol content and high temperature of fermentation (Berry and Ramsay, 1983).

The economic losses due to infections have been stressed by Dolan (1976, 1979). Even 1–10 million bacteria per ml in the fermenting mash can cause observable losses and a contaminant level of over 10^8 per ml can be serious. Some lactobacilli can convert the glycerol excreted by yeast into β-hydroxypropionaldehyde, which is converted to acrolein by the heat of distillation, and some bacteria can utilise fatty acids and their esters,

Table 4.4 Distiller's yeast specifications.

Requirement	*Alko Ltd** *(unpublished)*	*Chivas Brothers Ltd* *(Sharp and Watson, 1979)*	*Highland Distilleries Ltd* *(Dolan, 1979)*
1 Strain	Pure, genetically stable distiller's yeast	Acceptable fermentation capacity	*Saccharomyces cerevisiae*
2 Flavour	—	No organoleptically unacceptable aromas	—
3 Dry matter %	26–27	25 (min. 23)	$\geqslant 24$
4 Viability	> 90	> 75	> 95
5 Total bacterial count per gram yeast cake	$< 10^4$ very good 10^4–10^6 good $> 10^6$ poor	$< 10^6$ (acidophilic)	$< 1.5 \times 10^7$†
6 Lactic acid bacteria per gram yeast cake	$< 10^3$ very good 10^3–10^5 good $> 10^5$ poor	$< 5 \times 10^5$	$< 10^5$†
7 Coliform bacteria per gram yeast cake	fecal: < 10 very good 10–50 good > 50 poor	$< 5 \times 10^5$ (total)	—
8 Total Gram-positive bacteria per gram yeast cake	—	—	$< 5 \times 10^5$†

Table 4.4 (*cont.*)

Requirement	*Alko Ltd** (*unpublished*)	*Chivas Brothers Ltd* (*Sharp and Watson, 1979*)	*Highland Distilleries Ltd* (*Dohlan, 1979*)
9 Total *Acetobacter* per gram yeast cake	—	$< 5 \times 10^4$	—
10 Total wild yeast/mould per gram yeast cake	$< 10^4$ very good 10^4–10^6 good $> 10^6$ poor	$< 10^4$	—
11 Storage/Transportation	2–6°C	1–4°C Must not be frozen	1–4°C
12 Consistency	Cake	Cake	Cake
13 Presence of additives (enzymes or fermentables)	—	None allowed	—

— = means no information specified or available.
*Used for internal quality control.
†The original specification of less than 1500 bacteria per million yeast cells on aerobic or anaerobic incubation has been converted to bacteria per gram yeast cake by assuming that one gram of yeast cake consists of 10^{10} yeast cells.
145 245 345

thus reducing the flavour (Watson, 1981). It should be pointed out that some acrolein is always formed from glycerol at pot-still distillation.

Lactic acid bacteria are by far the worst contaminants in alcoholic fermentations. This is due to their ability to utilise sugars like pentoses, which are not readily utilised by yeast, and their abilities to grow well at pH 4–5 and tolerate high (10–12 vol%) concentrations of ethanol (Bryan-Jones, 1975; Lafon-Lafourcade *et al.*, 1983; Makanjuola and Springham, 1984).

Earlier reports (Bryan-Jones, 1975; Dolan, 1979) named *Lactobacillus brevis* and especially *L. fermentum* as causing the most problems. Dolan (1979) also stressed the importance of sanitation and hygienic control in keeping up the quality of the final product. Later studies have shown that a great diversity of lactobacilli are found, most of which do not comply with the present taxonomic classification but may represent new species (Bryan-Jones, 1975; Barbour and Priest, 1983). An extensive study of contamination levels of different stages of malt whisky manufacture and identification of the lactic acid bacteria found was made by Makanjuola and Springham (1984). They found that three genera of lactic acid bacteria are involved: *Lactobacillus, Leuconostoc* and *Pediococcus*. The identification was based on rather few characteristics, however. *Leuconostoc* was implicated as being the most frequent contaminant, but dying off at later stages of fermentation. At various production phases, from yeast and wort, to the end of fermentation, contaminant levels of zero to 10^7 bacteria per ml were found. Yeast and its handling were found to be the most significant sources of contaminants.

Even though malt may contain high levels of bacteria, e.g. *Erwinia herbicola*, 2×10^8 per gram (Haikara *et al.*, 1977) they probably do not survive the fermentation conditions too well, given adequate plant hygiene.

Partly because of difficulties encountered in identification (naming) of genera and species of contaminants, we have recently started to use a computer for API 50 identification and other data storage, comparison and processing (Tanner and Korhola, 1985).

The control of contaminants to minimum levels facilitates efficient and clean fermentations without undesirable flavours and possibly with increased yields. These factors improve the economics of production by, for example, shortening the maturing period required to obtain a palatable, balanced product.

References

AHVENAINEN J 1982 Lipid composition of aerobically and anaerobically propagated brewer's bottom yeast. *Journal of the Institute of Brewing* **88**: 367–70

BARBOUR E A, PRIEST F G 1983 Numerical classification of lactobacilli from Scotch whisky distilleries. In Priest F G, Campbell I (eds) *Current Developments in Malting, Brewing and Distilling, Aviemore Conference 1982*. Institute of Brewing, London, pp 289–91

BARNETT J A 1976 The utilization of sugars by yeast. *Advances in Carbohydrate Chemistry and Biochemistry* **32**: 125–234

BARNETT J A 1981 The utilization of disaccharides and some other sugars by yeasts. *Advances in Carbohydrate Chemistry and Biochemistry* **39**: 347–404

BARNETT J A, PAYNE R W, YARROW D 1983 *Yeast characteristics and identification*. Cambridge University Press, Cambridge, pp 467–9

DE BECZE G I 1964 Reproduction of distiller's yeast. *Biotechnology and Bioengineering* **6**: 191–221

BERRY D R, RAMSAY C M 1983 The whisky fermentation: past, present and future. In Priest F G, Campbell I (eds) *Current Developments in Malting, Brewing and Distilling, Aviemore Conference 1982*. The Institute of Brewing, London, pp 45–58

BRANDT D A 1982 Distilled beverage alcohol. In Reed G (ed) *Prescott & Dunn's Industrial Microbiology*. The AVI Publishing Co Inc, Westport, Connecticut, pp 468–91

BRYAN-JONES G 1975 Lactic acid bacteria in distillery fermentations. In Carr J G, Cutting C V, Whiting G C (eds) *Lactic Acid Bacteria in Beverages and Food*. Academic Press, London – New York – San Francisco, pp 165–75

BUSSEY H, MEADEN P 1985 Selection and stability of yeast transformants expressing cDNA of an M1 killer toxin-immunity gene. *Current Genetics* **9**: 285–91

COOPER T G 1982 Transport in *Saccharomyces cerevisiae*. In Strathern J N, Jones E W, Broach J R (eds) *The Molecular Biology of the Yeast Saccharomyces, Metabolism and Gene Expression*. Cold Spring Harbor Laboratory, New York, pp 399–461

DOLAN T C S 1976 Some aspects of the impact of brewing science on Scotch malt whisky production. *Journal of the Institute of Brewing* **82**: 177–81

DOLAN T C S 1979 Bacteria in whisky production. *The Brewer* **2**: 60–4

ERRATT J A, STEWART G G 1978 Genetic and biochemical studies on yeast strains able to utilize dextrins. *ASBC Journal* **37**: 151–61

GUYMON J F, INGRAHAM J L, CROWELL E A 1961 The formation of *n*-propyl alcohol by *Saccharomyces cerevisiae*. *Archives of Biochemistry and Biophysics* **95**: 163–8

HAIKARA A, MÄKINEN V, HAKULINEN R 1977 On the microflora of barley after harvesting, during storage and in malting. In *EBC Congress*, Amsterdam, pp 35–46

HARRISON J S, GRAHAM J C J 1971 Yeast in distillery practice. In Rose A H, Harrison J S (eds) *The Yeasts*. Academic Press, London, vol 3 *Yeast Technology*, pp 283–348

JONES R P, GREENFIELD P F 1984 A review of yeast ionic nutrition. Part I: Growth and fermentation requirements. *Process Biochemistry* **19**(2): 48–60

JONES R P, PAMMENT N, GREENFIELD P F 1981 Alcohol fermentation by yeast – the effect of environmental and other variables. *Process Biochemistry* **16**(3): 42–9

KEIDING A K 1985 Genetic and molecular characterization of a distiller's yeast. *Carlsberg Research Communication* **50**: 95–125

KIRSOP B H 1981 Aeration in fermentation for ethanol production. *Enzyme and Microbial Technology* **3**: 375

KORHOLA M 1983 Improvement of yeast strains for added ethanol tolerance. In Korhola M, Väisänen E (eds) *Gene Expression in Yeast*. Foundation for Biotechnical and Industrial Fermentation Research, Helsinki, vol 1, pp 231–42

KRAUSS G, FORCH M 1975 Influence of different fermentation methods on the formation of lower free fatty acids. *Proceedings of the American Society of Brewing Chemists* **33**: 37–41

KUNKEE R E, GOSWELL R W 1977 Table wines. In Rose A H (ed) *Economic Microbiology*. Academic Press, London, vol 1 *Alcoholic Beverages*, pp 315–16

LAFON-LAFOURCADE S, CARRE E, RIBÉREAU-GAYON P 1983 Occurrence of lactic acid bacteria during the different stages of vinification and conservation of wines. *Applied Environmental Microbiology* **46**: 874–80

LAMPEN J O 1968 External enzymes of yeast: their nature and formation. *Antonie van Leeuwenhoek* **34**: 1–18

LAVERY M, BERRY D R 1983 Amylolytic activity during the Scotch malt whisky fermentation. In Priest F G, Campbell I (eds) *Current Developments in Malting, Brewing and Distilling, Aviemore Conference 1982*. Institute of Brewing, London, pp 293–6

LEHTONEN M, SUOMALAINEN H 1977 Rum. In Rose A H (ed) *Economic Microbiology*. Academic Press, London, vol 1, *Alcoholic Beverages*, pp 595–633

LEHTONEN M, SUOMALAINEN H 1979 The analytical profile of some whisky brands. *Process Biochemistry* **14**(2): 5–6, 8–9, 26

LYNEN F 1967 The role of biotin-dependent carboxylations in biosynthetic reactions. *Biochemistry Journal* **102**: 381–400

MAKANJUOLA D B, SPRINGHAM D G 1984 Identification of lactic acid bacteria isolated from different stages of malt whisky distillery fermentation. *Journal of the Institute of Brewing* **90**: 374–83

Merritt N R 1966 The influence of temperature on some properties of yeast. *Journal of the Institute of Brewing* **72**: 374–83

Nordström K 1964 Formation of esters from alcohols by brewer's yeast. *Journal of the Institute of Brewing* **70**: 328–36

Nykänen L, Nykänen I 1977 Production of esters by different yeast strains in sugar fermentations. *Journal of the Institute of Brewing* **83**: 30–1

Nykänen L, Nykänen I, Suomalainen H 1977 Distribution of esters produced during sugar fermentation between the yeast cell and the medium. *Journal of the Institute of Brewing* **83**: 32–4

Nykänen L, Suomalainen H 1983 *Aroma of Beer, Wine and Distilled Alcoholic Beverages*. Akademie-Verlag, Berlin, pp 3–15

Ohno T, Takahashi R 1983 Role of wort aeration in the brewing process, (1) Biosynthesis of lipids in brewer's yeast. *Report of the Research Laboratories of Kirin Brewery Co Ltd* **26**: 15–23

Oura E 1977 Reaction products of yeast fermentations. *Process Biochemistry* **12**(3): 19–21

Oura E, Suomalainen H, Viskari R 1982 Breadmaking. In Rose A H (ed) *Economic Microbiology*. Academic Press, Oxford, vol 7, *Fermented Foods*, pp 87–146

Parfait A, Jouret C 1975 Formation of higher alcohols in rum. *Annales de Technologie Agricole* **24**: 421–36

Pyke M 1965 The manufacture of Scotch grain whisky. *Journal of the Institute of Brewing* **71**: 209–18

Ramsay C M 1982 Physiological control of volatile formation by yeast in the Scotch malt whisky fermentation. PhD thesis, University of Strathclyde, Glasgow

Ramsay C M, Berry D R 1981 Physiological control of higher alcohol formation in the Scotch whisky fermentation. In Stewart G G, Russell I (eds) *Current Developments in Yeast Research*. Pergamon, Oxford, pp 63–8

Ramsay C M, Berry D R 1983a Development of a small scale mashing and fermentation system for studies on malt whisky production. *European Journal of Applied Microbiology Biotechnology* **18**: 207–13

Ramsay C M, Berry D R 1983b Factors influencing the aroma and flavour of whisky. *Brewing & Distilling International* **13**: 34–7

Ramsay C M, Berry D R 1984a The effect of inoculum level on the formation of higher alcohols, fatty acids and esters in the malt whisky fermentation. *Food Microbiology* **1**: 111–15

Ramsay C M, Berry D R 1984b Effect of temperature and pH on the formation of higher alcohols, fatty acids and esters in the malt whisky fermentation. *Food Microbiology* **1**: 117–21

Reed G, Peppler H J 1973 *Yeast Technology*. The AVI Publishing Co Inc, Westport, Connecticut, pp 304–16

Sharp R, Watson D C 1979 Improving yeast strains for whisky

production. *Brewing & Distilling International* **9**(6): 50–1, 53

SIMPSON A C 1968 Manufacture of Scotch malt whisky. *Process Biochemistry* **3**(1): 9–12

SPONHOLZ W-R, DITTRICH H H 1974 Die Bildung von SO_2-bindenden Gärungs-Nebenprodukten, höheren Alkoholen und Estern bei einigen Reinzuchthefestämmen und bei einigen für die Weinbereitung wichtigen 'wilden' Hefen. *Die Wein-Wissenschaft* **29**: 301–14

STEWART G G, RUSSELL I (eds) 1981 *Current Developments in Yeast Research*. Pergamon, Oxford

SUOMALAINEN H, LEHTONEN M 1979 The production of aroma compounds by yeast. *Journal of the Institute of Brewing* **85**: 149–56

SUOMALAINEN H, NYKÄNEN L 1966 The aroma components produced by yeast in nitrogen-free sugar solution. *Journal of the Institute of Brewing* **72**: 469–74

SUOMALAINEN H, OURA E 1971 Yeast nutrition and solute uptake. In Rose A H, Harrison J S (eds) *The Yeasts*. Academic Press, London – New York, vol 2 *Physiology and Biochemistry of Yeasts*, pp 3–74

TANNER R, KORHOLA M 1985 Contaminants in fermentation industry: identification and data handling by computer. *Abstracts of the Seventh International Conference on Global Impacts of Applied Microbiology (GIAM)*, Helsinki 1985, p 124

THURSTON P A, QUAIN D E, TUBB R S 1982 Lipid metabolism and the regulation of volatile ester synthesis in *Saccharomyces cerevisiae*. *Journal of the Institute of Brewing* **88**: 90–4

TUBB R S 1983 Genetic pathways to super-attenuating (amylolytic) yeasts. In Priest F G, Campbell I (eds) *Current Developments in Malting, Brewing and Distilling, Aviemore Conference 1982*. Institute of Brewing, London, pp 67–82

WATSON D C 1981 The development of specialised yeast strains for use in Scotch malt whisky fermentations. In Stewart G G, Russell I (eds) *Current Developments in Yeast Research*. Pergamon, Oxford, pp 57–62

WATSON D C 1984 Distilling yeast. In *Developments in Industrial Microbiology*, vol 25. Proceedings of the Fortieth General Meeting of the Society for Industrial Microbiology, Sarasota, Florida 1983. Society for Industrial Microbiology, Arlington, Virginia, pp 213–20

WEBB A D, INGRAHAM J L 1963 Fusel oil: a review. *Advances in Applied Microbiology* **5**: 317–53

WEBB A D, KEPNER R E 1961 Fusel oil analysis by means of gas-liquid partition chromatography. *American Journal of Enology and Viticulture* **12**: 51–9

YARROW D 1984 *Saccharomyces* Meyen ex Reess. In Kreger-van Rij N J W (ed) *The Yeasts, a Taxonomic Study*. Elsevier Science Publishers, Amsterdam, pp 379–95

YOSHIOKA K, HASHIMOTO N 1981 Ester formation by alcohol acetyltransferase from brewer's yeast. *Agricultural and Biological Chemistry* **45**: 2183–90

Yoshioka K, Hashimoto N 1983 Cellular fatty acid and ester formation by brewer's yeast. *Agricultural and Biological Chemistry* **47**: 2287–94

Yoshioka K, Hashimoto N 1984a Acetyl-CoA of brewer's yeast and formation of acetate esters. *Agricultural and Biological Chemistry* **48**: 207–9

Yoshioka K, Hashimoto N 1984b Ester formation by brewer's yeast during sugar fermentation. *Agricultural and Biological Chemistry* **48**: 333–40

CHAPTER 5

Batch distillation

D NICOL

Introduction

Until the introduction of the continuous Stein still in 1827 and its improvement, shortly after, by Aeneas Coffey in 1830 (Moss and Hume, 1981), all distillations had been carried out in pot stills (Nettleton, 1893) using the batch distillation technique. That pot-still distillation has survived in Scotland and elsewhere is a reflection of the unique product emanating from such stills, a product which commands demand from blenders and the palate of those not averse to a single malt.

Design and operation of stills

Traditionally, the actual distilling apparatus (Fig. 5.1), the still and condensing equipment, have been manufactured from copper (Underwood, 1935). It is no accident that this element has continued at least to the present time as the only suitable material for distilling high-quality malt whisky. Malleable, a good conductor of heat and showing good resistance to wear, it has been associated with distilling over many centuries. It is only comparatively recently that another attribute, its ability to influence flavour, has been more fully appreciated. That copper should be the material from which pot-still equipment is manufactured is a testimony to its undying role as an essential ingredient in the whisky recipe, as other areas have succumbed to the intrusion of stainless steel.

A still consists of three distinct, separate parts: the pot, the swan neck and the lye pipe or lyne arm, and the condenser (worm, formerly). The pot, as its name suggests, is that part of

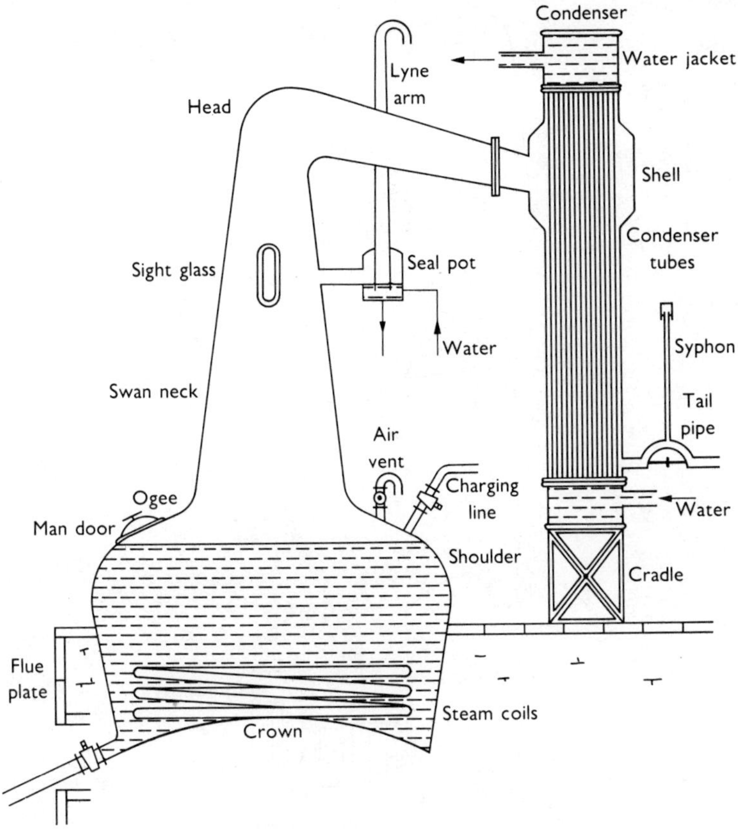

Fig. 5.1 Plain wash still.

the still which holds the alcoholic liquor to be distilled. It is provided with ancillary equipment – a man door, an anti-collapse valve, an air vent, charging and discharging cocks. These fitments are usually made of brass. The application of heat to the pot is by means of solid fuel, steam jacket, coils, kettles or gas, depending on the distillery.

The swan neck, attached vertically to the pot, may take several configurations – short, long, straight or tapering upwards; it may be provided with a ball shape at the base, be swept in at the base or just plain (Fig. 5.2). The point at which the neck is fitted to the body of the pot, above the shoulder, is referred to as the ogee. An anti-collapse valve, or sight glass, may be added in the case of wash stills. The lyne arm, a continuation of the swan neck, tapers towards the condenser. Like the swan neck, the

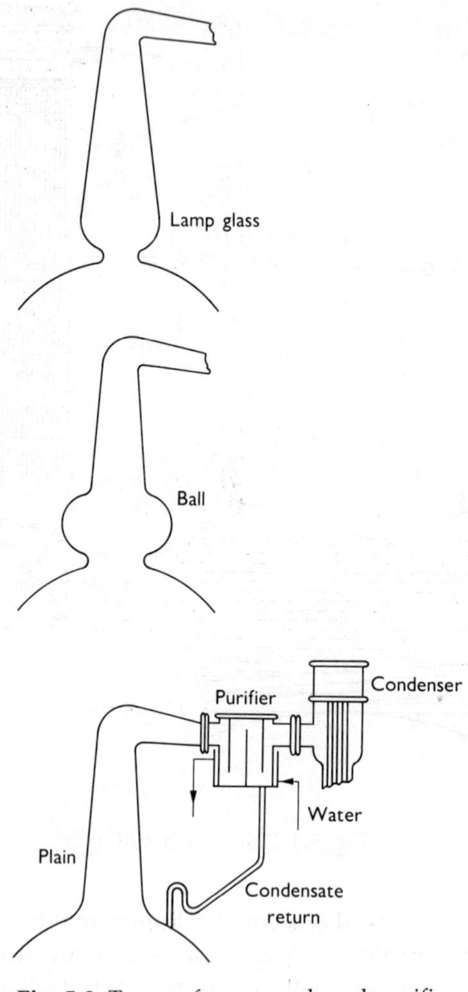

Fig. 5.2 Types of swan neck and purifier.

lyne arm may be short or long, be rising or falling gently, or be horizontal. A purifier may interrupt the lyne arm before it reaches the condenser.

In the earlier part of the twentieth century and before, it was customary to rivet the various components of the still together, but this has now been superseded by welding.

As would be expected, the gauge or thickness of copper used in still construction varies according to its particular location in the still. Heavy gauge will be found in the crown and the flue plates of the pot, thinner gauges being associated with the parts of the still on which vapours impinge. All copper material used must be arsenic-free.

The condenser, a derivative of the traditional worm, is a shell and tube heat exchanger designed to remove the latent heat of condensation of the vapours. It is normally made of copper and consists of a bundle of copper tubes sandwiched between two cooling water jackets; the tube bundle is encased in a copper shell into which the lyne arm is connected. The number of tubes and the overall length are governed by the worst conditions which may be met in terms of water supply and temperature during the distillation season.

A recent advance on the shell and tube condenser is the plate heat exchanger, which is now being introduced to some distilleries. Aesthetically not so appealing, it offers maintenance advantages over tubes, and as time goes on it may gain wider acceptance. It has the double advantage of using less water, thus producing higher-grade heat, which offers greater potential for reuse in the distillery.

PIPEWORK, SAFES AND RECEIVERS

Without the accompaniment of ancillary pipework, safes and vessels, no distillery would be complete (Fig. 5.3). Condensing apparatus is connected, via tail pipes, to the corresponding safe. Low wines distillates connect with the wash safe; foreshots, middle cut and feints with the spirit safe.

Low wines, derived from the wash, flow via a syphon, which retains a small sub-pool of distillate in the base of the condenser, to the wash safe by gravity. A small-bore sample pipe feeds distillate from the main flow via a three-way valve to a hydrometer jar in the safe for alcoholic strength determination. The jar is equipped with a hydrometer and thermometer. If necessary, the safe, vented to atmosphere through flame arresters can be close fastened by padlock to protect the distillate. The low wines distillate flows from a glass bowl in the wash safe directly by gravity to the low wines and feints receiver. The glass bowl is fed by a fixed spout fitted with a flap or tongue. The wash safe is usually constructed of brass.

Distillate derived from low wines follows a similar path from the condenser to the separate spirit safe. The spirit safe is the nerve centre of the production of good potable spirit and is

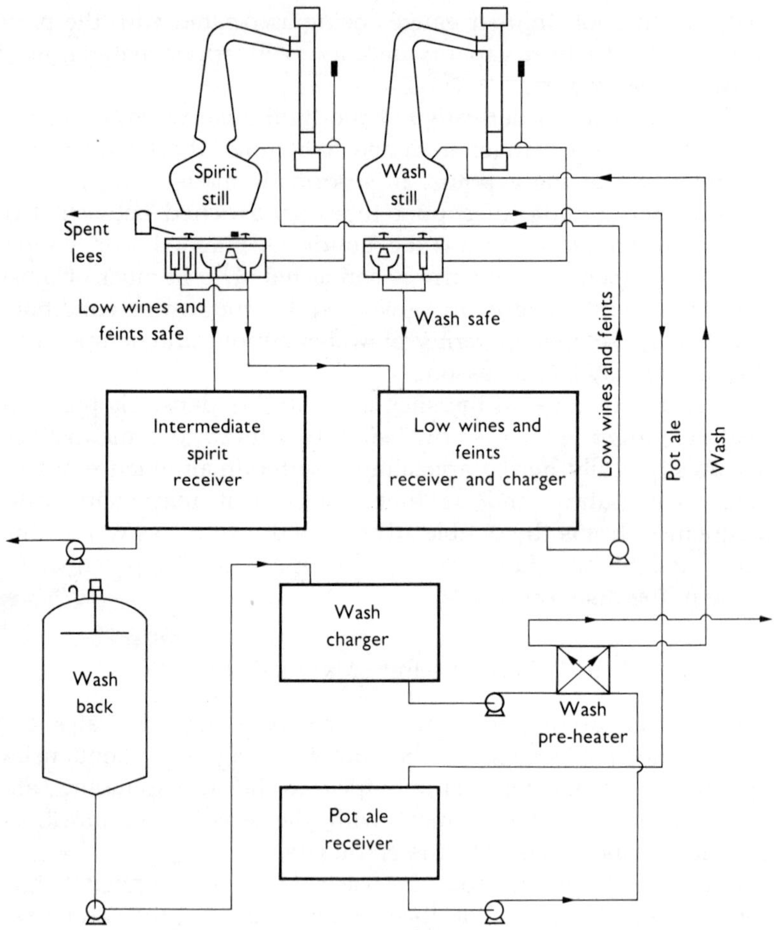

Fig. 5.3 Typical malt distillery flow diagram.

slightly more complicated than the wash safe. Similar to the wash safe, the spirit safe is divided into two sections: one for sampling and the other for observing and indicating heating adjustments for the correct flow of the main distillates. Unlike the wash safe, the spirit safe is provided with a horizontally rotating spout to separate the different fractions encountered during low wines and feints distillation. Each spirit still requires two sampling jars, one for general evaluation and the other for the demisting test. By means of the swivelling spout, distillate can be directed from the corresponding still to the low wines and feints receiver or the spirit receiver when desired. Distillates are gravity fed through the safe and sampled in a manner similar to that for wash distillations. The spirit safe is supplied with a small supply of mains water for direct use or it possesses a small cistern tank which is topped up with fresh water for the demisting test. At all times, the spirit safe is close fastened.

Receivers into which distillates flow are constructed of wood, glass-lined steel, epoxy resin-lined 'Corten' steel or preferably stainless steel. For low wines distillates the receiver should be capable of holding in excess of two complete wash distillations or the low wines equivalent obtained from a whole wash back. This is dependent on the number of wash stills, which should together be capable of accepting the contents of a complete wash back, at one time. Where the low wines receiver acts as a combined feints receiver, its size should be increased accordingly – approximately double that of a separate low wines receiver, with allowances upwards for rousing. Receivers are equipped with man doors, check dip facilities and are vented via flame arresters to atmosphere.

Prior to wash distillation, wash may be fed to a wash charger which should accept the total contents of a wash back. Preferably fabricated in stainless steel, such a vessel should be totally enclosed to prevent evaporation losses. It should also be equipped with an agitator to keep solids in suspension. A wash pre-heating system may be incorporated.

An intermediate spirit receiver is provided for collecting several new spirit distillations. Its dimensions should be such that it can accept at least three days (weekend) spirit production. The final spirit receiver and warehouse vat should be large enough to hold the total content of the intermediate spirit receiver. The spirit receiver is equipped with a close fastened man door, dip facility and flame arrester venting to atmosphere. Provision is also allowed for the addition of reducing water when necessary.

As the presence of any form of iron, especially post-distillation, is detrimental to the maturation of new spirit, due to the formation of blue-black iron complexes with the tannins extracted from the wood, on no account should new spirit be contained in vessels which are constructed in this metal.

WASH DISTILLATION

The wash still, capable of holding either a whole wash back charge or equal proportions of a charge derived from the wash back, is used to effect the separation of the volatiles – ethanol and accompanying congeners – from the wash. The strength of the charge in the still will be dependent on the original gravity of the worts and will approximate to 7.0% v/v ethanol. The volume of low wines distilled over will be slightly over one-third of the original volume of the charge to the still. The strength of the low wines collected will be about 21% v/v ethanol.

In the case of direct fired wash stills heated by coal or gas where the internal heating surfaces accumulate charred material, a rummager is necessary to prevent this accumulation, which would otherwise adversely affect heat transfer. The rummager consists of a rotating device from which copper chains or webbing are suspended, the webbing impinging on the internal surface of the pot. Rummagers are not necessary in indirectly fired wash stills, heated by internal steam coils.

Wash presented for distillation should be fully attenuated, and with properly maintained distillation plants full recovery of ethanol, within the limits of fuel efficiency, should be realisable. In some instances, wash may appear to be prematurely attenuated, within 24 hours, say, but in our experience it is preferable to distil wash over 44 hours old. Wash, by its very nature, contains unfermented dextrins, yeast residues, some husk and dissolved gases. When charged to the still, on initial heating, it can behave unpredictably, and as it approaches boiling point and immediately thereafter it has a tendency to froth. To control this frothing, a sight glass is situated two-thirds of the way up the swan neck – should froth advance beyond this point as seen in the sight glass, then a reduction in heat is necessary. The advance of uncontrolled frothing into the safe and hence the receiver is to be avoided. If this occurs it is referred to as a foul distillation. Prior to the introduction of sight glasses, stills were sounded with a wooden ball on a chain. The advance of the frothing up the swan neck resulted in a dull note when the neck was sounded with the ball, which was an indication to reduce the heat.

In addition to sighting or sounding, the state of the initial distillation can be observed in the wash safe. The imminent arrival of distillate is announced by advancing expanding gases and vapours which cause the tongue on the spout in the wash safe to flap vigorously. Again, this is a signal to adjust the heating.

Overcharging a wash still is to be avoided as an overfilled still will be difficult to control with respect to frothing. Ideally, the still should be filled to two-thirds pot capacity. At this level, once the low wines have been recovered, the residual pot ale in the pot should cover the flue plates or steam heaters. At no time should the heating surfaces, during and after distillation, be exposed.

With cold wash, the over-enthusiastic use of steam or solid fuel will exacerbate fouling of heat transfer surfaces, as cold solids will more readily bake on to the hot heating surfaces. It is therefore judicious to apply gentle heat initially until the wash is sufficiently heated to avoid this effect. Where wash pre-heating is used, this bake-on effect is minimised. From time to time it will be necessary physically to clean the internal heating systems where distillation times are being affected due to poor heat transfer.

Wash solids riding on the wave of initial froth and vapours will tend to accumulate in the swan neck and even in the lyne arm. As this is deterimental to reflux, it will be necessary to scrape off these deposits.

In most distilleries the stills are now equipped with steam pressure gauges and temperature recorders for recording wash and vapour temperatures. Level controls may also be incorporated and in a few distilleries wash distillation is automatically controlled through programmable logic controllers linked to densitometers and steam valves.

A plentiful supply of cold water to the condensers is necessary at all times. Failure of condenser water will result in a still literally boiling over; this is disastrous for the glassware in the particular safe and also contaminates the low wines with wash. Should this occur the heat must be stopped and safety observed. The distillate should flow smoothly into the safe and be maintained at 15–20°C. Collection of warm distillates, e.g. in excess of 30°C, will lead to evaporation losses, apart from being well outside the calibration range for hydrometers. Distillate temperatures which vary considerably from the instrument calibration temperatures will result in erroneous spirit indications and it will be necessary to make temperature corrections.

At all times, a record is maintained of the charges, the charge number, volume (dip), time charged, time in safe, the strength, temperature, time complete and the dip indication in the low wines receiver. During distillation regular checks are kept on the strength of the distillate until it approaches the cut-off point. For the conservation of fuel it is now normal practice to cease wash distillation when the strength of the distillate is about 1% v/v ethanol.

As the need for fuel conservation (Malkin, 1983) continues to grow, several techniques, in addition to wash pre-heating, are infiltrating the still house. Special attention is being paid to condenser water, which can be controlled to provide hot water at 85°C. The condensate may require further sub-cooling, as only the latent heat has been extracted, but the hot water may be used for mashing, malt drying, secondary industries or it may be reused within the distillation process. This latter use hinges on the incorporation of a thermocompressor or mechanical vapour recompression unit, with the flash steam generated being used as the driving force for the distillation. Where steam coils or kettles are used, condensate is returned to the boiler feed water tank. With direct fired stills using solid fuel, LPG or natural gas, heat can be recovered from the flue gases. As a further aid to energy conservation, as well as for safety, all charging or discharging lines conveying hot liquor or water and all steam lines should be lagged.

Residual pot ale (Black, 1984), being a valuable food resource for ruminants as a feed supplement when dried, is normally collected for further processing into distillers' dried solubles or distillers' dark grains.

LOW WINES AND FEINTS DISTILLATION OR SPIRIT DISTILLATION

Unlike wash distillation, where care must be exercised at the first sign of distillate appearing in the safe to avoid frothing, spirit distillations are not beset by this problem. The spirit distillation is, however, critical, as it is governed by the recovery of three fractions – foreshots, middle cut (new spirit) and feints. The separation of these fractions by cut points peculiar to the distillery in which they are used, determines the quality and consistency of the final product (Fig. 5.4).

The charge to a spirit still is usually a mixture of low wines, foreshots and feints at a strength of between 25–30% v/v ethanol. In distilleries where low wines, foreshots and feints are separately collected, a charge is blended from each before being fed to the still. This charge, regardless of its source, should be maintained at a constant strength. Where the receiver acts as a vessel for collecting low wines, foreshots and feints together, the charge will comprise a mixture of the three components, with variations in strength. The variation should lie within 25–30% v/v ethanol.

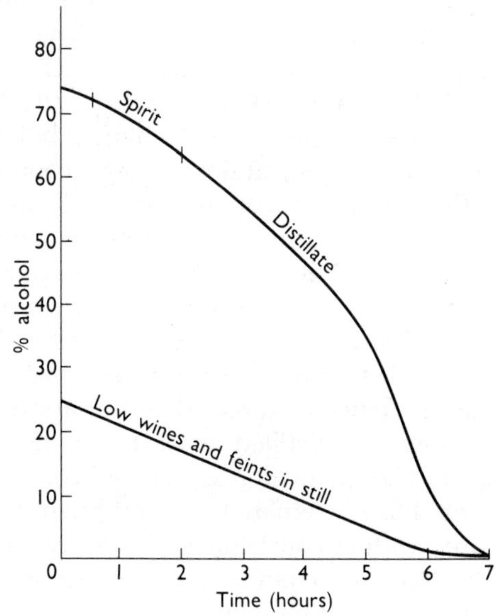

Fig. 5.4 Variation in composition of still contents and distillate during spirit distillation.

As with wash, the charge may be pre-heated with residual spent lees or from hot condenser water. On no account should the spirit still be overcharged, as the reflux ability of the still will be adversely affected, with detriment to the final product.

The spirit still can be brought to the boil quickly, but heat must be reduced to ensure a slow to moderate flow of foreshots, to effect reflux. The foreshots, responsible for cleansing or purging out the heavy oils clinging to the internal surfaces from the previous distillation, are collected between 15–45 min, depending on the distillery. They contain water-insoluble oils, readily dissolved at 75–80% v/v ethanol, and produce a turbid, milky solution when diluted with water to 45.7% alcohol. This is the basis of the demisting test, the test which determines the point in time and strength when the foreshots become potable spirit. As the foreshots continue to be collected, they are subjected to reduction to 45.7% alcohol with water from the small cistern tank at the safe, to check the clarity of the mixture. As time elapses the milky appearance on the addition of water will diminish until, as mentioned above, in the worst case, the foreshots plus water will remain clear after 45 min collection. At this point, the spirit is deemed potable and the flow of distillate is directed to the intermediate spirit receiver. The strength of the distillate at this point may vary from charge to charge but will be in the range 70–75% ethanol. This is the first cut, dependent on the alcoholic strength of the original charge, but varying little from distillation to distillation, under constant conditions.

The flow of the middle cut or new spirit is maintained at a slow pace to ensure maximum reflux. Strength and temperature of the distillate are carefully monitored. The collection of the middle cut continues until the second predetermined and fixed cut point is reached. The strength may be 62–64% depending on the distillery. Once this second cut point has been reached, the flow of distillate, as feints, is directed to the appropriate receiver. Feints can now be distilled at a faster rate than either foreshots or spirit. The distillation is deemed complete when the distillate contains 1% v/v ethanol, as indicated by the hydrometer. To ensure product consistency and to avoid undue spirit losses, reference to correction tables for spirit indication at different temperatures may be required.

Similar to wash distillation, a record of each distillation is kept, recording date, time charged, time in safe, temperature, strength, time on spirit and strength, time off spirit and strength, time of completion of distillation. An initial and final receiver dip is also recorded. A typical malt distillery production cycle is shown in Fig. 5.5.

Modelling of batch distillation

The ethanol and its accompanying congeners, which eventually give rise to the characteristic aroma of a whisky or any other similarly distilled product, are derived from the raw materials, the changes wrought on the raw materials by the various process stages, the fermentation and the effects of distillation. Some aroma compounds are selectively concentrated by distillation while others are reduced. At the fermentation stage most flavour and aroma compounds are at their most dilute. Batch

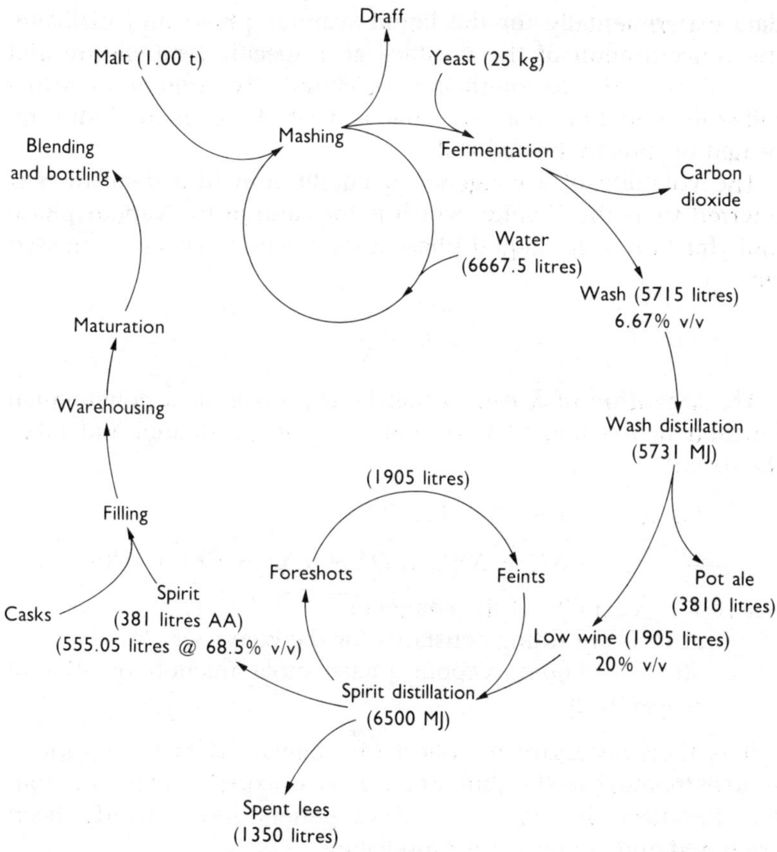

Fig. 5.5 Typical malt distillery production cycle. All figures are standardised to 1 t malt.

distillation seeks to produce a consistent balanced distillate reflecting in its overall aroma the interplay of several hundred compounds in various concentrations, which contribute individually and cumulatively to the final bouquet.

The volatility of the various congeners is governed by vapour liquid phase equilibria. As their concentration relative to ethanol is low, ethanol plays an important role in the study of such phase equilibria. The modelling (Williams and Knuttel, 1983) of batch distillation has, so far, been limited to studies on ethanol and water phase equilibria, and ethanol, water and aroma compound phase equilibria.

Before a model can be constructed, it is necessary to collect data experimentally for the liquid, vapour phase and distillate, the concentration of the volatiles at a specific temperature and time during distillation being measured. To adhere to actual full-scale distilling practice, this is best done using indirectly heated or directly heated stills.

The volatility of a congener at equilibrium in a distillation is referred to as the K value which is the ratio of the vapour phase mole fraction to the liquid phase mole fraction. This is expressed as:

$$K = \frac{Y}{X}$$

The logarithm of K can further be expressed as a polynominal function of the liquid phase mole fraction of ethanol and takes the form:

$$\log K = A(1 - X) + BX + X(1 - X)(C + DX + EX^2 + FX^3 + GX^4)$$

where K = volatility of the congener
A to G are fitting constants for the congener
X, Y = liquid, vapour phase mole fraction of ethanol respectively

It is then necessary to collect the relevant data from various sources relating to the different volatiles, experimentally or from the literature. In excess of 33 congeners have already been examined and submitted for modelling.

A constructed model, based on specified criteria, should enable the percentage of the charge, the final ethanol content of the residual charge, the instantaneous ethanol content or the cumulative ethanol content of the distillate to be determined. From the above equations the vapour phase mole fraction of any selected congener and the corresponding material balance can be calculated by the program. Pot liquid and corresponding distillate mole fractions, the instantaneous K values, the cumulative mole fraction in the distillate, the percentage recovery of the initial congener present, can all be calculated by the program as a function of the percentage of the charge distilled over.

By constructing a model and designing a computer program, which ideally should incorporate calculations for reflux, it is possible to simulate the dynamics of batch distillation. Along with ethanol, the more common esters, higher alcohols and volatile acids have been subjected to this technique. In the long term, it will enhance the understanding of a process which has been regarded more as an art than as a science.

Optimisation of product and yield

PRODUCT OPTIMISATION

It has often been said that 'there is no such thing as a bad whisky – some are better than others'. That this is so, is more than likely due to the subjectivity surrounding flavour and aroma, to the different psychological responses that individual whiskies elicit from those engaged in nosing or tasting the product. Such is the complexity of the different bouquets, that there will be as numerous and complex a response to the impact of specific whiskies.

It is the duty of every distiller to ensure that the individual product of the pot still should maintain its own identity and profile, recognisable as not just coming from a vague geographical location but from its actual place of origin. That this statement no longer carries universal weight is a reflection of how, in the interests of economy, the raw material from which malt whisky is made is procured from large malting units; the distillery equipment – from the mashtun to wash backs, from the wash backs to the stills – has acquired an air of uniformity. Fortunately, to counter-balance this move to overall uniformity or standardisation, the location of the still with its own peculiar environment and water supply remain unchanged. In this 'graduation' to uniformity a few elements within each distillery remain unsullied – the shape and size of the still, the cut points which give the distillate its unique characteristic bouquet.

It is the fusion of the four states of matter – earth, water, air and fire, recalling the alchemy from which present distillation (Underwood, 1935) is derived – that has projected this ancient art into the science of the present time. Before the advent of chemical and instrumental analysis (Swan *et al.*, 1981), the early distiller had to rely, as he still does, on the senses of smell and taste as the means of classifying the quality of his product. 'Proof' of the distillate was determined by soaking gunpowder in raw spirit; if it ignited when lit by a flame, it was deemed to be overproof and if there was no ignition then the spirit was underproof. The advent of the spirit hydrometer injected a greater degree of control over the whole distillation process.

Means of controlling and monitoring the process were not confined to the distiller, but were also available to HM Customs & Excise. In fact the origin (Clow, 1952) and state of the present distilling scene in Scotland owe much to the Excise officers. It was the latter, as Preventative Officers engaged in the seizure and destruction of illicit stills, and finally the passing of the Excise Act of 1823, which founded the conditions under which the manufacture of malt whisky has developed into the industry which is known today. With the requirements that all stills should have a minimum size, not compatible with illicit operations, and that each still should be licensed, the first moves to uniformity had begun. More scientific control (originally hydrometry); the introduction of the spirit and wash safes; crown locks on all vessels; declaration of amounts of malt used and volumes of spirit produced from a given weight of raw material; provision of secure Customs & Excise warehouses in which to store the filled casks; all ensured that at every stage of the process both the distiller and the Revenue knew what was happening. Failure to exercise control and produce less than the malt's potential yield of alcohol led to penalties levied on the distiller. It was the close scrutiny to which the distiller was subjected by the Revenue that inevitably led to the demand for optimisation of both product quality and yield, each stage of the process being minutely examined in an attempt to extract every last drop of the alcohol from the raw material.

Without a product of good quality, the reputation of a distiller and his still would be sorely dented. It was therefore in the interest of the distiller to produce a good product in order to maintain and increase his market share. That the reputation of the product of most distilleries is the result more of accident than design is no fatuous statement.

The size and shape of stills tend to be fixed; the length of the swan neck, the length and attitude of the lyne arm, whether gently rising or falling towards the condenser; the means of heating the pot, whether directly fired externally by solid fuel or gas or internally heated by steam, using coils, kettles or panel heaters – such fixed physical elements impinge directly and indelibly on the flavour of the final product, before racking.

The quality of the final distillate is governed by the type of yeast used for fermentation; the age and strength of the wash to be distilled; the rate of distillation; the selection of the cut points, from foreshots to middle cut and from the middle cut to feints, during low wines and feints distillation. The consistency of the charge to the spirit still will also affect the quality of the final product; variations in strength invariably lead to fluctuations in congener levels from distillate to distillate, with very low strengths leading to the occasional expensive blank run when no spirit is taken off.

It has recently been demonstrated that immature wash (Geddes, 1985) produces a distillate with a pronounced 'green note'. Such a product is derived from wash about 35 hours old. Although yeast fermentation, to all intents and purposes, may appear complete, the effects of clean wash appear to be detrimental to the final product quality. It is believed that the action of lactobacilli on the cell contents of autolysed yeast cells in long and prolonged fermentations, without detriment to yield, produces a more acceptable spirit.

Distillation of fresh wash is not to be encouraged, as it leads to the potential bumping over of the wash still contents in a foul distillation. Such a distillation adds contaminants to the contents of the low wines and feints receiver, which will produce off-notes in the final distillate.

Length of distillation is governed by the condenser efficiency (cooling water availability), the heat impact, and volume to be distilled. Ideally distillation length should be governed by the mashing cycle and as far as possible be equated with this for both wash and spirit distillations. Distillations in warm summer weather are affected by cooling water and ambient still house air temperatures – high temperatures inevitably slow down the collection of distillates and adversely affect reflux. The spectrum of congeners within the final distillates in summer may therefore show marked variations from distillates collected during winter conditions.

A wash distillation can be fraught with problems associated with the age of the wash, as previously mentioned. Initially, if wash pre-heating is not available, heat should be applied gently to avoid fouling of the heating surfaces, but once initial frothing has subsided the rate can be increased without detriment to low wines quality. The temperature of the distillate in the wash safe should not exceed 20°C and, if possible, should equate with the calibration temperatures of the safe hydrometers, otherwise temperature correction will have to be applied.

With low wines and feints distillations, as expected, much greater vigilance is required, as there are two cut points to be considered. Cut points (one for the foreshots to middle cut, with the demisting test, and the other for middle cut to feints), are usually decided and fixed for all time when a distillery first commences production. Like the stills from which the distillates flow, they have a marked influence on the final aroma and hence consistency of product. Unless drastic problems are encountered in distillation, e.g. instability of charge phases, on no account should they be altered, as to do so will alter the character of the product. The cut points affect the strength of the final distillate, which, depending on the distillery, will range from 66–75% v/v ethanol. Examples of chromatograms of foreshots, feints and mid-range volatiles are shown in Figs 5.6, 5.7 and 5.8 and Table 5.1 represents analyses of three different spirit charges from a malt distillery.

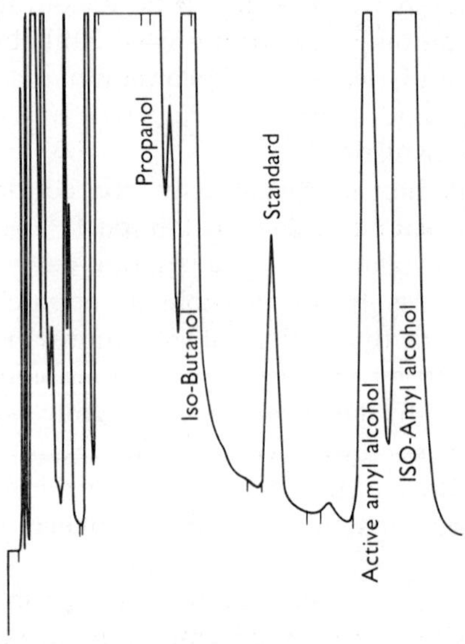

Fig. 5.6 Chromatogram of foreshots.

The demisting test ensures clarity of the final distillate – a distillate with low concentrations of higher fatty acid esters C_{16} upwards. These esters inevitably contribute to chill haze problems on the reduction to 40% v/v alcohol on the bottling of the

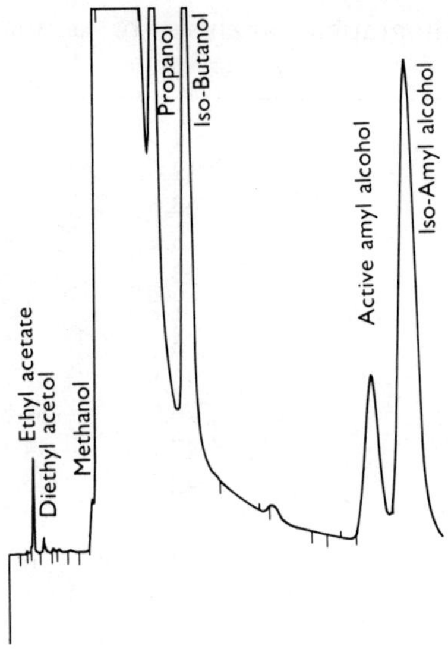

Fig. 5.7 Chromatogram of feints.

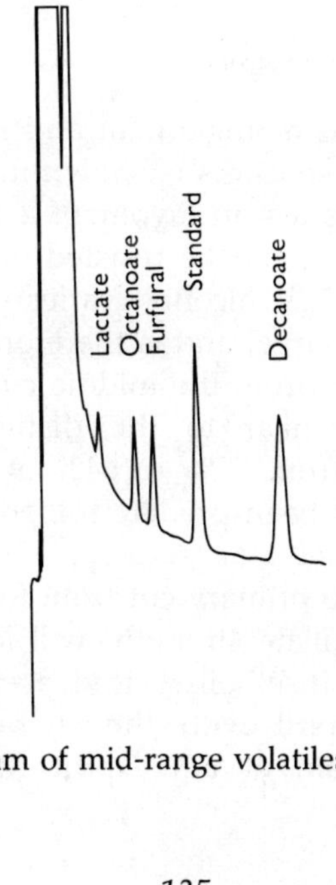

Fig. 5.8 Chromatogram of mid-range volatiles, showing ethyl esters.

Table 5.1 Chromatographic analysis of a malt whisky new make.

Period	4	4	4
Charge	12	14	16
Strength (% v/v)	63.5	71.5	69.4
Acetaldehyde	3.2	3.8	6.8
Ethyl acetate	23.7	25.5	27.0
Diethyl acetal	1.7	1.2	2.2
Methanol	5.1	4.6	5.3
Propanol	40.8	42.7	41.9
iso-Butanol	79.8	80.8	80.5
o.a. Amyl alcohol	47.7	44.7	49.5
iso-Amyl alcohol	142.5	145.5	142.5
Total higher alcohols	331.1	313.7	314.4
Ethyl lactate	4.7	2.5	4.1
Ethyl octanoate	1.6	1.9	1.7
Furfural	3.3	3.9	4.2
Ethyl decanoate	5.7	5.6	4.5
β-Phenethyl acetate	5.7	7.5	5.9
Ethyl laurate	2.1	2.6	2.1
β-Phenylethanol	3.8	0.6	0.6
Ethyl myristate	0.6	1.1	0.6
Ethyl palmitate	2.7	3.3	2.6
Ethyl palmitoleate	1.5	1.9	1.4

o.a. = optically active
Concentrations are g per 100 litre alcohol

mature product, both as a single malt and in a blended whisky. A chromatogram of these esters is shown in Fig. 5.9. Pressure to eliminate the demisting test in favour of a fixed time period for collection of foreshots should be resisted, as the point at which foreshots clear at 45.7% alcohol varies with time and the strength of the initial charge, and is therefore not fixed.

The second cut point, from the middle cut to feints, is fixed at a known strength peculiar to the distillery in which it is operated. It can vary from 57% to 64% alcohol, depending on the distillery. As it has been predetermined, any deviation from it will affect spirit quality.

Failure to observe the primary cut from foreshots to spirit, e.g. early cut at higher distillate strength, will lead to a spirit which is contaminated with fusel oils, ethyl acetate and short-chain fatty acid esters, a spirit with the characteristic 'pear drop' aroma. A chromatogram of new spirit fusel oil is shown in

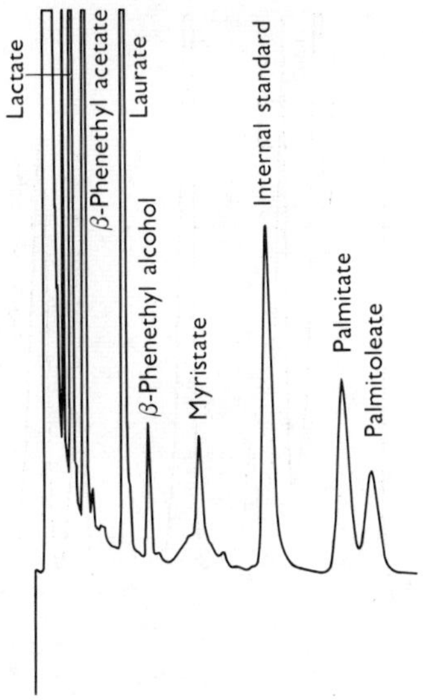

Fig. 5.9 Chromatogram of chill haze precursors, showing ethyl esters.

Fig. 5.10. The strength of the spirit will increase while that of the low wines and feints in a common receiver will drop, upsetting the equilibrium of the concentration of the congeners in the low wines and feints pool, with alteration of congener concentrations in the subsequent spirit. The opposite effect will occur with an over-run of foreshots, with increasing alcohol strength of the common pool with concomitant decrease in spirit strength. Should this state be allowed to continue in subsequent distillations, it will eventually lead to blank spirit runs as more and more water-insoluble congeners dissolve in the alcohol-enriched aqueous layer in the low wines and feints receiver. The combined strength of the low wines and feints should not exceed 30% v/v alcohol.

The second cut, predetermined and fixed, ensures that feints-related compounds do not contaminate the final product. Feints are characterised by a distinctive stale note with a metallic after-taste. Over-run of the middle cut will introduce this note and taste to the spirit, will result in a weakening of the strength of the low wines and feints, and if continued unchecked in subsequent distillations will cause a drop in strength of the final product. Although facilitating the clearing of foreshots in subsequent distillations, introduction of feints to spirit by failing to

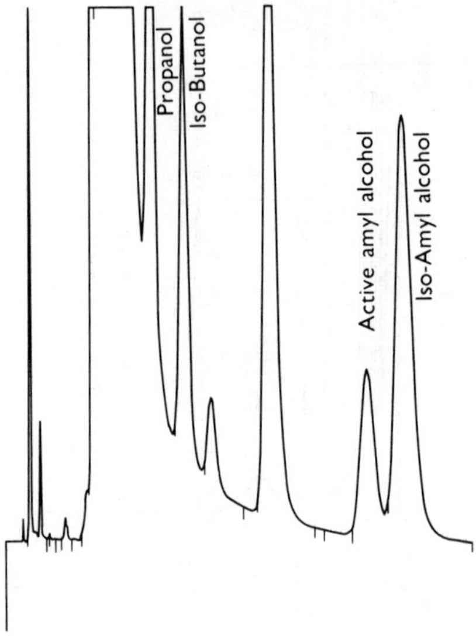

Fig. 5.10 Chromatogram of new spirit fusel oil.

observe the cut point will most decidedly produce a spirit of inferior quality.

Early cutting from spirit to feints will produce an effect similar to that of over-running foreshots, that is, lighten the spirit bouquet, increase the strength of the common pool and lead to potential non-clearing problems.

It is therefore essential that cut points should be strictly observed to avoid instability in the phase equilibrium within the low wines and feints receiver, with its common pool. Such problems can be reduced by the use of separate receivers for collecting low wines, foreshots and feints; spirit charges (to spirit still) are then blended together from the individual receivers to produce a consistent charge.

Inevitably, slight variations in charge composition do occur, which are reflected in the final distillates; by collecting several spirit distillations in a large intermediate spirit receiver such differences will be smoothed out.

The quality and consistency of final spirit charges are determined against an agreed standard and the congener levels are determined by gas chromatography.

Foreshots can be brought in quickly (no frothing, as with wash distillation) and collected slowly, ensuring adequate reflux. Once cleared using the demisting test (which governs the point at which the distillate is deemed potable), to ensure a balanced spirit a constant steady flow of middle cut should be maintained at a rate conducive to good reflux; once off spirits, at the second cut, the distillation rate can be increased. The total cycle should parallel that of a wash distillation, again, determined by the mashing cycle. Failure to observe distillation rates by over-generous application of heat will produce an unbalanced fiery spirit containing an excess of congeners derived from the fore-shots, feints and low wines, and will potentially overload the condenser system, adversely affecting any reflux.

Where several stills are being simultaneously operated, their operation should be staggered so that no two or more stills of the same type are running in parallel. This will tend to produce a consistent pool of low wines and feints, rather than peaks and troughs in the strength of charges being fed to spirit stills. Should a build-up of feints occur due to improper timing of distillations, it may be necessary to cease wash distillation in order to clear the excesses. It is inadvisable to stop spirit distillations, as a restart will adversely affect congener levels. A practice called 'eking a still' may be used to clear excesses, but this should be done only as a last resort. This technique applies only to spirit distillations. Any deviations in distilling practice in an attempt to alleviate imbalances will lead to departures from a consistent distillate.

It should be remembered that the copper patina which has formed on the copper surfaces exposed to spirit vapours, plays a marked role in the fermentation of congeners, especially the esterification of organic acids by alcohols. The residence time and interplay of the various volatiles exposed to the copper will affect the formation of the different reaction compounds which eventually emanate from the still. In this respect, it has been demonstrated that the volatile compounds of sulphur (Leppänen *et al.*, 1983) derived from the yeast (sulphides and mercaptans) are reduced and removed by the interaction with copper, especially in the wash still. Copper is therefore an essential element in producing a clean spirit. Any misguided attempts to introduce stainless steel as an alternative to copper, especially in areas where reflux and condensation occur, is fraught with danger.

Although some spirits are noted for their distinctive peaty aroma, derived from peat fired kilns at the malting stage (Howie and Swan, 1984), the carry over of phenolic characteristics, so distinctive in these spirits, will be affected by the alcoholic strength of the low wines and feints constituting the charge to the spirit still. The greatest recoverable concentration of steam volatile phenols distils over towards the end of the middle cut as the strength of the middle cut reduces and the vapour phase is enriched with water vapour in which the phenols are entrained. It is in the feints that the greatest preponderance of phenols is congregated. Any imbalance within the feints in terms of increasing strength will adversely affect the ability to impart the correct phenolic characteristics to the spirit.

As oil prices hardened in the 1970–80s, the cost of distillation has increased. Attempts to assuage the cost penalty incurred by distillation have meant that several distillers have introduced measures to counteract this. In this respect, there has been a move towards high-gravity brewing (Rutherford, 1981) or mashing. Such a move has to be taken with caution, as worts of increased strength inevitably result in the production of higher strength low wines. Depending on the configuration of the distillation equipment, especially the spirit stills, difficulties may be encountered in the ability of the foreshots to clear, as the low wines and feints pool will tend to contain a higher proportion of ethanol soluble congeners, e.g. fusel oil and higher fatty acids and esters. Spirit stills possessing long swan necks and lyne arms, with their existing enhanced reflux ability, will not encounter this problem unless distillation rates are unusually high.

Examination of the contents of a low wines and feints receiver reveals two distinctive layers: a thin ethanol soluble top layer of low density, floating on the surface of the aqueous ethanol lower layer. Most spirit charges are derived from the lower aqueous layer. It is debatable whether or not both layers should be emulsified, or left and allowed to separate before charging to the spirit still. Experience has demonstrated that in some cases, depending on whether more than one low wines and feints receiver exists, it is somewhat advantageous to allow phase separation to occur in the low wines and feints receiver and charger. As the low wines and feints consist of a mixture of water-insoluble and ethanol-soluble compounds, an aqueous ethanol-insoluble layer exists on the surface of the aqueous ethanol layer. To discharge the total contents of the receiver/charger into a spirit still means that at least one still will receive a disproportionate preponderance of water-insoluble oils. These compounds, fusel oils and higher fatty acid esters, will most certainly result in a blank distillation, when the demisting test fails and no spirit is collected. To avoid this problem and its knock-on effect on immediate and subsequent distillations, it is advisable to withdraw charges only from the lower aqueous layer, ensuring that the insoluble layer remains in the receiver – in effect the low wines and feints receiver is being used as a phase separation vessel. Contrary to certain beliefs, an equilibrium develops between the two phases and a thin molecular layer of fusel oils, esters, fatty acids etc., remains on top of the aqueous layer. Disturbance of this layer once it has developed, by emulsifying or discharging it into a still, will have serious quality implications. Where more than one low wines and feints receiver and charger exists it is prudent to use one of the receivers as the primary collecting vessel with an underbalance system to a second or third vessel. Charges to spirit stills are drawn from the latter vessels which will contain a clearer aqueous ethanol layer than the primary vessel. Blank distillations will then be very uncommon unless distillations become unbalanced, resulting in increasing strength of the low wines and feints. Should this unfortunate condition develop, water may be added to the receiver to effect phase separation. In the reconstruction of a distillery, cognisance of the layout and configuration of low wines and feints receivers should not be ignored, as alterations to their size, shape or number can adversely affect the congener levels in the final product.

The use of hot condenser water (Malkin, 1983) as a source of flash steam, produced by either mechanical vapour recompression or steam ejectors and heat pumps, is gaining momentum in distilleries. The effects of running condensers at higher than traditional temperatures does not, at first, appear to have affected spirit quality; but where changes to the process are being envisaged, potential quality problems should always be borne in mind.

OPTIMISATION OF YIELD

With the development of barley strains possessing improved malting quality and increased starch reserves; as the various facets of the processes within the distillery have become better understood; with the introduction of improved mashing systems and improved yeast strains; and with fastidious attention to hygiene, distillery yields have, over the years, continued to increase towards the theoretical maximum. The distillation equipment, in itself, does not play a marked part in this improved yield – it is more the careful manipulation of the equipment with specific reference to final end-points at the completion of the distillation, which has a direct bearing on yield.

Yield, expressed in litres of absolute alcohol per tonne of grain mashed, is derived from the original barley. Responsibility for yield rests with the seed merchant, the farmer, the grain merchant, the maltster, the yeast supplier and the end-user, the distiller. A good distiller will pay minute attention to detail in an attempt to extract every available molecule of fermentable sugar from the malt at the conversion stage in the mashtun; he will nurse the fermentations in the wash backs and coax every litre of alcohol from the yeast; finally, he will recover, as far as is economically feasible, as many litres of alcohol from the fermented wash during distillation. It is the distiller, alone, who is in complete communion with the process of transforming the fruit of the earth to the water of life. The litres of alcohol collected after distillation reflect the integrity, efficiency and judgement of all those involved in the various stages of supply, purchase and use of the raw materials. In this way unit costs are kept low.

At the present time, it is not without the bounds of expectation to exceed 400 litres per tonne. This figure can be much reduced where problems are encountered in the stages prior to distillation. It is at the stage of purchase that quality assurance and control functions (Duncan and Philp, 1966; Dolan, 1976; Sharp, 1982) can lend invaluable expertise and assistance in the selection of the best available varieties, maintaining an unbroken monitoring chain between supplier, purchaser and end-user. A distiller's production overheads remain constant, all things being equal, regardless of whether or not good or bad quality malt is processed in the search for economy. Poor malt will increase the unit costs. Poor quality yeast will also have a detrimental effect on yield.

Given that the integrity of distillation plant and equipment is well maintained, total recovery of available spirit will be ensured. This, however, is an ideal situation and is not always the case. The most effective means of identifying a loss of product during distillation is obtained by calculating 'the percentage over attenuation'. For a given amount of wash distilled, derived from a known quantity of malt, the distiller will expect a corresponding volume of spirit equivalent, within a certain percentage, to the alcohol contained in the original wash. Any large deviation outside the bounds of this percentage calls for a critical examination of the process, from worts declaration in the wash back to the condition of the distilling equipment.

Percentage over attenuation is calculated from a formula based on attenuation and the original volume of the declared worts. For every 100 litres of worts collected and fermented, attenuating through 8° specific gravity, 1 litre of absolute alcohol can be expected. For example, a wash back holding 41 000 litres of wash, which was declared at 55° of an original gravity, and which reaches a final gravity of 998 or $\overline{02}$, will have attenuated 57°. The attenuation charge is calculated as follows:

$$\text{Attenuation charge} = \frac{\text{litres of wash} \times \text{attenuation}}{8 \times 100}$$

$$= (41\,000 \times 57)/800 = 2921 \text{ litres AA}$$

In practice this figure is a slight underestimation, and given good fermentation a distiller might expect to recover, say, 3023 litres AA in the example above. To calculate the percentage over attenuation, the following formula is used:

percentage over attenuation

$$= \frac{\text{actual volume of spirit collected in litres AA} \times 100}{\text{attenuation charge}} - 100$$

$$= \frac{3023 \times 100}{2921} - 100 = +\,3.49\%$$

This figure reflects approximately true statements of declaration both in original and final gravities. Departure from the above figure, either negative or high positive values, demands investigation.

A negative figure points to either over-enthusiasm in declaring the OG (over-declaration) or a loss of wash between wash back and wash still, or losses in distillation. A high value, in excess of 3–4%, will be exclusively due to under-declaration, i.e. reporting a lower OG. Where wash back declaration is suspected in reflected percentage values, the accuracy of the tun room hydrometers will be implicated and perhaps the allowances for temperature and work (fermentation). In such cases, accuracy of declaration can be determined in the laboratory where facilities should exist for calculating OGs from samples of wash. Potential spirit yields from fully fermented wash will give an indication of the volume of alcohol to be derived from a specific wash back.

Loss of wash or distillation losses or even high evaporation losses can be the result of plant failure, e.g. leaking pumps, flanges, vessels etc. Unfortunately, sometimes the human element is involved, and losses of great magnitude can occur – very depressed percentages over attenuation with high negative values will result. Distillation losses can occur due to product losses via leaks in the pot, to incomplete distillation via pot ale or spent lees, or as vapour losses in the neck or lyne arm, and even in the condensers. In fact, losses can occur, unseen and seen, anywhere within the equipment. It is essential that the distillation apparatus is examined regularly, checking for passing valves, the suspected vapour leaks or the tell-tale sign of condenser cooling water appearing in the safe when a distillation has been completed. Where indirect steam heating is used, it is not impossible for the contents of the pot to appear in the condensate return to the boiler feedwater tank. Substantial insidious losses of product will therefore continue undetected by the unwary distiller, for example from still discharge valves, invisible vapour leaks from swan necks or lye pipes and faulty condenser tubes.

To confirm suspected product loss, the services of the laboratory should be called upon. Excessive alcohol in pot ale or spent lees implicates wash safe and spirit safe hydrometers respectively, with a premature ceasing of distillation. Alcohol present in condensate suggests leaking flanges in the steam coils. As a matter of routine the laboratory should receive regular samples of condensate return, pot ale and spent lees for alcohol detection.

Arbitrary limits for losses are:

	%
pot ale	<0.03
spent lees	<0.05
condensate	<0.0001
vapour loss	0.00
condenser water	<0.0001

A major calamity which can occur, with loss of product, time and plant, is the collapse of a still. Invariably caused by a combination of circumstances, both human and mechanical, it is avoided by paying strict attention to charging and discharging procedures and safety procedures, should a distillation require to be stopped in mid-stream for any reason. Further insurance against such an occurrence is the provision of a water seal pot as opposed to the normal vacuum relief valve. The latter type of valve requires constant examination, especially the seat and the lever, to ensure that solids are not impairing freedom of movement and that there are no leaks. Such vacuum relief valves are more vulnerable on wash stills than spirit stills.

Strict surveillance and attention to detail at every stage of the process, close co-operation between distiller and the laboratory control function (Duncan and Philp, 1966; Dolan, 1976; Sharp, 1982) will ensure that the awaiting casks will be delivered of a charge which, in volume, will reflect the maximum alcohol yield possible. In the pursuit of volume, however, quality remains paramount.

The universal scene

Scotch whisky, with its world-wide reputation, has attracted imitators – not least Japan, where various companies have established distilleries modelled on Scotch malt distilleries. Grain spirit, simple to produce, can be distilled almost anywhere in the world, as it is less affected by climatic conditions than malt whisky. It is therefore possible to produce a Scotch malt-type whisky, locally distilled and matured, and blend it with the mature product of an indigenous grain distillery.

It is not surprising that, given a source of grain and malt, blended whiskies produced locally have sprung up not only in Japan, but in Korea, Argentina, New Zealand, Spain and even Mexico and the sub-continent of India. The expertise for such activity may have originated in Scotland, but fortunately, peculiar to Scotland alone is its location and climate. These latter attributes cannot be exported and remain the prerogative of the Scotch malt whisky distiller.

In a chapter on batch distillation, mention must be made of the Irish distillers, who would claim that the origins of Scotch whisky came from that island. As opposed to Scotch malt whisky production, Irish whiskey differs from malt whisky in that the raw material is a combination of unmalted cereals and malt, and the distillation process is taken one step further by three distillations instead of two. The overall effect is a distillate of higher strength and imbued with a higher proportion of congeners than would be obtained in Scotch malt whisky.

Any fruit or grain, or starch source capable of releasing fermentable sugar, can be processed and converted to a distillate. Hence apples, apricots, cherries, coconuts, dates, gentian, hogweed, milk, molasses, plums, potatoes, rice, cane sugar, water melons, grapes, grape skins and cactus have all been used as the raw material for producing a distillate (Johnson, 1977; Rose, 1977). Of these distillates, cognac, brandy, rum, gin and vodka vie in the market place with whisky. The first three follow a distillation pattern similar to malt whisky but the raw material is different, and hence the final bouquet.

Were it not for the influence of the various Acts passed by Parliament in an attempt to discourage and prevent illicit distillation (Macdonald, 1914) over the centuries; the cunning attempts by distillers to interpret these Acts to their advantage; their final succumbing to the Act of 1823 and subsequent Acts; the outcome of the Royal Commission of 1909 and the resulting definitions of Scotch whisky; the universal protection of the Scotch Whisky Association; the product would have failed to develop into the major force on the world alcoholic beverage scene that it is today.

References

Black H 1984 *Distillery Byproducts as Feeds for Livestock*. North of Scotland College of Agriculture

Clow A 1952 The fermentation industries. In *The Chemical Revolution*. Batchworth Press, London

Dolan T C S 1976 Some aspects of brewing science on Scotch malt whisky production. *Journal of the Institute of Brewing* **82**: 177–81

Duncan R E B, Philp J M 1966 Methods for the analysis of Scotch whisky. *Journal of the Science of Food and Agriculture* **17**: 208–14

Geddes P 1985 Bacteriology in the Scotch whisky industry. *Journal of the Institute of Brewing* **91**: 56–7

Howie D, Swan J S 1984 Compounds influencing peatiness in Scotch malt whisky flavour. In Nykänen L, Lehtonen P (eds) *Flavour Research of Alcoholic Beverages*. Foundation for Biotechnical and Industrial Fermentation Research, Helsinki

Johnson H 1977 *The World Atlas of Wine*. Mitchell Beazley, London

Leppänen O, Ronkainen P, Denslow J, Laakso R, Lindeman A,

Nykänen I 1983 Polysulphides and thiophenes in whisky. In Piggott J R (ed) *Flavour of Distilled Beverages*. Ellis Horwood, Chichester

Macdonald I 1914 *Smuggling in the Highlands*. Eneas Mackay, Stirling

Malkin L S 1983 Energy saving in malting and malt processing. In Priest F G, Campbell I (eds) *Current Developments in Malting, Brewing and Distilling, Aviemore Conference 1982*. Institute of Brewing, London, pp 171–92

Moss M S, Hume J R 1981 *The Making of Scotch Whisky*. James and James, Ashburton

Nettleton J A 1893 *The Manufacture of Spirit*. Marcus Ward, Belfast and New York

Rose A H 1977 *Economic Microbiology*. Academic Press, London, vol 1, *Alcoholic Beverages*

Rutherford A G 1981 Mashing for malt whisky production. *The Brewer* **67**(797): 82–6

Sharp R 1982 Optimising the secrets of Scotch. *Brewing & Distilling International* **12**(3): 27, 32, 37

Swan J S, Howie D, Burtles S M 1981 Sensory and instrumental studies of Scotch whisky flavour. In Charalambous G, Inglett G E (eds) *The Quality of Foods and Beverages*. Academic Press, New York and London

Underwood A J V 1935 The historical development of distilling plant. *Transactions of the Institute of Chemical Engineers* **13**: 34–62

Williams L A, Knuttel W P 1983 Computer modelling of aroma compound behaviour during batch distillation. In Piggott J R (ed) *Flavour of Distilled Beverages*. Ellis Horwood, Chichester

APPENDIX *Guide to troubleshooting*

Problem	Effect	Cause	Action
Excessive frothing Boil over (wash)	Contamination of LW Contamination of LW Damage to safe with product loss	Fresh wash Condenser water failure Blocked condenser tubes Excessive heat Overloaded still	Check fermentation time Check water system Examine tubes Improve heating regime Avoid, if possible
Feinty spirits	Organoleptic	Over-run of middle cut	Check hydrometers Watch temperature of distillate – apply temperature correction, if necessary
Blank runs	No spirit collected Increased fuel usage	Weak charge Excess fusel oil Unbalanced charges	Check strength of LW & F Avoid charging fusel oil layer Distillation balance
Weak spirit	Altered congener balance Water in safe	Poor reflux Leaking condenser Poor cooling	Slow distillation rate Check tubes, pressure test
Increased distilling time (wash)	Distillation timing	Fouled heating surfaces	Check surfaces, clean if necessary

Increasing distilling time (wash & LW & F)	General slowing down of all operations	Poor cooling – water and air temps high – summer conditions	Adjust mashing programme
Product loss	Low yield – low percentage over attenuation	Extraction efficiency	Check sparge/malt ratios, mashing temperatures Avoid overloading mashtun
		Infection	Improve hygiene
		High wash temperatures	Lower setting temperatures
		Physical losses of worts, wash, LW & F, spirit	Investigate whole process – personnel and plant
Distillation losses	Water in safe	Leaking condenser tubes	Pressure test tubes
	Unfamiliar, high-pitched noise in still	Leaking steam coils/pans	Check flanges, etc.
	Staining on surfaces	Passing discharge, air or safety valves	Check valves
		Leaking pot	Fill pot to man door rim with water
	Adverse analysis	Spirit in pot ale, spent lees	Check hydrometers
Abnormal percentage over attenuation	Apparent losses/gains	Wrong declaration	Check hydrometers in tun room and correct allowances, if necessary

LW = low wines; F = feints

CHAPTER 6

Continuous distillation

R J PANEK and A R BOUCHER

This chapter is intended to be used as an introduction to the fundamentals of distillation. It discusses the techniques commonly employed in a modern distillery but is not intended to be a conclusive treatise on still design, since there are many books and articles that are devoted solely to this task. The intention of the authors is to present the fundamental theories and concepts in such a fashion that the reader can combine these theories and concepts with deductive reasoning for the comprehension of continuous distillation and for the understanding and solution of practical problems encountered in a distillery.

The first section is a general discussion on evaporation and volatility, which is necessary to understand the basic principles behind distillation theory. The next section describes the major parts and terminology of a typical continuous distillation unit. The final section is a description of the various types of continuous distillation columns and how they are combined in practice to perform a variety of functions.

Physical behaviour of boiling liquids

It is important to understand what physical changes occur when a liquid is processed in a distillation column. This is best exemplified by starting with a pure component system and then expanding to the binary ethyl alcohol-water system.

When a pure liquid is placed in an open container, the liquid molecules are constantly in motion. Some of the molecules at the liquid surface will pick up enough energy to break the surface tension of the liquid and escape as a vapour to the air space above the liquid. This process is called evaporation. If left unchecked, this process will continue until all of the liquid has vaporised. If the evaporation process were repeated in a closed vessel that contains no air or other impurities, the vapour that is formed would be trapped inside the container. The motion of the vapour molecules exerts a force on the walls of the vessel. This force can be measured and is referred to as the vapour pressure of the substance. If the temperature of this experimental system is changed, the vapour pressure increases as the temperature increases, and, conversely, the vapour pressure decreases as the temperature decreases. The boiling point of a liquid is defined as the temperature at which the vapour pressure of the liquid equals the pressure of the surroundings.

Evaporation is a surface phenomenon, since the molecules at the liquid surface are the most likely to overcome the surface tension of the liquid and transform into a vapour. On the other hand, boiling occurs when there is enough energy in the system to transform liquid into vapour within the body of the liquid. In a distillation system, the energy is added directly by sparging steam into the base of a column or indirectly by using a steam fed heat exchanger or calandria. The amount of energy needed to boil a liquid can be substantially reduced by operating the column under a vacuum. The vacuum reduces the vapour pressure of the liquid and reduces the kinetic energy necessary to vaporise the liquid.

Because of differences in molecular structure, the pure component vapour pressure and boiling temperature is usually different for different compounds. When more than one compound is combined in a mixture, the intermolecular forces can cause deviation from pure component behaviour.

Examination of the vapour pressure data for pure water and pure ethyl alcohol reveals that the vapour pressure for ethyl alcohol is greater than the vapour pressure for water at all temperatures. (See Table 6.1). It is this difference in vapour pressure that is responsible for the separation that occurs during the distillation of alcohol-water mixtures.

In a pure, single component boiling liquid, the composition of the vapour is always equal to the composition of the liquid and is always 100% pure component. A binary system such as the alcohol-water system is much more complex. Since the vapour pressure of ethyl alcohol is greater than the vapour pressure of water at any given temperature, alcohol is the more volatile substance and is present in the vapour at a higher concentration

Table 6.1 Vapour pressures of ethyl alcohol and water.

Temperature (°C)	(°F)	*Ethanol* (PSIA)	*Water* (PSIA)
76	168.8	13.40	5.83
80	176.0	15.70	6.87
84	183.2	18.37	8.06
88	190.4	21.31	9.42
92	197.6	24.75	10.96
96	204.8	28.48	12.72
100	212.0	32.74	14.9

than water. Attractive forces between the alcohol and water molecules due to similar polarity and mutual solubility prevent the complete separation of the alcohol from the water when it is vaporised. That is, there is always some water in the vapour state which is carried into the alcohol product.

In the closed system, equilibrium is defined as the point when the alcohol-water concentration of the vapour and the alcohol-water concentration of the liquid no longer change with respect to time. If the molecular concentration of the alcohol in the vapour and the molecular concentration of the alcohol in the liquid are measured at equilibrium, the volatility of the alcohol can be calculated.

$$\text{Volatility } (K) = \frac{\text{mole fraction of alcohol in vapour}}{\text{mole fraction of alcohol in liquid}}$$

Equilibrium data have been determined for all concentrations of ethyl alcohol-water mixtures and are summarised in Table 6.2 and graphically represented in Fig. 6.1. Note that the concentration of ethyl alcohol in the vapour is greater than the concentration in the liquid ($K > 1$) for all alcohol concentrations from 0 to 194.4° US proof. At this point the alcohol concentration in the vapour equals the concentration in the liquid ($K = 1$). This point is known as the azeotropic point and is the upper limit for conventional distillation. If the concentration of alcohol in a solution is greater than 194.4° US proof (95.57 wt%) the concentration of ethyl alcohol in the vapour is less than the concentration in the liquid ($K < 1$). Since 100% pure ethyl alcohol is not used by the beverage alcohol distiller we are concerned only with the equilibrium data in the range where $K > 1$.

Table 6.2 Equilibrium data for ethyl alcohol-water system.

Temperature		*Ethyl alcohol in liquid phase*		*Ethyl alcohol in vapour phase*	
(°F)	*(°C)*	*°US proof*	*Mole fraction*	*°US proof*	*Mole fraction*
212.0	100.0	0.0	0.0000	0.0	0.0000
203.4	95.2	12.50	0.0202	85.16	0.1782
197.2	91.8	24.83	0.0417	117.74	0.2894
192.6	89.2	36.97	0.0646	134.43	0.3650
189.1	87.2	48.95	0.0891	144.62	0.4201
186.1	85.6	60.80	0.1154	151.46	0.4619
183.0	83.9	83.69	0.1740	158.73	0.5114
180.9	82.7	105.35	0.2425	163.58	0.5477
179.2	81.8	125.71	0.3235	167.76	0.5817
176.7	80.4	144.76	0.4209	172.28	0.6219
175.1	79.5	162.58	0.5400	177.17	0.6702
173.5	78.6	178.95	0.6892	183.96	0.7471
172.8	78.2	193.56	0.8814	193.67	0.8830
172.4	78.0	194.37	0.8943	194.37	0.8943

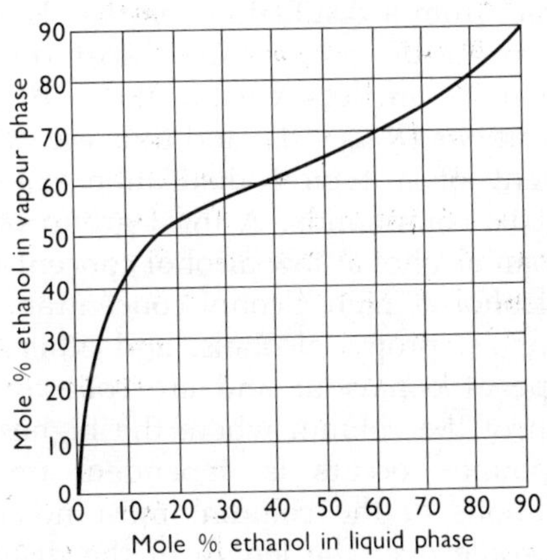

Fig. 6.1 Equilibrium data for ethyl alcohol-water system.

THE FLAVOUR CONTRIBUTION OF CONGENERS

The taste and odour of all foods and beverages is determined by the flavour contribution of the individual components that comprise that food or beverage. In alcoholic products, the flavour is determined by minute quantities of volatile congeners which are formed as by-products during the fermentation process and distilled into the product. Many of these by-products have been identified by modern analytical techniques. Even though these compounds are present in very small quantities, they have a significant impact on product flavour and quality.

If the volatility of the congener is compared with the volatility of ethyl alcohol, we can determine the relative volatility of the congener in a multi-component system.

$$K_{(\text{rel})} = \frac{\text{mole fraction of congener in the vapour}}{\text{mole fraction of alcohol in the vapour}}$$

In order to understand how these compounds behave in a distillation column they can be broadly classified into three groups. These groups are determined by thermodynamic properties, such as Dalton's law of partial pressures and Raoult's law. Some of these congeners, e.g. most esters and many aldehydes, are more volatile than alcohol over the entire range of ethyl alcohol concentration: $K_{(\text{rel})} > 1$. These compounds are more volatile, relative to alcohol, and are most abundant in the overhead product from a distillation column. They are commonly referred to as heads compounds. Most organic acids and phenolic compounds are less volatile than ethyl alcohol at all alcohol concentrations ($K_{(\text{rel})} < 1$), and thus are most abundant in the base effluent of a typical distillation column. They are referred to as tails compounds. A third group of compounds is more volatile than alcohol at low alcohol concentrations, and less volatile than alcohol at high alcohol concentrations. Amyl alcohols, butyl alcohols, propyl alcohols, and other higher alcohols exhibit this type of behaviour and are collectively called fusel oils. The region of the column where the highest concentration of these compounds occurs is dependent on the particular operating conditions of the column, most notably the alcohol concentration profile over the length of the distillation column. These fundamental relationships form the basis for column operation and design.

Terminology of continuous distillation

A typical distillation column consists of a cylindrical tower arranged vertically, which is subdivided into a series of stages by

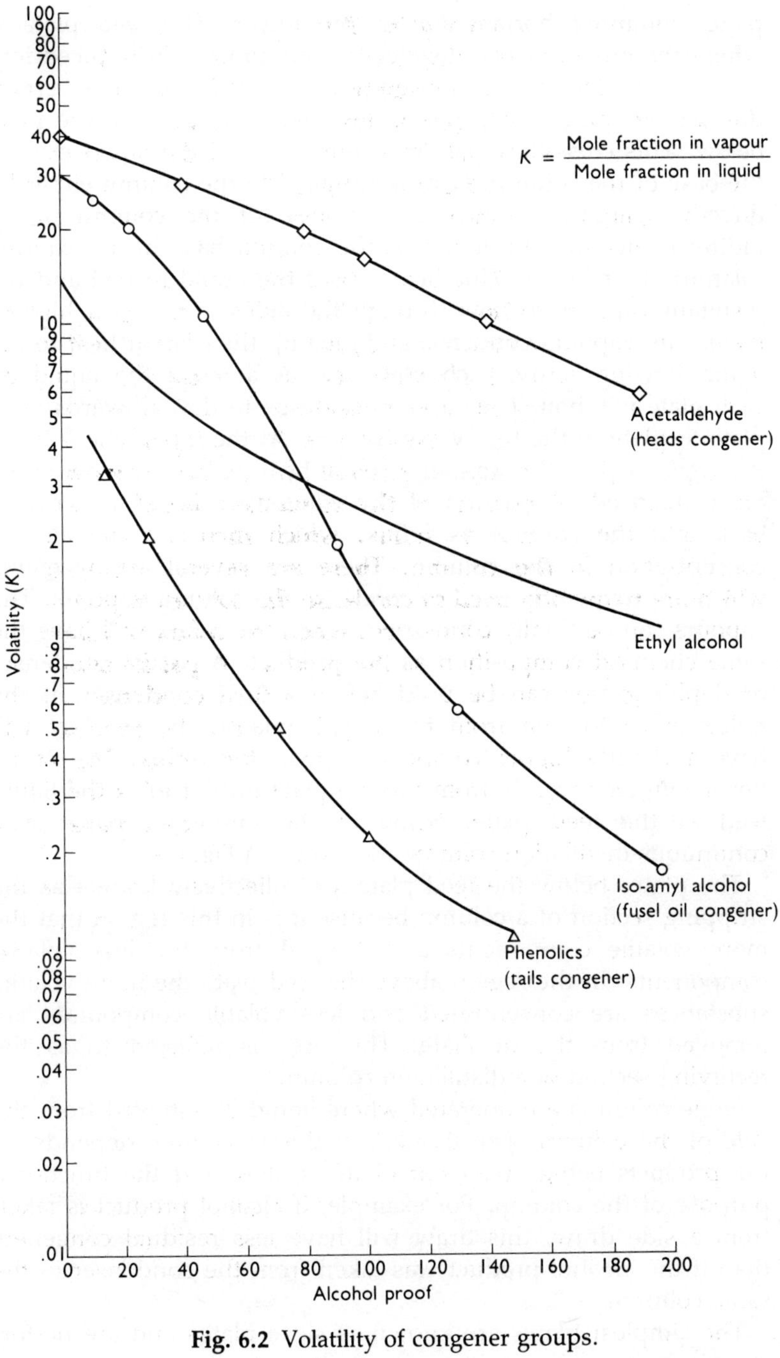

Fig. 6.2 Volatility of congener groups.

plates mounted horizontally in the tower. The feed plate is where the mixture to be distilled is continuously introduced into the column. This liquid flows across the feed plate to a vertical downcomer pipe which permits the liquid to flow to the plate below. This cross-flow pattern continues until the liquid exits at the base of the column. Heat is supplied to the column either by directly sparging steam into the base of the column or by indirectly heating the liquid at the column base in an external calandria or reboiler. This heat causes the liquid to boil and the resultant vapours to flow through the holes in the next highest plate. The vapours condense and give up their latent heat to the liquid flowing across each plate and as a result the liquid on each plate will boil. The vapours continue to flow upwards from plate to plate to the top of the column. At the top of the column the vapours pass through an external heat exchanger where they are condensed. A portion of the condensed liquid is recycled back into the column as reflux, which increases the alcohol concentration in the column. There are several arrangements which are frequently used to condense the column vapours. The vapours can be totally condensed, when the reflux will have the same chemical composition as the product. A partial condenser or dephlegmator can be used before a final condenser. If the reflux is withdrawn from the dephlegmator, the product will have a slightly higher composition than the reflux. The reflux flows countercurrently from the top plate until it joins the liquid feed at the feed plate. Some of the important parts of a continuous distillation column are shown in Fig. 6.3.

The plates below the feed plate are collectively known as the stripping section of a column because it is in this region that the more volatile components are stripped from the less volatile components. In the region above the feed plate the more volatile substances are concentrated and less volatile compounds are removed from the distillate. This area is referred to as the rectifying section of a distillation column.

Some columns are operated where liquid is removed from the side of the column. The location of this side draw depends on the products being taken out of the system and the functional purpose of the column. For example, if alcohol product is taken from a side draw, this draw will have less residual congeners than if the alcohol product was taken from the condenser in the same column.

The simplest plates are known as sieve plates and are perfor-

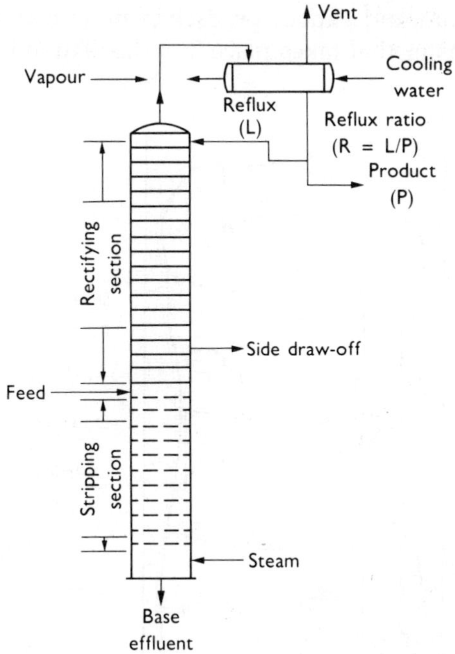

Fig. 6.3 Schematic diagram of a typical continuous distillation column.

ated with small holes which allow the vapour to pass from the underside of the plate to the upper portion. Each plate has an appropriately sized downcomer on one side to allow liquid to flow to the plate below. The downcomer is sealed in an overflow cup or weir in order to prevent vapours from flowing up the downcomer. The plate area near the downcomer and the plate area on the side opposite the downcomer where liquid enters the plate do not contain any holes and are inactive areas of the plate. The cross-flow pattern of the liquid is created by the liquid flowing across each plate from downcomer to downcomer (Fig. 6.4). A weir is used to regulate the flow on each plate and maintain a liquid level on each plate. The vapour from the plate below passes through the perforations and mixes with the liquid on the plate. These vapours actually give up their latent heat and condense. This causes the slightly more volatile compounds in the liquid to vaporise and the process continues to the plate above. It is very important that a 3.8–5.0 cm (1.5–2 in.) liquid level is maintained on all plates at all times. It is the intimate mixing of liquid and vapour on each plate that is responsible for the mass transfer that takes place in a distillation column.

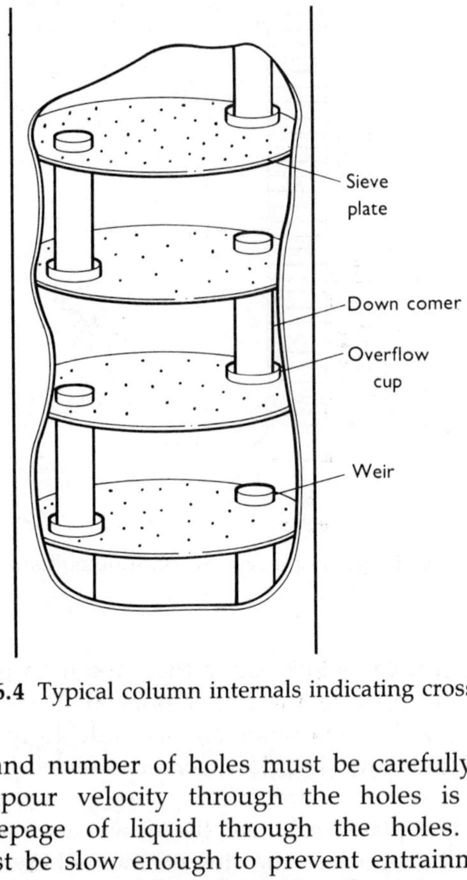

Fig. 6.4 Typical column internals indicating cross-flow.

The size and number of holes must be carefully designed so that the vapour velocity through the holes is sufficient to prevent weepage of liquid through the holes. The vapour velocity must be slow enough to prevent entrainment of liquid and ensure adequate mixing of liquid and vapour on each plate. There are several other plate designs that are frequently used in a beverage alcohol distillery. Bubble caps are constructed with a chimney which prevents liquid weepage and the vapour must pass through slots in the caps which create small bubbles and thereby increase the surface area of the vapour as it passes through the liquid. The primary advantage of devices such as bubble caps and tunnel caps is that the distillation column can be operated over a much wider range of feed rates than the sieve plates. However, these contacting devices significantly increase the capital cost of constructing a column.

Recent advances in the mathematical modelling of plate hydraulics and the increased use of computers has enabled us to design sieve plates more precisely, increasing their efficiency. We are now able to modify accurately the size of the holes in the different plates of a single column to compensate for the vapour density as it changes within a column.

The design of a distillation column begins with an evaluation of the intended function of the column. This evaluation includes determining composition and range of flow rates of the feed, the desired product composition, and the separation which will be required in the column. There is a vast difference between columns designed for the production of fuel alcohol and columns designed for the production of alcohol in beverage, pharmaceutical or cosmetics usage.

Continuous distillation systems

The design of the individual columns in a distillation system will differ depending on the number of columns in the system and the desired product neutrality. The columns can be sequenced in different orders to produce a wide variety of products. The columns typically used in the beverage alcohol distilling industry are described below. This is followed by a description of the most common systems used for the production of a wide variety of alcoholic products.

COLUMN FOR THE PRODUCTION OF HIGH WINE

A fermented mash is first distilled in a unit which separates ethyl alcohol and volatile congeners from the grain solids, non-volatile compounds, and most of the water which is present in the fermented mash. This alcoholic product is called a high wine or primary distillate. The high wine product can either be entered into barrels to be aged and used as whisky or can be further distilled into light whisky or spirits.

Beer still

A beer still typically has the feed plate located near the top of the column and product is collected from the condenser at 105–145° US proof. The base effluent or stillage contains the non-fermentable solids, yeast cells and water. A portion of the stillage is used as backset to adjust pH and buffer in subsequent fermentations and the remainder is dried to about 10% moisture in the dryer house.

The choice of product has an impact on the design and operation of the still. When the desired end-use for the distillate is aged whisky, emphasis is on product quality. Thermal degradation can be a source of undesirable quality characteristics in the distillate. In order to reduce the adverse effects of thermal degradation of the solids in fermented mash, the number of stripping plates is kept to a minimum, typically 12–14 plates. In addition, the beer is fed to the system at fermentation temperature, 29–32°C or 85–90°F, and the alcohol loss from the base of the column is relatively high, 0.05–0.10 vol%. These precautions minimise the residence time of the solids in the still and also minimise the amount of heat used in the still to separate the volatile product.

The beer still high wine which is used for whisky is sometimes processed in a doubler. A doubler is a large kettle where the high wine is fed at about 125° US proof. Heat is added via steam coils in the kettle of the doubler. The distillate from the doubler is usually 135–140° US proof. The liquid in the doubler is maintained at a constant level and the base effluent is fed back into the beer still one or two plates below the feed. Doubling removes some acid and phenolic compounds from the distillate and improves the distillate quality. This process is called continuous doubling.

If the high wine is to be further rectified into light whisky or spirit, the beer still is designed to operate at maximum efficiency. The stripping section of the column contains a greater number of plates (18–22), to reduce base losses to about 0.02–0.04%. It is also common practice to use a shell and tube heat exchanger to pre-heat the beer feed to approximately 85°C (185°F). The beer is passed through the tube side of the exchanger and heated, with the vapour from the column being passed through the shell side of the unit. The increased residence time increases the thermal breakdown in the still; however, these economic practices are possible because any undesirable quality characteristics from grain decomposition will be removed in subsequent processing.

Most beer stills are heated by directly sparging steam into the side of the column. If a calandria is used, the stillage is circulated through the tubes and steam is fed to the shell side of the unit. The use of a calandria instead of sparged steam reduces the amount of water which must be removed from the stillage in the dryer house. Calandria design is critical because this increases the residence time of the stillage in the still and the organic material in the stillage has a tendency to foul the calandria and plug the tubes, which will then require an extensive cleaning procedure.

In beer stills used for the production of fuel alcohol, since the primary concern is production cost and not quality, it is common practice to use a feed pre-heater and then pass the feed through another heat exchanger which raises the feed temperature to 100°C (212°F) by using base effluent in the heat exchanger.

GENERAL BEER STILL DESIGN AND OPERATION

The beer still stripping plates are typically sieve plates. The holes must be large enough so that the grain solids will not plug and small enough to prevent excessive weepage. Past experiments and experience have demonstrated that 0.95 cm ($\frac{3}{8}$ in.) holes are adequate and the total surface area of the holes should equal about 7–9% of the total cross-sectional area of the column. Typical vapour velocity through the holes in a sieve plate is 6–12 m sec^{-1} (20–40 ft sec^{-1}). Bubble caps cannot be used in the stripping section of the beer still because of the plugging problems associated with solids. Since some fermented mashes have a tendency to foam, the downcomers must be sized large enough and plate spacing must be high enough to prevent column flooding and excessive build-up across the column.

The rectifying section of a beer still is usually only 3–6 plates and can be constructed of either sieve plates or bubble cap plates. An entrainment plate is sometimes employed just above the feed plate. This plate has a series of large nozzles which propel the vapour from the feed plate in a tangential direction which uses centrifugal force to separate entrained solids from the vapour.

Copper is used in the construction of most beer stills. Soft copper is used for the shell and the plates are cold rolled to prevent bending and buckling. It is very important that the plates remain in a horizontal position in the column in order to maintain a constant level across the plates and prevent dry spots. Fermented grain products contain volatile sulphur compounds which have a negative effect on distillate quality, such as a gassy, cabbage-like odour. These compounds undergo a chemical reaction in the presence of copper, which acts as a catalyst, renders them insoluble and removes them from the product vapour, thereby improving distillate quality. Stainless steel has been used in still construction because it is cheaper and wears better than copper. Because of the problem with sulphur compounds, however, it is advisable to have some copper in the system, either in the rectifying section or in the vapour pipe leading to the condensers.

A demister is a 10 cm (4 in.) thick pad of fine copper mesh which is placed in the top of the column to provide a large exposed copper surface area for the distillate vapours to pass through. The demister serves to remove sulphurous compounds and also any solids particles which have been entrained in the vapours.

After extended operation, a beer still must be cleaned to remove from the still any accumulation of grain deposits and residue from the compounds which reacted with the copper surfaces. This accumulation is evidenced by an increase in the pressure drop across the column. The cleaning is also necessary to remove any film caused by the organic compounds in the mash or the precipitation of sulphide compounds. When cleaning is necessary, the beer feed is replaced with hot water and the still is operated for about 30 min to flush away most of the grain deposits. A mild 0.5% caustic solution is fed to the top plate of the column after the water feed is shut off and this solution is boiled through the column for 30 min. The boiling caustic should remove all the remaining grain deposits and organic films from the still. The next step is to feed a weak 0.3% citric acid solution to the top plate of the column to dissolve the sulphide compounds that have been deposited on the copper surfaces of the still. A thorough water rinse will complete the cleaning procedure. Demisters should be replaced twice a week and should be cleaned in a vat or tub. The cleaning sequence is similar to the procedure used for the still.

COLUMNS FOR CONGENER REMOVAL AND ALCOHOL CONCENTRATION

There are many variations of the basic distillation column which are used to further process alcoholic products. Columns are used to concentrate the ethyl alcohol and remove undesirable congeners from the distillate. The congener distribution throughout each column is determined by the volatility of the congener and the alcohol concentration profile at all points within the column.

Each of the following distillation columns was designed for a specific purpose. Most are combined with one or more other columns to form a complete distillation system. By examining the column operating parameters, especially overhead proof and base proof, and keeping in mind the volatile behaviour of the three groups of congeners (heads, tails and fusel oils), the intended purpose of the column can be determined. Understanding these concepts will assist in optimising column operation and ensure the production of a superior quality distillate. Trouble-shooting operational difficulties is also simplified when these concepts are understood. The nomenclature used to describe these columns is by no means universal. It is a common occurrence to have more than one name for columns that have similar design and accomplish the same functional purpose.

Rectifying column

The primary purpose of a rectifying column is to remove water from the product distillate. A rectifying column usually contains 50–75 plates. The feed plate is situated between plates 20–25. The column has enough stripping plates to remove all of the alcohol from the base effluent. A heads draw which contains 2–5% of the feed flow rate is drawn from the condenser and the remaining condensed overhead vapours are returned to the column as reflux. The product is taken from a side draw about 5–7 plates from the top of the column at a 192° product US proof. Since the proof at the top of the column is very high and the base proof is negligible, the fusel oil congeners accumulate 2–4 plates above the column feed on the plate where the liquid proof is 130–140°. The reflux ratio compares the amount of condensed vapour returned to the column with the amount withdrawn from the column. The magnitude of the reflux ratio has an impact on the alcohol concentration and purity of the product distillate.

The rectifying column will produce a product that is very close to the azeotropic composition. Because the volatility of ethyl alcohol approaches unity at high concentration, there is only a very small increase of proof on the plates at the top of the column and a large number of plates is required to get any closer to the azeotrope.

When this column is combined in a system with other columns which remove more effectively the heads and fusel oil congeners, a very neutral product can be produced. The water from the base of this column is frequently recycled to other columns in the system where dilution water is needed or the energy in this stream can be recovered by using it in a heat exchanger to pre-heat other process flow streams.

Super rectifying column

This column is sometimes used after a rectifying column when a very neutral product spirit is desired. It is a very large column with 60–80 plates. A 2–4% heads draw is taken from the condenser and the product at 194° US proof is withdrawn 5–7 plates from the top of the column. The base effluent is 182–185° US proof. Since the alcohol concentration of the base effluent is so high, the heat to the column is supplied by an external calandria to prevent dilution.

The high-proof environment in the super rectifying column causes all congeners which are less volatile than ethyl alcohol at very high alcohol concentrations, including the fusel oils, to exit the base of the still.

Aldehyde or heads purifying column

This column is responsible for removing the highly volatile aldehydes and esters from the alcohol product. The unit contains from 24–50 plates. The feed is at or below the column mid-point and is comprised of a mixture of high wine, recycle streams from other columns, and dilution water. The alcohol product exits the column base at 30–45° US proof.

The low proof at the base of the column takes advantage of the high volatility of aldehydes and esters at low alcohol concentrations. These congeners are separated from the alcohol product and are removed from the condenser at 192° US proof in a stream which contains 3–5% of the feed flow rate. An esters draw can be taken about 4 plates above the feed and will contain intermediate boiling esters which are at their highest concentration at about 175–185° US proof. If an esters draw is taken it will contain about 3–5% of the column feed flow rate.

Extractive distillation column

An extractive distillation column has 30–50 plates. High wine and other recycled streams enter the column at plate 15–20. The alcohol product exits the base of the column at 10–20° US proof. This unit differs from the aldehyde column in that the dilution water is added to the top rectifying plate in the column. If the column operation is controlled very strictly, the vapour to the condenser contains only 11–12% of the alcohol in the feed and is comprised of the highly volatile aldehydes and esters and also the fusel oils. These vapours are cooled and sent to a fusel oil concentrating column.

Purifying column

If a column of similar construction to a rectifying column is fed with high wine and recyle streams it is referred to as a purifying column. The acid and non-volatile congeners are removed in the alcohol-free base effluent, fusel oils in a side draw at 130° US proof, and aldehydes and esters from the condenser at very high proof. Both the fusel oil draw and the heads draw contain 5–10% of the column feed rate. The 190–193° US proof product, which is removed from the side of the column, has a heavy flavour because of the relatively high concentration of aldehydes and esters in the rectifying section of the column.

COLUMNS FOR CONGENER CONCENTRATION

Many of the columns described have congener streams which are withdrawn either from the condenser or a side draw. These congener streams contain 2–15% of the alcohol which is fed to the column. Rather than dispose of this alcohol with the congener by-products, it is economically desirable to recover this alcohol as product. This is the function of the following two columns.

Fusel oil concentrating column

This column contains 50–60 plates. The various fusel oil streams from the rectifying and purifying columns and the esters draw from the aldehyde column are fed to the unit at or around plate 17–20. The combined overhead vapours at 192° US proof or more are recycled back to the aldehyde or purifying column as second-grade alcohol for repurification. The base effluent is free of alcohol and is usually discharged to the sewer with undesirable high-boiling compounds.

Under these conditions the high volatility of the fusel oils at the base of the column forces them up the column, while the less volatile nature of the fusel oils at the high alcohol concentration at the top of the column forces them towards the base. This will cause the highest concentration of fusel oils to be at the plate in the column where the liquid is about 130–140° US proof. This is usually 2–3 plates above the feed. The fusel oils are drawn off at this point and sent to a decanter where cool dilution water is added to reduce the proof to 30–40°, where the fusel oils become insoluble and immiscible. The concentrated fusel oils form an upper liquid layer and the alcohol-water mixture forms a lower layer. The fusel oils are withdrawn from the system and the alcohol-water layer is recycled to the column feed.

Heads concentrating column

This column has 40–50 plates and operates with parameters similar to an aldehyde column. The feed consists of the heads draws from the aldehyde, purifying, and rectifying columns. Dilution water is added to the feed to control the alcohol concentration at the base of the column at 30–40° US proof. The highly volatile aldehydes and esters are concentrated at the top of the column, where they are removed from the system. Congener concentration at this point can be as high as 1% aldehydes and 10% esters, with the remainder being mostly alcohol with some water.

DISTILLATION SYSTEMS FOR HIGH-PROOF PRODUCT

These distillation columns can be combined and operated in a number of different configurations to produce distillates with a wide variety of organoleptic characteristics. Modifications and changes are constantly being implemented in these systems to reduce operating costs and to take advantage of advances in process control and column design. Since each column requires a significant capital investment, the size of the distillery and the volume throughput are important factors when determining the number of columns and size of the distillation system. The greater the throughput, the easier it is to justify a large system. However, the desired quality characteristics of the distillate are still the most important factors. The design of the system will determine the quality characteristics of the distillate and the flexibility of the system.

The volume flow rate of the congener draws in a multi-column distillation system and their destination play an important part in determining the congener level in the product distillate and in determining the alcohol losses from the system. It is possible to concentrate the fusel oils in the rectifying column without the use of a fusel oil concentrating column to a concentration that is high enough for them to separate in a decanter. However, when this is done, the fusel oils will accumulate in the column and some will exit with the alcohol product. For this reason, multi-column distillation systems with only three columns or less are not capable of producing a distillate that is vodka quality and the alcohol loss from the system is usually significant.

Coffey still

The Coffey still (Fig. 6.5) is constructed of two columns which are located side by side. The first unit operates functionally as a beer still. Mash is fed into one of the upper plates and heat is supplied by sparging steam into the base of the unit. The alcohol-free stillage is withdrawn from the column base. The primary difference between this still and the beer still is that the vapours from the overhead of this unit are fed directly to the base of the second unit and act as a heat source for that unit. The second column is referred to as the rectifier or analyser. Product is withdrawn from the side of the analyser column 5–10 plates from the top, heads are withdrawn from the condenser, and fusel oil is withdrawn 7–10 plates above the bottom of the analyser.

This unit looks like a two-column system but there is only one liquid feed stream to the stripping column and only one external source of heat. The primary advantage of splitting the columns is that this reduces column height. A major disadvantage is that the liquid which discharges from the base of the analyser must be pumped to the top of the stripping column where it serves as reflux.

This still is traditionally built in a rectangular shape. Although earliest construction was of copper plates within a wooden shell, the shell is now constructed of copper. The feed line is built so that it crosses through the vapour space on each plate in the analysing section. This not only pre-heats the feed, but also increases the internal reflux of the system.

The Coffey still produces a spirit (189–191° US proof) that has a high congener concentration and a very heavy flavour. This unit was the first commercially successful continuous still for the production of a high-proof spirit. The first unit was built in Scotland in the early 1830s and is the traditional method for the production of grain whisky. Stills similar to this are used in California for the production of brandy and in Jamaica for the production of light-bodied rum.

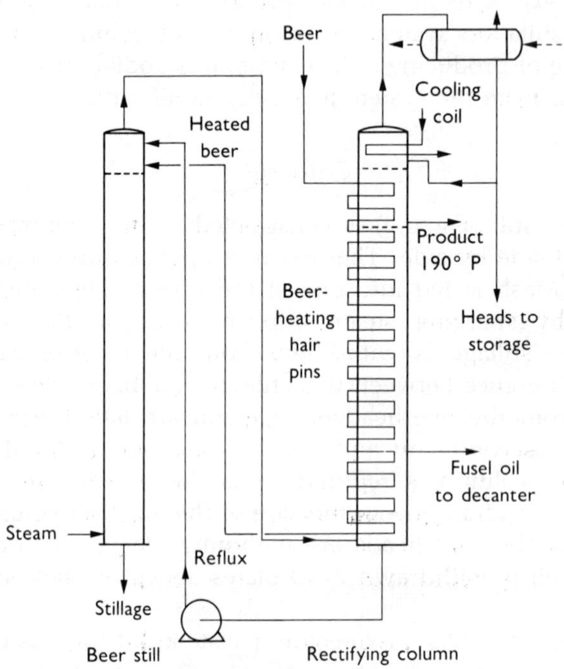

Fig. 6.5 Coffey still.

Aldehyde-rectifying column system

This system (Fig. 6.6) is used for the production of a heavy-bodied product. Heads congeners are removed from the aldehyde and rectifying column condensers. Fusel oils are removed from a side draw in the rectifying column. This process is known as the Barbet process. The primary disadvantage of this system is that to produce a distillate with low congener levels, the alcohol losses in the fusel oil and heads draws are high.

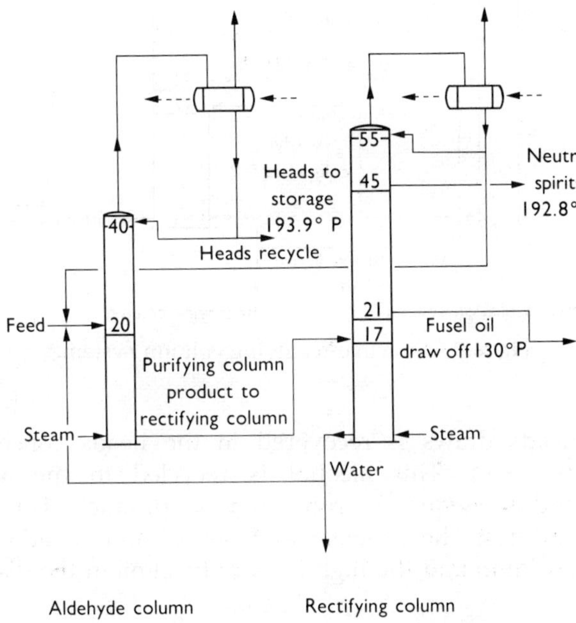

Fig. 6.6 Aldehyde-rectifying column system.

Extractive-rectifying column system

In this system, the fusel oils and heads are removed from the extractive column condenser (Fig. 6.7). The fusel oils, however, tend to accumulate in the extractive column and pass with the base effluent into the rectifying column where they must be removed in a side draw. Alcohol losses in this system are also high.

Aldehyde-rectifying-heads concentrating column system

This represents an improvement over the two-column systems because the alcohol that is lost in the aldehyde and rectifying

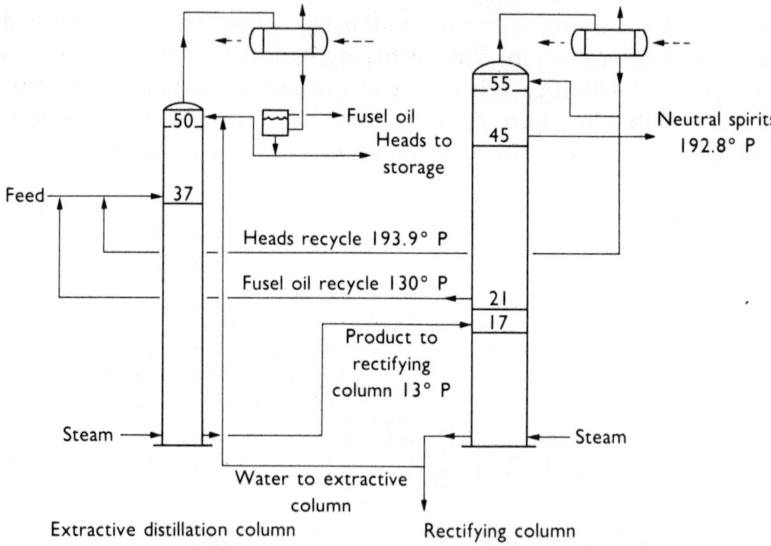

Fig. 6.7 Extractive-rectifying column system.

column heads draws is recovered in the heads concentrating column (Fig. 6.8). This alcohol is recycled to the aldehyde column and subsequently recovered as product. This system does not address the problem of fusel oil accumulation in the rectifying column and the high level of fusel oil in the distillate.

Aldehyde-rectifying-fusel oil concentrating column system

Since the fusel oil draw from the rectifying column is reprocessed, the volume flow rate of this stream can be increased to prevent the accumulation of fusel oils in the rectifying column (Fig. 6.9). This provides a more neutral product. An esters draw may be taken from the side of the aldehyde column. The alcohol losses in the heads draws from this system are significant.

Aldehyde-rectifying-extractive column system

In this system (Fig. 6.10) the heads from the aldehyde column and the rectifying column fusel oil are fed to an extractive column. The primary advantage of this system is that there is a high congener concentration in the feed to the extractive column and a subsequent high concentration in the overhead vapours.

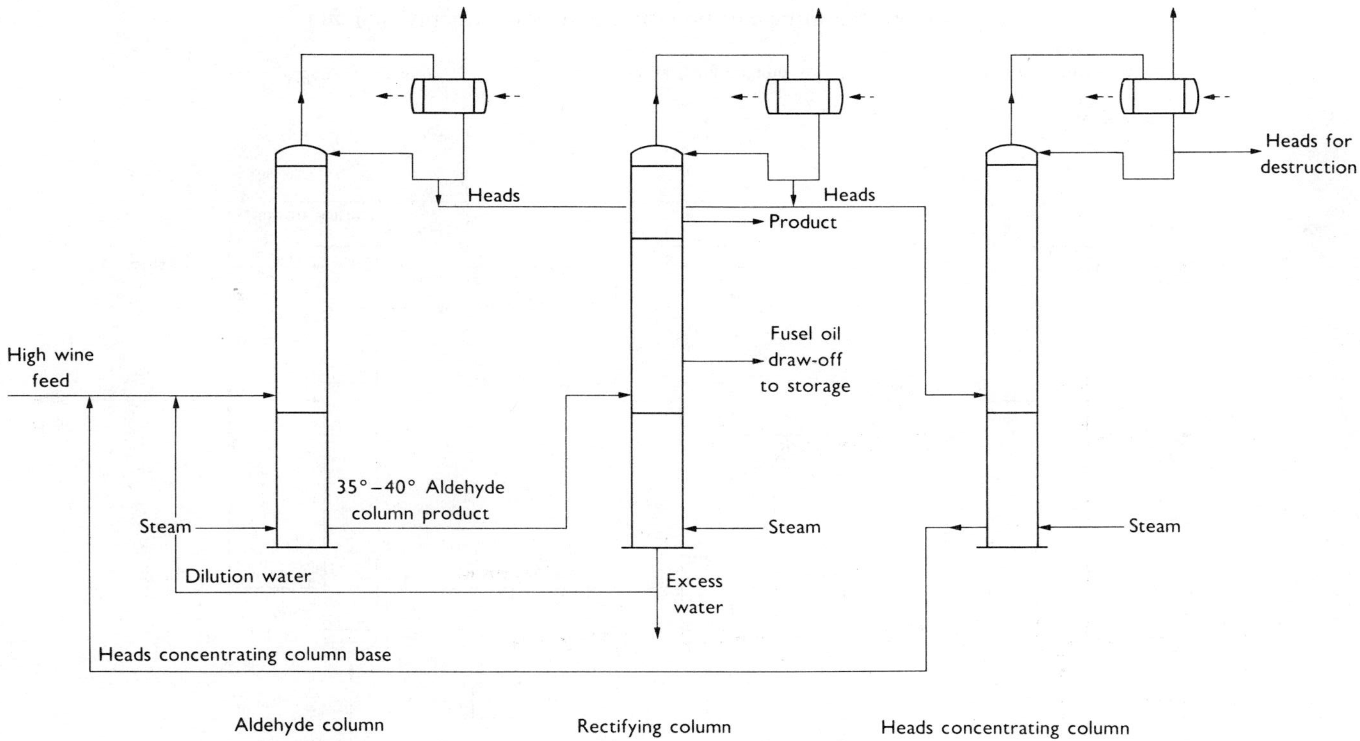

Fig. 6.8 Aldehyde-rectifying-heads concentrating column system.

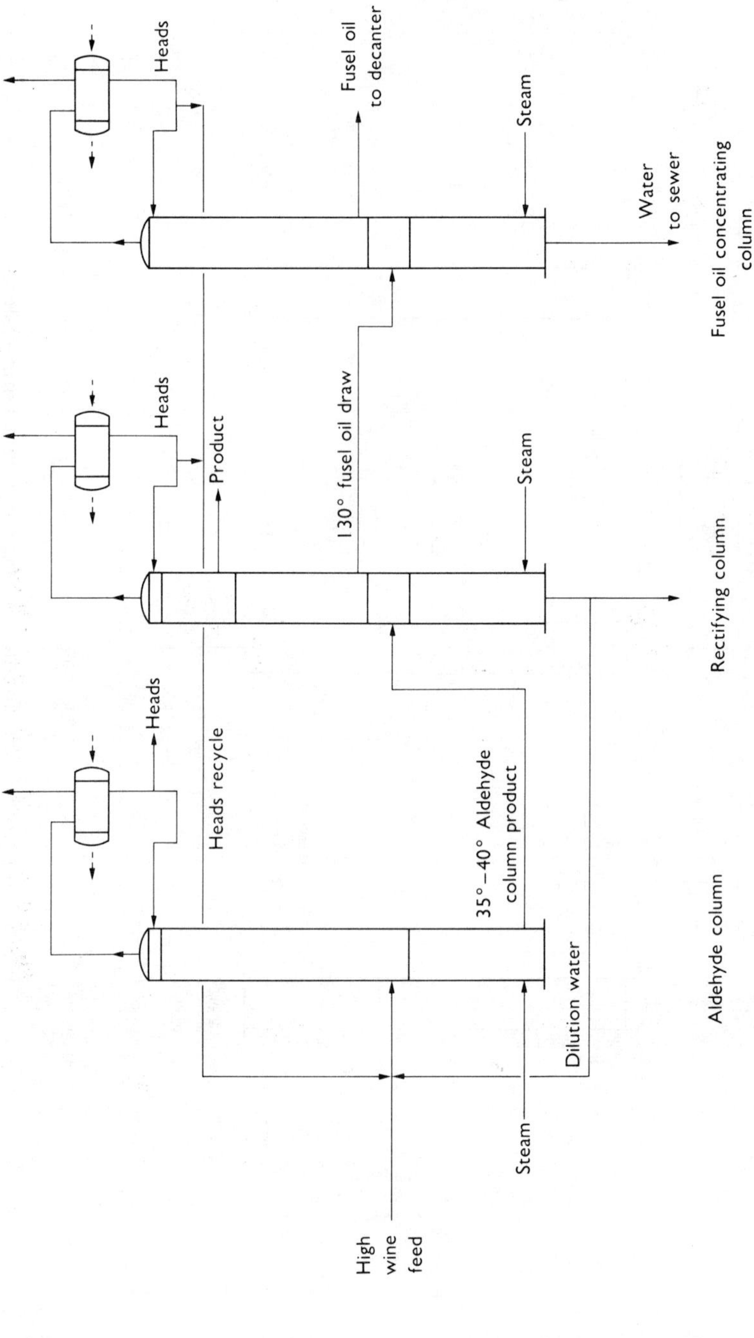

Fig. 6.9 Aldehyde-rectifying-fusel oil concentrating column system.

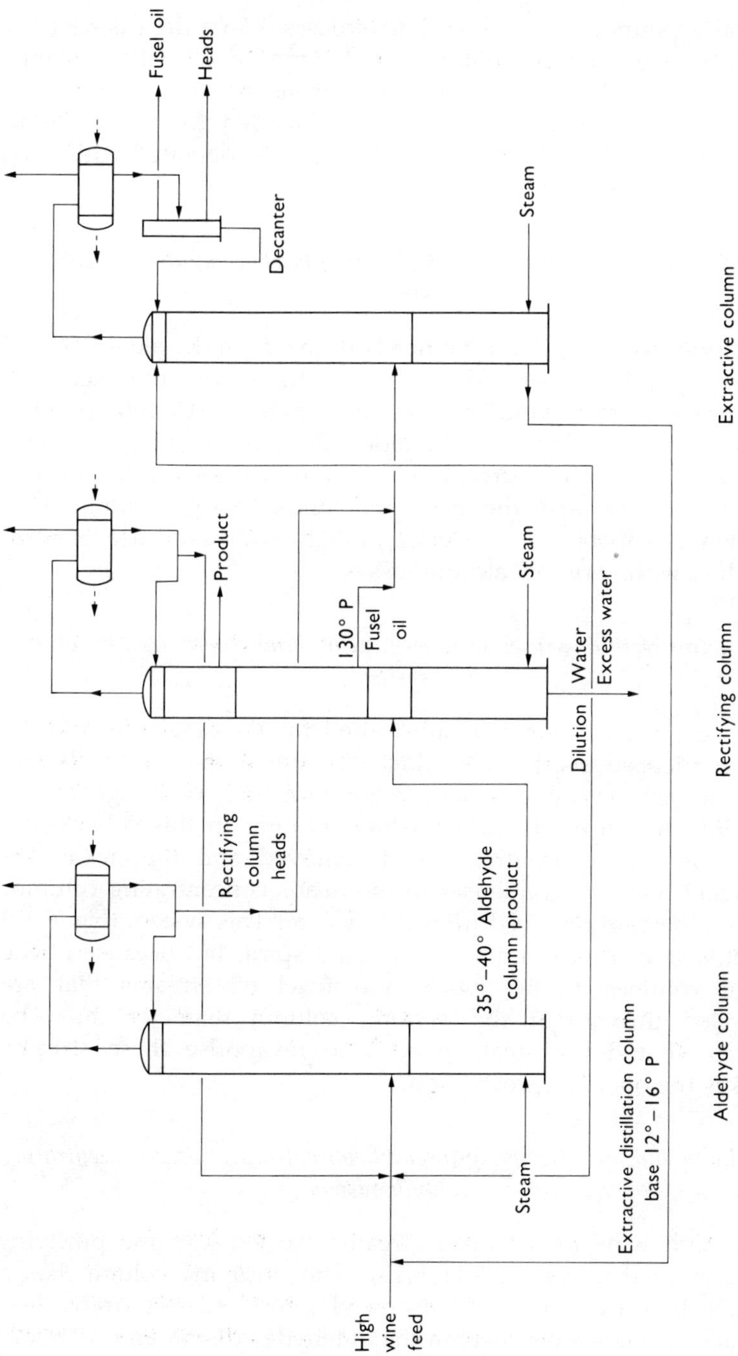

Fig. 6.10 Aldehyde-rectifying-extractive column system.

These vapours can be cooled and processed in a decanter where the congeners are concentrated and removed from the system. The ethyl alcohol in the extractive distillation column is recovered in the base effluent, which is recycled to the aldehyde column feed. This distillation system can be operated with very low alcohol losses.

Aldehyde-rectifying-fusel oil concentrating-heads concentrating column system

In this system (Fig. 6.11) the heads draws from the aldehyde and rectifying columns and the fusel oil draw from the rectifying column can have volume flow rates high enough to prevent congener accumulation in the aldehyde and rectifying columns. The alcohol in these streams is recovered in the heads concentrating column and the fusel oil concentrating column. This system is capable of producing a light-bodied or near-neutral product with very low alcohol losses.

Extractive-rectifying-fusel oil concentrating-heads concentrating column system

If an extractive column is substituted for the aldehyde column, the condensed overhead product from the extractive distillation column, which contains both heads and fusel oil congeners, is fed into the fusel oil concentrating column. In this column the fusel oils are concentrated and removed and the heads are recycled from the condenser to the heads concentrating column, where they are removed from the system. This system (Fig. 6.12) is capable of producing a very neutral spirit, but because of the large volumes of the heads and fusel oil streams that are recycled throughout the system, column diameter must be increased and the steam needed to revaporise these streams makes the operating costs high.

Purifying-aldehyde-rectifying-fusel oil concentrating-heads concentrating column system

The high wine plus recycle streams are fed into the purifying column in this system (Fig. 6.13). The purifying column has a heads draw from the condenser and a fusel oil side draw. This reduces the congener level in the aldehyde column and therefore

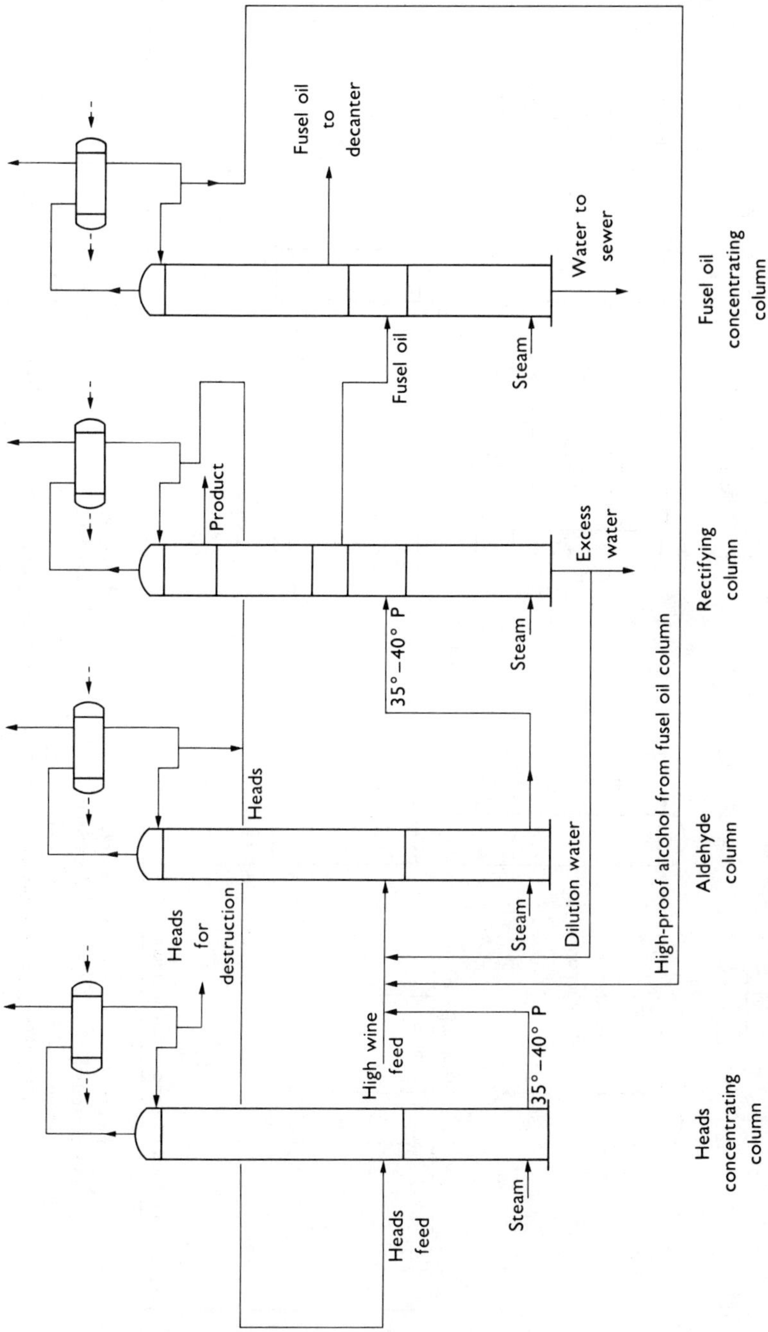

Fig. 6.11 Aldehyde-rectifying-fusel oil concentrating-heads concentrating column system.

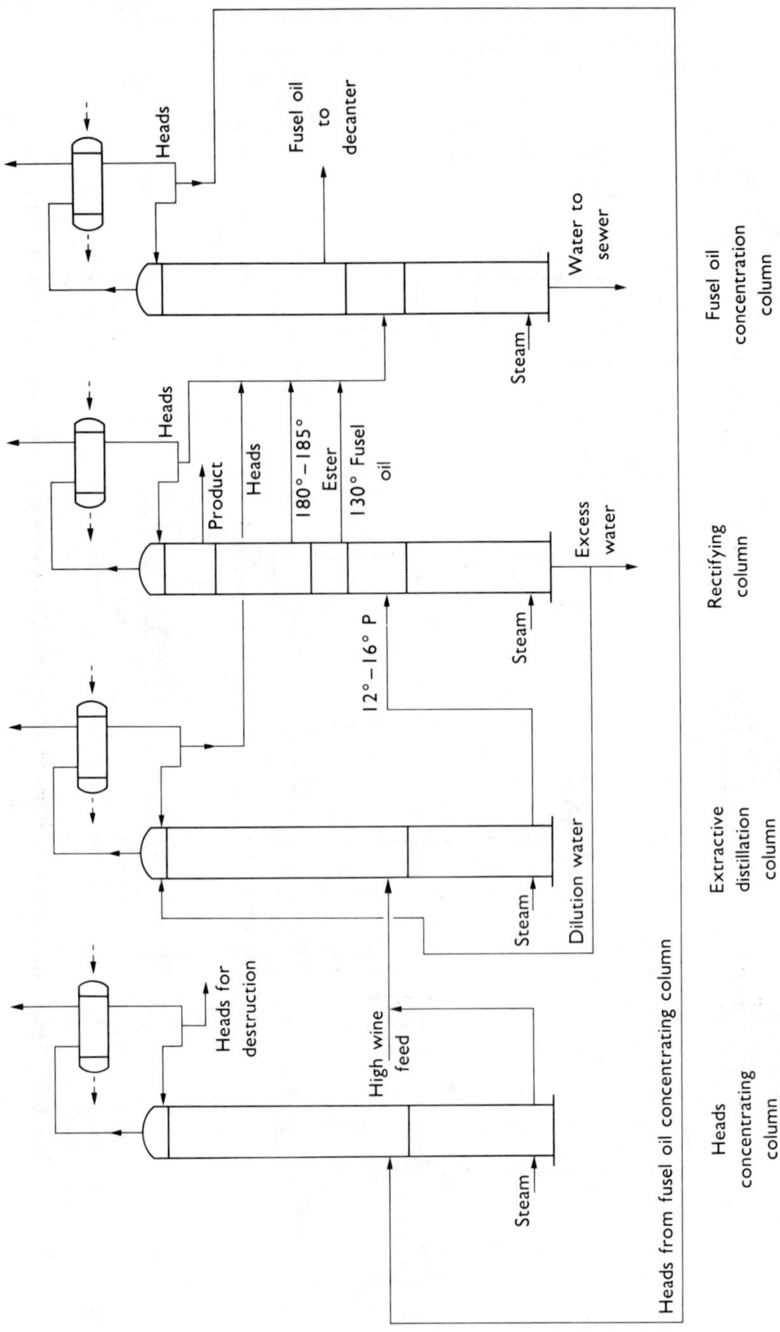

Fig. 6.12 Extractive-rectifying-fusel oil concentrating-heads concentrating column system.

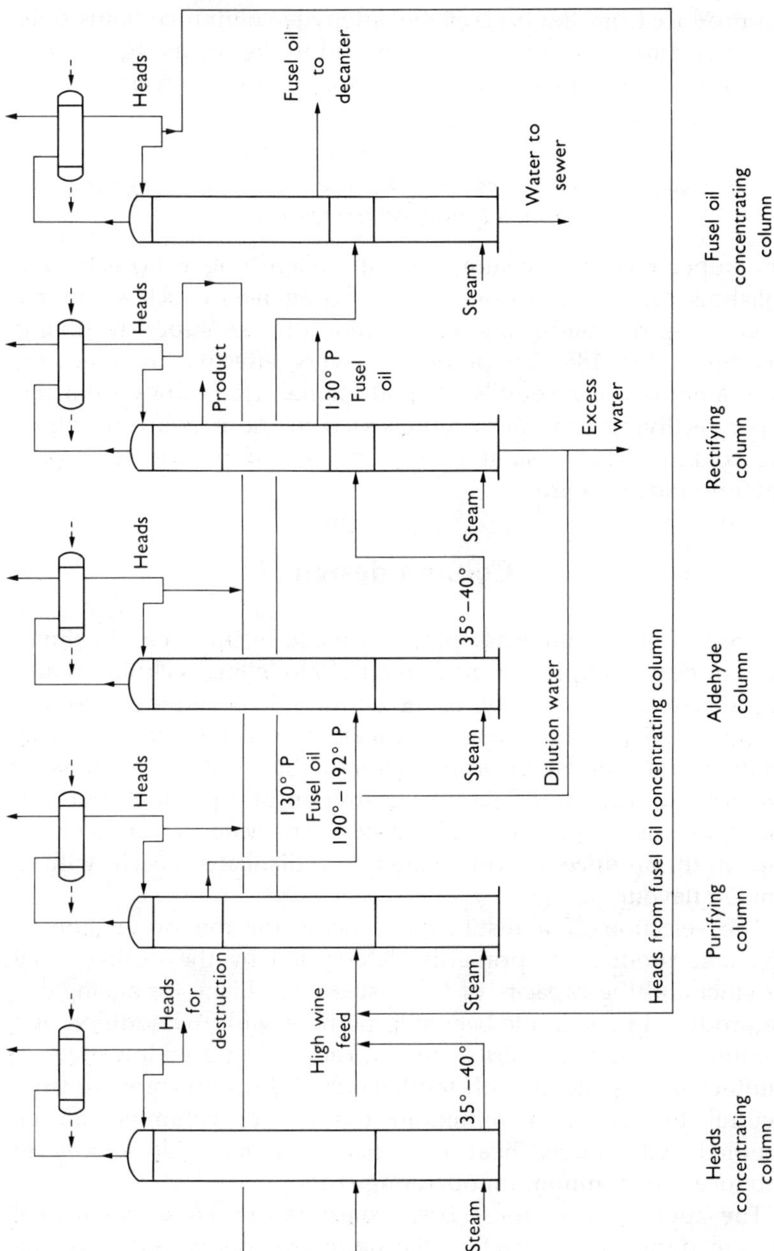

Fig. 6.13 Purifying-aldehyde-rectifying-fusel oil concentrating-heads concentrating column system.

the product from the base of the aldehyde column contains only traces of congeners, which are removed in the rectifying column. This type of column arrangement will produce a very neutral type of distillate.

Aldehyde-rectifying-super rectifying-fusel oil concentrating-heads concentrating column system

The super rectifying column in this system (Fig. 6.14) acts as a polishing column to remove traces of congeners which are in the product spirit. Since the base proof of the super rectifying column is 182–185° US proof, it is very effective in removing trace amounts of fusel oils. The additional heads draw from the super rectifying column is reprocessed in the heads concentrating column. This system is also capable of producing a very neutral grade of spirit.

Column design

There has been significant progress made in the area of column design, due to improved mathematical modelling techniques and the increased availability of powerful microcomputers to the design engineer. The volatile behaviour of congeners has been studied and can be incorporated into the design equations to predict not only the alcohol concentration of a product, but also the congener profile as well. Distillation systems can be conceived to produce a wide variety of distillates, each with a unique flavour.

The selection of a distillation process for the production of alcoholic products is primarily determined by the desired final product and the capacity of the system. A whisky for ageing can be produced in a simple beer still, perhaps with the addition of a continuous doubler. Spirits can be produced in a wide variety of multi-column systems with the flavour of the spirit ranging from neutral to heavy. A minimum number of columns can be coupled with novel heat exchange systems and devices to produce fuel at minimum operating costs.

The application of these basic concepts can assist the alcohol system designer in selecting the optimum configuration for the production of the desired end-product. These principles can also be utilised by operating personnel to maintain optimum operations and assist in trouble-shooting plant problems.

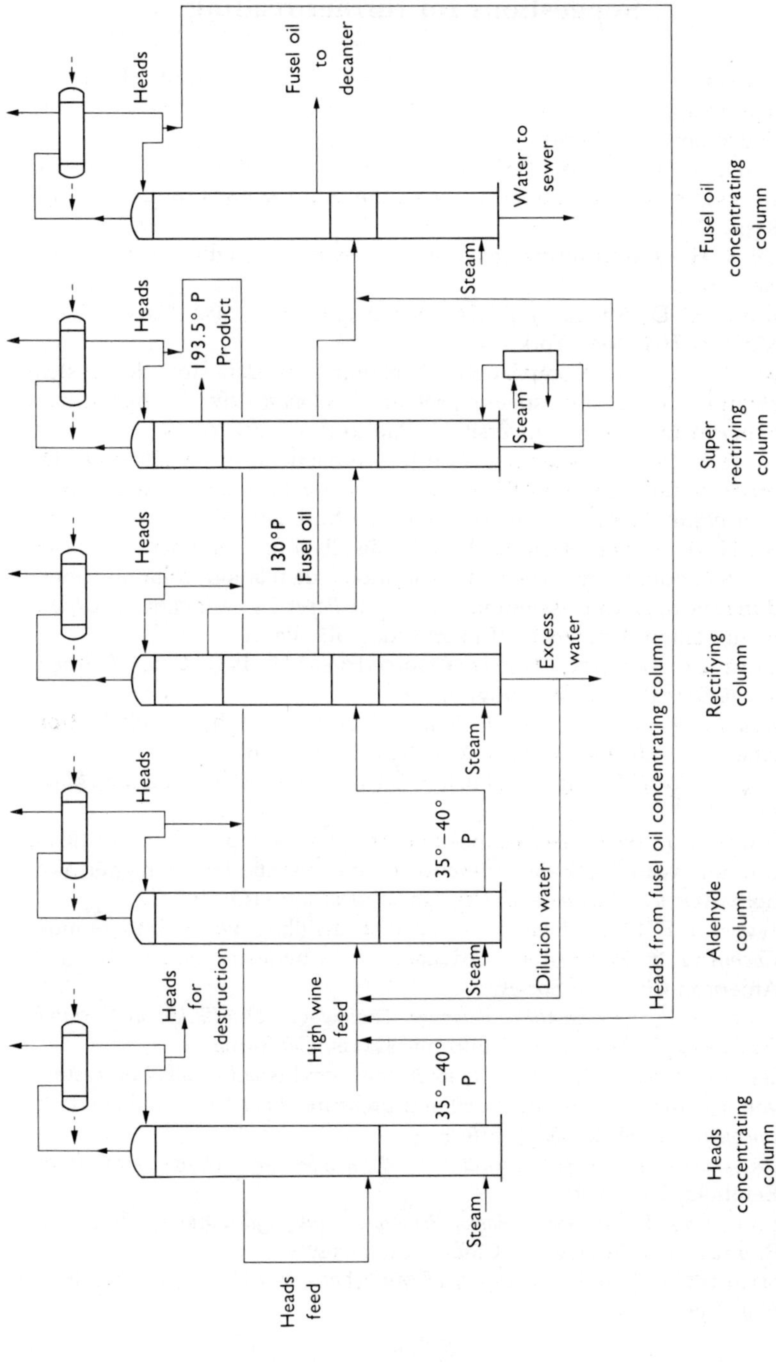

Fig. 6.14 Aldeyde-rectifying-super rectifying-fusel oil concentrating-heads concentrating column system.

Suggestions for further reading

ALTSHELER W B, UNGER E D, KOLACHOV P 1951 Improved still for liquid-vapor equilibria. *Industrial and Engineering Chemistry* XLIII (November): 2259–564

ANDREASEN A A (ed) 1982 *Beverage Alcohol Production*. Research & Development Department, Joseph E Seagram & Sons, Inc, Louisville, Kentucky

BARRON H 1944 *Distillation of Alcohol*. Joseph E. Seagram & Sons, Inc, New York

BENNETT C D, MEYERS J E 1962 *Momentum, Heat, and Mass Transfer*. McGraw-Hill, New York

BRAU H M 1957 Compilation of equilibrium data for the system ethanol-water at atmospheric pressure. *Technical Paper 19*, Agricultural Experiment Station, University of Puerto Rico, Rio Padres

BRAU H M 1958 Considerations on the convenience of distilling at high proof in continuous alcoholic distillation for beverages. *The Journal of Agriculture of the University of Puerto Rico* XLII: 248–57

BRAU H M 1958 Fusel oil in alcoholic distillation: experimental data on its distribution by plates of continuous distillation columns under different operating conditions. *Technical Paper 24*, Agricultural Experiment Station, University of Puerto Rico, Rio Padres

BUREAU OF ALCOHOL, TOBACCO, AND FIREARMS 1981 *Code of Federal Regulations* Part 27. Washington, DC

COLES H W, TOURNAY W E 1947 Removal of higher alcohols from ethanol. *Analytical Chemistry* XIX (November): 936–7

ELLERBE R W 1973 Batch distillation basics. *Chemical Engineering* (May): 110–16

GUYMON J F 1960 The composition of plate samples from distilling columns with particular reference to the distribution of higher alcohols. *American Journal of Enology and Viticulture* **11**(3): 105–12

GUYMON J F 1974 Chemical aspects of distilling wines into brandy. *Chemistry of Winemaking*, Advances in Chemistry Series, No. 137. American Chemical Society

GUYMON J F 1976 *Distilled Beverage Technology: Distillation and Brandy Production*. University of California, Davis, California

HAFSLUND E R, LOVELL C L 1946 Active amyl-isoamyl alcohol system: vapor-liquid equilibrium at 760 mm pressure. *Industrial and Engineering Chemistry* XXXVIII (May): 556–9

HENGSTEBECK R J 1961 *Distillation: Principles and Design Procedures*. Reinhold, New York

HIMMELBLAU D M 1967 *Basic Principles and Calculations*, 2nd edn. Prentice-Hall, Englewood Cliffs, New Jersey

HOLLAND C D 1981 *Fundamentals of Multicomponent Distillation*. McGraw-Hill, New York

KIRSCHBAUM E 1948 *Distillation and Rectification*. Chemical Publishing Co, Brooklyn, New York

LORD A 1979 *The World Guide to Spirits, Aperitifs, and Cocktails*. Sovereign Books, Gulf and Western, New York

MCCABE W L, SMITH J C 1967 *Unit Operations of Chemical Engineering*, 2nd edn. McGraw-Hill, New York, chs 17, 18, 19

MCCABE W L, THIELE E W 1925 Graphical design of fractionating columns. *Industrial and Engineering Chemisty* XVII (June): 605–11

NISENFELD A E, SEEMANN R C 1981 *Distillation Columns*. Instrument Society of America, Research Triangle Park, North Carolina

PACKOWSKI G W 1978 Beverage spirits, distilled. *Kirk-Othmer Encyclopedia of Chemical Technology* 3rd edn. Wiley, New York, Chichester, vol III, p 830

PERRY R H, CHILTON C H (eds) 1973 *Chemical Engineers' Handbook*, 5th edn. McGraw-Hill, New York, chs 13, 18

ROBINSON C S, GILLILAND E R 1950 *Elements of Fractional Distillation*, 4th edn. McGraw-Hill, New York

SHAH G C 1978 Troubleshooting distillation columns. *Chemical Engineering* (July): 70–8

SHINSKY F G 1977 *Distillation Control; for Productivity and Energy Conservation*. McGraw-Hill, New York

A Source Book of Technical Literature on Fractional Distillation. Gulf Research and Development Co., undated

SMITH J M, VAN NESS H C 1959 *Introduction to Chemical Engineering Thermodynamics*. 2nd edn. McGraw-Hill, New York

UNGER E D 1980 *Beverage Alcohol Distillation*. Research & Development Department, Joseph E Seagram & Sons, Inc, Louisville, Kentucky

CHAPTER 7

Process control

P J HARDY and J H BROWN

Introduction

'Process control' in distilleries refers to the steps that are taken to maximise yields, economy, and, in times of high production requirement, throughput, whilst making whisky with the desired qualities. This is achieved by monitoring each stage of the process using both on-line and laboratory measurements from which decisions can be made regarding the adjustment of process conditions. Process control can also mean the use of automatic systems to carry out process operations with the additional objective of making savings in labour costs. The legal definition of Scotch whisky (Finance Act 1969, Schedule 7, and amendments of 1979 and 1980, which is administered by HM Customs & Excise) states that:

> the expression 'whisky' or 'whiskey' shall mean spirits which have been distilled from a mash of cereals which have been—
>
> (i) saccharified by the diastase of malt contained therein with or without other natural diastases approved by the Commissioners; and
>
> (ii) fermented by the action of yeast; and
>
> (iii) distilled at an alcoholic strength (computed in accordance with section 2 of the Alcoholic Liquor Duties Act 1979) less than 94.8 percent in such a way that the distillate has an aroma and flavour derived from the materials used.

This imposes constraints on the processing materials and conditions which can be used, and restricts the use of added enzymes or chemicals that could increase yields and reduce costs. In these respects there are fewer options open to process control scientists than in most other distilling and brewing operations. In addition, Customs & Excise regulations (Statutory Instrument 1982 No. 611) control the type of plant, conduct of processes, and records to be kept by distillers. The approval of the Customs is also required for the use of any additives in the process, such as yeast foods and enzymes.

Information about the efficiency of making whisky may be obtained by physical measurements or by chemical and microbiological analyses. Some information, particularly about the quality of raw materials, may be obtained in time to allow action to save costs and enhance yields. Other information, typically that concerning fermentation efficiency, may only be obtained sufficiently early to stop a process fault being perpetuated, but often not until it has been repeated several times over a series of fermentations. The value of information must always be balanced against the cost of acquiring it.

Information collected after processing events have taken place does not allow action for the production batches concerned but forms a base of knowledge of the milling or cooking, mashing and fermentation processes. Interpretation of this information database should lead to better yields and companies operating several distilleries should be in a position to acquire new knowledge sooner than those operating only one. Process control operations are now being greatly assisted by the use of computers for data acquisition and interpretation.

It is an essential part of process control that education of all personnel who can affect process efficiency is carried out. Automated processes used in some distilleries take this controlling influence away from certain operators, for example mashmen, and transfer it to others, for example instrument engineers.

Sampling

The provision of accurate and relevant analytical information for process control is only possible if representative samples are available for analysis. Whereas this may be reasonably easy for samples of homogeneous liquids it is difficult to achieve for raw materials and process intermediates such as worts and wash. For example, if grain sampling is being carried out to provide a weekly yield target based on raw material analysis, then a representative sample is required from the week's throughput of perhaps 1000 t in a medium-sized grain distillery. To rely on sampling by distillery personnel leaves room for doubt as to the representative nature of the sample. In-line trickle or grab samplers followed by sample division are to be preferred. Commercially available samplers are not always suited to the task, and systems designed in the distillery are generally best. Much use is made of the sample of wash taken at the end of fermentation. Generally, a few litres of sample are taken from a fermenter which may contain up to 6000 hectolitres. Efforts are made to render the liquid in the wash back homogeneous by stirring or rousing with air prior to sampling. This may be supplemented by the use of samplers allowing withdrawal from various depths in the fermenter. This latter course is particularly necessary for sampling 'all grains in' fermentations. The installation of in-line sensors for parameters such as ethanol, lactate and sugars may alleviate misgivings presently held about measurements based on small 'representative' samples taken from large bulks.

Raw materials and the prediction of distillery yield

Raw materials represent the greatest part of distilling costs, accounting for about two-thirds of the cost of making new malt and grain whisky. It is normal practice each year for distilleries to obtain samples of grain or malt from the various suppliers in the market before drawing up contracts. The distiller must have facilities for analysing the samples in order to decide for himself which represent the best value for money. Any process control operation must monitor carefully the malt or grain received at the distillery, both to check that it is as good as the specifications against which it has been purchased, and also to provide a predicted or target yield in litres of alcohol per tonne.

In the Scotch whisky industry, trading in malt and cereals is generally on the basis of analyses carried out under the *Recommended Methods of Analysis* of the Institute of Brewing (1977) but more recently, EEC regulations have been introduced for some purposes. These methods are continually being revised and extended to new analyses, such improvements normally appearing first in the Institute journal before being collated in a revised edition of the *Recommended Methods*.

GRAIN FOR COOKING

Unmalted cereal for grain distilling is traded on qualities which are relatively easily measured, such as moisture content, specific weight, sieve analysis and nitrogen content. Grain infested with insects will invariably be rejected. The emphasis here lies on speed of obtaining results at distillery intakes, thus allowing deliveries to be rejected when necessary. Many reliable instruments are available presently for carrying out rapid determinations of these parameters. The relatively high cost of such instruments compared with that of the slower, standard methods of analysis can be justified easily when the raw material costs are considered. Equipment costing £20,000–£30,000 can represent money well spent when it is considered that a medium-sized grain distillery can process 50 000 t grain costing some £6–7 million each year.

Distillers may require their grain to conform to quality standards other than those used in normal trading. For example, maximum levels for pesticide residues and mycotoxin contents may be specified. Such specifications, however desirable, require time-consuming analyses to be carried out and only infrequent analysis can be justified.

Since the grain used in cooking is the main cereal starch source for grain whisky distilling, it should be analysed to determine its alcohol yielding potential. Starch measurement can indicate the approximate potential, and polarimetric (Association of Official Analytical Chemists, 1980) and enzymic (Beutler, 1978; Ahluwalia and Ellis, 1984) methods are used. However, these methods may not be sufficiently accurate for the full range of cereals now in use. Trial fermentations in the laboratory, with the cereals to be used in the distilleries, may be carried out if adequate facilities are available for miniaturising the operations of cooking, mashing and fermentation. If accurate process control is to be carried out then regular (weekly perhaps) estimation of the alcohol-yielding potential of cereals should be made on representative samples of the cereal in use.

In this way the yield of alcohol achieved in the distillery may be compared with a 'target' yield calculated from laboratory fermentations, which are carried out to mimic the distillery process as far as practicable. However, lower standard fermentation gravities and temperatures may be used, in as much as they are less likely to strain the fermentation. Conversion of starch to fermentable sugars may be carried out using fungal enzymes in the laboratory test in order to remove the effect of variable malt quality.

The choice of raw material will depend on prices prevailing at the time of purchase, but comparative analyses and yields are shown in Table 7.1.

Table 7.1 Composition and yield potentials of cereals used in grain distillery cooking (percentage results calculated on dry basis).

	Maize	*Wheat*	*Barley*
Starch (%)	71.9	71.0	62.1
Protein (%)	10.3	11.2	12.5
Oil (%)	4.2	1.6	2.0
1000 corn weight (g)	264.2	43.8	35.8
Specific weight (kg per hl)	70.5	76.8	66.8
Yield of alcohol (litres of alcohol per tonne dry grain)	465.0	452.0	385.0
Glucan (%)	< 0.1	0.6	4.0
Pentosan (%)	< 0.1	6.0	3.0

Techniques for setting target yields are particularly useful for comparing the potential of different cereals and of the same cereal from different sources, for example African white versus French or American yellow maize. Such techniques have also indicated the week-to-week variations in yield potential of a cereal in regular use caused by varying moisture, starch or other qualities. If non-malt enzymes are used for starch conversion in the laboratory test it must be remembered that they may have activities which are either missing from the distillery malt source or they may be deficient in activities which may be present in the distillery malt.

MALT FOR GRAIN DISTILLING

The cost of dried or green (unkilned) malt used for saccharification in grain distilleries is high relative to that of the cereals used in cooking, and therefore distillers look first and foremost at the diastatic activity of their malt. Malt having relatively high α- and β-amylase contents can be used at a lesser percentage in the mash. It should not be forgotten, however, that malt also supplies modified starch to the process and the relative amounts of this in malts of different enzyme quality must be considered in calculating the least cost option. Since malt is often the lower alcohol-yielding source of starch, inefficient saccharification brought about by using slightly less than the optimum percentage of malt might be the best cost option. Such judgements are risky, however, both because of the imprecision of the enzyme analyses that are presently available and also in view of the effects of processing conditions on saccharification.

The important qualities to be assessed are enzyme content, particularly α-amylase, and alcohol-yielding potential. The latter assessment may be carried out in similar fashion to that for cereals used in cooking, i.e. by trial fermentation. The α- and β-amylase contents are analysed by measurement of dextrinising units (DU) and diastatic power (DP), using the *Recommended Methods of Analysis* of the Institute of Brewing (1977). From the knowledge of DU and DP values of the malt used in distilleries and of corresponding distillery yields, it is possible to build up a body of data from which the optimum percentage of malt of any given quality can be predicted. Different relationships may hold in cases where malts from certain varieties of barley have different ratios of α- and β-amylase, and distillery trials may be carried out to determine the optimum malt percentage.

For distilleries using green malt more rapid determinations of enzyme contents are required, and these have been reported by Henry (1984) and Mathewson and Pomeranz (1981). Rapid, accurate techniques for enzyme activities such as limit dextrinase (Lee and Pyler, 1982) and β-glucanase (Martin and Bamforth, 1983; Bamforth, 1985) may also be beneficial to the distiller, and advances are being made in this field.

Malt supplied to grain distilleries is also expected to conform to moisture and extract specifications and be free from insect infestation. The use of chemical additives in malting is not permitted.

MALT FOR MALT WHISKY

Extract

The most important analysis is that of hot-water extract, giving a measure of the extract which can be expected in the mashing operation at the distillery and hence the distillery yield. Extract is expressed as a percentage of the fresh weight, and in commercial malts is usually in the range 74–79% at a moisture content of 4.5–5.0%.

Prediction of spirit yield from malt hot-water extract

A prediction of spirit yield in litres of alcohol per tonne is obtained by multiplying per cent hot-water extract by 5.22. This factor is used widely in the distilling industry, as are other factors of apparently empirical origin, and it therefore deserves some explanation. On the introduction of metric methods of measurement in 1981 the factor 5.22 was obtained by direct conversion from the factor previously in use, 0.236 proof gallons per brewer's pound extract per bushel of malt. This factor seems to have been introduced in 1921 as a result of laboratory fermentation experiments. The following calculation shows that the 5.22 factor is an underestimate for predicting spirit in wash from per cent extract.

A typical malt distillery fermentation starts at a gravity of 55° and loses 47°, giving a residual gravity of 8°. Referring to the original gravity table for 'worts of beer' (a discussion of the validity of this table for use in distilleries is given later in this chapter) in the *Recommended Methods of Analysis* (1977):

Spirit indication for 47° gravity lost	= 10.538
Specific gravity of alcoholic distillate	$= 1 - \dfrac{10.538}{1000}$
	= 0.98946
Density of alcoholic distillate	= 0.98946 × 0.99715
	= 0.98664
From *Laboratory Alcohol Table* (1979)	
Alcohol concentration	= 7.589 % vol
Therefore for each degree of gravity lost the alcohol produced will be	0.1615%
or 0.001615 litres of alcohol per litre degree of extract fermented.	

Extract in percentage terms as used in the distilling industry is calculated from litre degrees per kg by dividing by the solution factor 3.86. Predicted yield of alcohol in wash per tonne will therefore be 0.001615 × 3.86 × 1000 × per cent fermentable extract, or 623.4 litres per tonne of fermentable extract. The factor 5.22 therefore corresponds to a fermentation of 83.7%, a figure easily exceeded in distilleries today. Hastie (1925) presents a vivid picture of the unsterile procedures used in the early part of the century to obtain supplies of brewer's yeast, the only source then available, when the target yield of 0.236 per pound per bushel must have been difficult to attain in commercial practice.

The extract figure used for predicting spirit yield is that obtained by the Institute method using the coarse grind (setting 7 on the Buhler-Miag laboratory disc mill). However, the grist fractions so obtained bear little resemblance to those in the distillery, although results are more reproducible than was the case when the old Boby roller laboratory mill was used. Furthermore, the distillery is carrying out a three- or four-water extraction with higher temperatures extracting substances not dissolved in the one-water laboratory mash.

The factor 5.22 does not take into account differences in malt fermentability, a method for determining which has now been accepted by the trade and by the Institute of Brewing (Dolan *et al.*, 1981). This method makes use of the sample of wort which has already been prepared in the laboratory for measurement of hot-water extract. The wort is subjected to a laboratory fermentation at a constant temperature of 33°C. During fermentation, a proportion of the solids present in the worts, usually about 86%, is used by the yeast, this figure being known as per cent fermentability. Although good correlations can be demonstrated between malt fermentability as measured in the laboratory and distillery fermentation results, the difference in fermentation conditions leads to the figure for the laboratory test being slightly higher than that achievable in the distillery.

The product of extract and fermentability is per cent fermentable extract, usually about 65%. Multiplying this by 6.06 (a factor currently accepted by the trade but which is, as shown above, on the low side), gives a figure for predicted, or potential, spirit yield in litres of alcohol per tonne. This figure is now coming into increasing use, both as a malt specification and as a target yield. Neither this nor the target figure obtained from extract alone takes account of the alcohol left in spent wash and spent lees after distillation, a fact which must be appreciated by the distiller when calculating overall distillery efficiency by comparing actual and predicted yields. For the reasons stated above, per cent fermentable extract is a more satisfactory basis for predicting spirit yields than hot-water extract, especially if determined using a fine grind wort.

Other malt analyses

Malt moisture and diastatic power are normally included in malt specifications. Moisture is usually between 4.0% and 5.5%, low moistures requiring extended kilning cycles and consequently being more expensive to produce. The advantage of low moisture malt to the distiller is that a cleaner separation of husk from endosperm is obtained during milling, which may lead to better extract recovery in the mashtun. Some distillers also feel that long kilning is desirable for the generation of flavour substances. Diastatic power is rarely a limiting factor in the production of fermentable sugars; however, low DP values can sometimes be an indication of faulty kilning and this is an analysis a distiller may carry out when comparing different suppliers' samples. Diastatic power as a means of assessing malt quality has now largely been overtaken by the measurement of malt fermentability.

A further aspect of malt quality which can be of importance to the distiller is the degree of modification. During germination the processes of cell wall breakdown and protein dissolution result in a grain which is friable and can be milled to give a grist without excessive flour content yet with complete separation of the endosperm from husks. Poorly modified malts, or malts in which a significant proportion of the grains have not germinated, contain 'hard ends' or whole grains which cannot be milled. This can be tested by means of the 'friabilimeter', which can be used to find both the overall per cent friability and the percentage of unmodified (whole) grains. The presence of unmodified material leads to extract losses, and in addition has been shown (Bathgate, 1983) to lead to increases in wort viscosity and impeded run-off. The friabilimeter can give a much more sensitive and reproducible measure of malt modification than the standard Institute method of coarse-fine extract difference. Although malts giving a result of 80% or less can fairly be said to be undermodified, malts giving a high result can sometimes give problems because unmodified sections of endosperm may be small enough to pass through the friabilimeter sieve. Bourne and Wheeler (1982) used another extract method which takes account not only of the increased coarse-fine extract difference that occurs with poorly modified malts, but also of the increased dependence on mash concentration. It gives a much wider range of values when the extract value from a concentrated mash with coarsely ground malt is subtracted from the extract of a dilute, fine-grind mash, ranging from 4–6% for a well-modified malt to over 10% for certain problem malts.

Other malt analyses which the distiller may wish to carry out because of possible effects on whisky quality and composition include volatile phenols, sulphur dioxide and nitrosodimethylamine (NDMA). Volatile phenols are an indication of peatiness and are important in distilleries where a medium or high level of peaty flavour is required. Sulphur dioxide is absorbed by malt during kilning from the gases produced by the combustion of fuels with a high sulphur content, and distillers who believe that it has an effect on whisky quality include a maximum sulphur dioxide content in their malt specifications. NDMA is also formed in malt during kilning under certain conditions and must be kept to less than 1 or 2 ppb. The Thermal Energy Analyser for carrying out NDMA analysis (Long, 1982) is expensive, but companies who are not able to carry out the analysis themselves are able to use the services of consulting laboratories.

The Scotch whisky industry, unlike the brewing industry, does not allow the use of malt that has been treated with chemicals, such as gibberellic acid. Although direct detection by chemical analysis may not be possible, evidence of such treatment can be obtained from unnatural values for worts colour and soluble nitrogen ratio.

YEAST

Whereas testing of grain and malt prior to use is generally possible, the same is not always true of yeast, because an important part of the process control is to use it as quickly and in as fresh a condition as possible. Distillers use relatively few sources of yeast since only a few strains demonstrate the qualities of attenuation required. For the last 20 years or so the 'M' strain yeast manufactured by the Distillers Company has been the major source of yeast throughout the Scotch whisky distilling industry. Prior to the introduction of pure culture yeasts distillers used the yeast removed as a by-product from brewery fermentations. Brewer's yeast is still used in malt distilleries by some companies but as a mixture with a pure culture distilling yeast.

Distilling yeast is supplied in pressed form at 25% solids content. The yeast is used once only. Viability is usually near 100% and though microbiological cleanliness is generally good, numbers of bacteria can rise rapidly if the yeast has to be kept for any length of time before use. Brewer's yeast is a much more variable commodity, coming as it does from different breweries where care in harvesting, storage and transport may not be adequate. Brewer's yeast may contain anything between 15% and 90% living cells, but because of the bactericidal properties of hop α-iso acids the content of bacteria is usually much less than that of distilling yeast. Now that brewers are tending towards the addition of hop extracts to beer after fermentation and removal of the yeast, the bactericidal value of brewer's yeast is becoming lost. Autolysis occurring during spoilage provides an ideal substrate for bacteria and is indicated by softness and the bags becoming 'blown'. Such yeast should be rejected.

Standard tests for yeast viability (*Recommended Methods of Analysis*, 1977) and counts of total bacteria can be used to monitor yeast quality. Dolan (1976) advises that only yeast with a viability of 85% or more on arrival should be accepted. Although the function of brewer's yeast is not entirely connected with its fermenting activity, viability is an index of condition and if the yeast is to be stored for any length of time its condition at the time of use should be examined.

Although many of the bacteria present in distiller's yeast will not grow under the anaerobic conditions of distillery fermentation, the bacteria count is a measure of the care that has been taken in the production and handling of the yeast at source, during transport and storage at the distillery. Very high counts can usually be correlated with reduced distillery yields and although populations should be well below 1 million per gram leaving the factory, numbers can double during each subsequent day. Dolan (1979) recommends that yeast should show fewer than 1500 total bacteria per million yeast cells and of these no more than ten should be lactobacilli.

Testing of distilling yeast by means of laboratory fermentations is carried out by the yeast-producing companies, and where samples can be tested with sufficient speed, distillers also may be able to screen yeast batches before use. Simple attenuation testing of laboratory fermentations carried out at different temperatures and after different periods of storage can help to identify yeast batches likely to give problems in distilleries. Despite this, results frequently arrive too late for action to be taken and it would be an advantage to have more rapid tests for identifying yeast of suspect quality.

Although yeast may arrive in good condition, deterioration will take place before use and special precautions should be taken to avoid this. These are: using the yeast as soon as possible after delivery; keeping it in a cold store at between 0 and 5°C; and mixing distilling yeast with brewer's yeast in a liquid suspension with approximately two parts of water to one of yeast. A brewer's yeast should contain sufficient hop α-iso acids to prevent the growth of bacteria in the mixture.

WATER

Distilleries require water for mashing, cooling heat exchangers, condensers or worms, spirit reduction, boiler feed make-up and washing purposes. The quality of mashing water used in grain distilleries is generally a stable factor, although perhaps being unique to a particular distillery. The complete range from very soft water to very hard water may be used successfully without modification in grain distillery processes, although a hard water would be preferred for reasons of enzyme stability in mashing (Briggs *et al.*, 1971; Bertoft *et al.*, 1984). Quality monitoring may be carried out off-line to ensure consistency of supply, especially at certain times of the year. Contamination by decaying vegetation may lead to undesirable off-odours which, while they may be detected by nosing, are mirrored in the ammoniacal nitrogen content. For distillers operating carbon dioxide recovery plant it can be appropriate to monitor the nitrate content of the supply so that remedial action may be taken to avoid nitrous gases in the product. Mashing or process waters for malt distilleries vary considerably in composition, coming from various sources such as springs, mountain streams, lochs and even town supplies. Diastatic activity is not a limiting factor in malt mashes and most process waters are soft (Rutherford, 1981) with a hardness of less than 100 ppm calcium carbonate ($CaCO_3$). No chemical treatment is necessary, but chemical and bacteriological analyses are occasionally carried out to ensure that no drastic changes have occurred. Cooling water is required in the largest amount; as long as its content of algae or other suspended material is not such as to clog up waterways in the cooling system, and steps are taken to avoid scale build-up, its composition is immaterial.

The purest source of water available is used for spirit reduction at both grain and malt whisky distilleries. Water with abnormally high iron content or significant hardness is unsuitable for this purpose. Discoloured or opaque spirit may be caused and the problem is worse in grain distilleries where reduction is from a much higher alcohol concentration. While some distilleries use reducing water supplies that are almost guaranteed safe, routine monitoring for iron and hardness is prudent. For supplies that have to be demineralised prior to use, on-line monitoring using a conductivity meter is essential.

The water for boiler feed make-up usually has to be treated using a cation exchange resin to avoid the dangerous conditions resulting from the build-up of scale. Insurance contracts typically specify a maximum of 5 ppm $CaCO_3$, which can only be achieved without treatment using the softest of natural raw supplies. Suspended solids must be removed prior to treatment to avoid clogging of water softener resin, and output of water softeners must be monitored.

Control of the extraction process in grain whisky distilleries

COOKING

Different temperature and pressure regimes may be used for cooking the different cereals used in grain distilleries. The gelatinisation temperatures of the starches in the commonly used cereals are shown in Table 7.2. The objectives of control remain the same. Whether the grain is presented ground or whole to the cooking process, cooked continuously (Wilkin, 1983) or in batches, the prime objective is that of achieving and maintaining the desired temperature and pressure conditions. Optimum cooking conditions vary for different cereals, whether ground or whole, for different varieties of a cereal and for different cooking equipment. These conditions may be indicated by laboratory cooking and fermentation trials if facilities are available. Such an approach should still be reinforced by adjustment of conditions in distillery trials and this latter approach may be the only option open to some distillers. Success in achieving the desired conditions of cooking may be readily monitored using temperature and pressure recorders. Deviations from the desired regime will be noted and collated with other process efficiency information. The direct assessment of the success of the cooking operation is difficult. The degree of gelatinisation and bursting open of whole grains may be assessed by visual inspection of cooked grains, but this is necessarily a crude measure. For whole grain cooking, inspection may provide the earliest warning of inefficient stirring in the grain cooker. The avoidance of starch lipid complexes is desirable (Murtagh, 1973) but their presence cannot readily be measured. The ratio of liquor to cereal used in the cooking process can affect cooking efficiency, in particular too concentrated a mix can be detrimental to efficient gelatinisation. Control of the desired ratio is readily achieved by cereal weighing and liquor metering. The degree of grinding of cereal may be important in some distilleries and this is readily monitored by sieve analysis. For ground cereals, the cooking regime must be such as to avoid balling during mixing of cereal and water. In continuous cooking it may prove necessary to employ malt to allow viscosity reduction of the grain slurry in the cooking tubes (Unger *et al.*, 1944).

Table 7.2 Gelatinisation temperatures of starches.

Starch source	*Gelatinisation temperatures (°C)*
Maize	70–75
Wheat	52–54
Barley	61–62

While a rapid, easy to use, preferably in-line measurement of cooking efficiency would be welcomed by the distiller, he is left basically with the indications of temperatures, pressures and times taken from recorders, and information which may be gained from residual starch analysis of worts (Schur *et al.*, 1978) and fermented wash. This latter information may be of a historical nature since many batches of cereal may have been cooked prior to a cooking fault being detected by such means.

MASHING

The process of saccharifying the cooked, gelatinised starch with, in the case of the Scotch whisky industry, the diastase of malt may take place either in batch fashion or continuously. Coarse cereal solids may be removed during the mashing operation, but are now predominantly transferred to the fermenters in an 'all grains in' process. To ensure the most efficient use of enzymes available in malt, optimum mashing conditions must be supplied. Mashing temperatures and pH are critical. Grain distillery mashing operations vary widely and different regimes are used depending on whether dried or green malt is used, on the cooked cereal in use and on whether coarse solids are to be separated prior to fermentation. The objective remains that of saccharifying most of the starch from the cookers and killing the bacteria present in the malt, whilst preserving satisfactory enzyme activity for completion of saccharification during fermentation ('secondary conversion'). Distillers using 'backset' or 'backslop', essentially reuse of process liquor recovered after distillation (Ronkainen *et al.*, 1978), must pay close attention to the pH of worts during mashing. Optimum mashing conditions are set either from laboratory trials, from controlled distillery trials, or a combination of both.

Malt should be ground as finely as possible for release of enzymes into solution. A compromise may be necessary, however, if coarse solids separation is to be carried out after mashing. Fineness of grind can be adequately and conveniently checked by sieving techniques.

Techniques of mashing together the cooked grain and malt slurry also vary widely. However, it is a general aim to achieve a final mash temperature of between 60 and 65°C; 62°C is commonly used. The temperature of the mash will vary during mixing and the profile of temperature required may be varied according to the raw materials used. Typically, low initial mashing temperatures (approximately 50°C) are used to ensure proteolytic and, when necessary, β-glucanase activity. The temperature is then raised to carry out saccharification and extraction of the starch, this also being sufficient to kill the majority of bacteria present in the dried or green malt (Dolan, 1979).

A major aspect of process control is ensuring that the correct proportions of malt and cooked cereal are mixed together. Accurate weighing and logging of weights is therefore essential, and mashing systems benefit greatly from automation of the admixing. Good control of temperature profiles can be achieved and the opportunity for controlled distillery trials to optimise mashing conditions is enhanced. Hydrometer measurements traditionally used in the mash-house for gravity determination are informative but do not indicate fully the progress of the mashing operation. This technique may be supplemented by other useful measurements, such as α-amino nitrogen content, pH, worts viscosity, sugar composition of worts, residual starch composition of worts and enzyme activity. The number of such parameters routinely monitored and the frequency with which they are monitored depend on the process used. It is relatively easy to monitor pH and this may be considered essential if recycling of spent liquor is carried out. A knowledge of worts viscosity may be crucial to the successful operation of an effluent evaporation plant if barley, particularly, is being used. In view of the importance of secondary conversion during fermentation, a knowledge of the residual enzymic activity of worts is extremely useful to grain distillers trying to optimise malt usage. Specific electrode development is a rapidly expanding field (Janata and Blackburn, 1984) and it is likely that such devices may be available in future.

The limitations under which the grain distiller works are nowhere more evident than at the mashing stage. Overcooking of cereal due to plant failure or accidental excessive recycle of backset both act to give non-optimal acidic conditions in the mashtun and subsequent fermentation. Adjustment of pH, whilst desirable, is only allowed by the express permission of HM Customs & Excise and is generally considered unacceptable in the Scotch whisky industry.

Control of the extraction process in malt whisky distilleries

Extraction differs from the process in grain distilleries in that dry malt is the only starch source, and whereas in many grain distilleries fermentations are carried out with all grains in, it is invariably the case in the Scotch malt whisky industry that worts are drained from malt mashes. The control of the mashing process is a compromise between efficient conversion of substrates to a soluble form and separation of the worts. Various tests carried out both on site and in the laboratory can be used to optimise the process. Assuming that the malt in use is well modified and has a moisture content as specified, failure to achieve an efficient recovery of the extract must be due to inadequacies in the milling or mashing operations.

MEASUREMENT OF EXTRACTION EFFICIENCY

In malt distilleries dry malt is milled prior to mashing, and weighing is by mechanical weighing machines which pivot and discharge after filling with a pre-set weight of malt. The tipping mechanism is connected to a counter which registers after every tip. These machines are set by means of counterweights and must be adjusted so that tipping occurs when the weight of malt is exactly the same as that of the counterweights.

The volume of wort recovered in the fermenting vessel or wash back is found using a dipstick and by reference to gauging tables, which were once provided by HM Customs & Excise but are now the responsibility of the distiller. The traditional instrument used for measuring wort concentration, as in the grain distillery, is the saccharometer, which reads specific gravity, correct at 20°C. The figure used in calculating mashtun extract is the degrees of excess gravity (or simply 'gravity'):

$$\text{excess gravity} = 1000\,(\text{specific gravity} - 1)$$

$$\text{mashtun extract} = \frac{\text{excess gravity} \times \text{litres wort}}{3.86 \times \text{kg malt}}$$

The factor 3.86 is the solution divisor which is used to calculate per cent sugar in a solution of known gravity. The solution factor was in use over a hundred years ago and some of the early work is discussed by Nettleton (1913).

Because yeast is added soon after the first runnings from the mashtun have entered the wash back, a certain amount of fermentation occurs as the back is being filled, so that it is not possible to obtain an accurate direct measurement of the original specific gravity of the wort. The figure for wort gravity that is used for calculation of mashtun extract in the distillery is an approximation achieved by adding a correction to the stem reading on the saccharometer. In order to obtain an accurate figure a laboratory determination of 'original gravity' is carried out.

ORIGINAL GRAVITY

The determination of original gravity, i.e. the degrees of excess gravity of worts as if no fermentation had taken place, from the analysis of fermented wash is possible because of the relationship between gravity lost and alcohol produced. In the traditional method of analysis (*Recommended Methods of Analysis*, 1977) a sample of wash is distilled and the distillate is carefully collected so that no alcohol is lost and is then made up to the same volume as that of the original sample. The specific gravity is found and from it the spirit indication:

$$\text{spirit indication} = 1000\ (1 - \text{specific gravity of distillate})$$

The gravity lost corresponding to the spirit indication is found by reference to tables. The non-volatile residue in the distilling flask is made up to the original volume also, and its degrees of excess gravity measured. Original gravity is the sum of gravity lost and residual gravity.

The original gravity tables now in use came into force on 1 January 1980 as a Statutory Instrument (1979 No. 1146), following the introduction of metric methods of measurement. This document contains two tables: Schedule 1, for the original gravity of 'wort or wash'; and Schedule 2, for 'worts of beer'. In both cases the temperature for measuring spirit indication is 20°C. The table which now appears in the *Recommended Methods of Analysis* is the Schedule 2 table with additional interpolated values, and is in use throughout the brewing industry. In view of the unsatisfactory situation where we have two tables – Schedule 1 ostensibly being for use in the distilling industry – it is worth while looking into the origins of both.

Both the Schedule 1 and Schedule 2 tables are the results of converting earlier tables for use at the new reference temperature of 20°C stipulated at metrication, this work being carried out by the Laboratory of the Government Chemist. The 'Mean Brewery Table' (Thorpe and Brown, 1914) was taken and the effects on both spirit indication and gravity lost that the change from 60°F to 20°C would entail were calculated (Savage, 1971). However, the relationship between the thermal expansion of worts and wort gravity had not been fully considered and a further correction was applied (Lisle and Savage, 1976).

The classic experimental work of Thorpe and Brown in 1914 was carried out because the tables then in use for worts of beer, contained in the first schedule to the Inland Revenue Act of 1880, had been found to give inaccurate results. Thorpe and Brown found that using filtered samples of fermented worts, the gravity lost corresponding to any spirit indication was greater than that shown in the earlier table. In Part IV of the work (Brown, 1914) there is an interesting historical account of the earlier tables and Brown goes on to discuss the reasons for the discrepancy between the old and new tables. From this work it is clear that the discrepancy is due to the use of unfiltered worts in the earlier work, which had been based on a small number of laboratory fermentations using very high yeast pitching rates. This had the effects of adding a significant amount of alcohol to the wort before fermentation and also of increasing the residual gravity by inclusion of yeast solids.

Unfiltered worts could be used for original gravity determination with appropriate tables, provided that either the whole of the fermentation was taken for the analysis, or a sample was taken from the bulk in which yeast cells were suspended uniformly throughout. In sampling commercial fermentations, whether in the brewery or distillery, this is difficult or impossible and there is no alternative but to use filtered samples. In spite of this fact the distilling industry seems to have continued to use the old table long after the Board of Customs and Excise had applied Thorpe and Brown's new table for revenue purposes in beer production. The table which had been in use prior to Thorpe and Brown's work was produced in 1852 and is shown in Brown's paper. Comparison of this with the table shown in the present Customs and Excise Schedule 1 shows that the two are virtually identical, adjustments at the higher spirit indications presumably having been made to conform with the new reference temperature of 20°C. A comparison of the various tables relating gravity lost to spirit indication is shown in Table 7.3.

Thorpe and Brown's work showed not only that their table gave accurate results using filtered samples but also that it could be used with any fermentation where starch was the main raw material. In the Scotch whisky industry, therefore, the original gravity table of Schedule 2 and the *Recommended Methods of Analysis* should be used for process control purposes, at least by those using filtered wash samples.

Mention has been made above of Thorpe and Brown's work on the effects of yeast solids. Even if yeast cells are removed by

Table 7.3 Summary of the various tables of gravity lost.

Spirit indication	*Degrees of gravity lost* *Graham, Hofmann and Redwood, 1852* (60°F)*	*Mean Brewery Table, 1909–10* (60°F)*	*Statutory Instrument 1979 No. 1146* *Schedule 1 Wort or wash (20°C)*	*Schedule 2 Worts of beer (20°C)*
1	3.0	4.25	3.0	4.24
2	6.6	8.50	6.6	8.47
3	10.7	12.90	10.7	12.84
4	15.1	17.30	15.0	17.22
5	19.5	21.85	19.4	21.70
6	24.1	26.40	23.9	26.21
7	28.8	31.00	28.6	30.75
8	33.7	35.65	33.4	35.31
9	39.1	40.30	38.6	39.88
10	44.2	45.00	43.7	44.48
11	49.0	49.85	48.4	49.20
12	54.3	54.85	53.5	54.06
13	59.4	59.95	58.4	58.99

*Data from Graham, Hofmann and Redwood and Mean Brewery Table taken from Brown (1914).

filtration the possible influences of solutes taken up by the yeast for growth and then released back into the solution during the later stages of fermentation (MacKenzie and Kenny, 1965) have to be considered. It can be shown in laboratory experiments that for fully fermented samples the original gravity calculated from wash analysis is slightly higher, perhaps by as much as 0.5°, than the gravity measured on the worts before fermentation. However, such samples remain stable for a considerable number of days, at least as far as gravity and alcohol measurements are concerned and samples do not have to be analysed immediately. A more serious problem exists with samples from fermentations which are used in the distillery before fermentation is complete. Unless such samples are analysed immediately or action taken, such as freezing, to prevent further alcohol production, the samples will continue to ferment, and subsequent analysis will give results not representative of the wash at the time of distillation.

Once original gravity has been determined, mashtun extract can be calculated as described above, original gravity being equivalent to the degrees of excess gravity measured with a hydrometer. Mashtun extract should be similar to the hot-water extract of the malt mashed, but it should be recognised that the mashing methods employed in the distillery and in the laboratory are not the same. The main difference is that the distillery employs a three- or four-water extraction with higher temperatures for each succeeding water. Thus it is possible for protein and higher molecular weight polysaccharides to be rendered soluble and to achieve an extract slightly higher than the laboratory figure. On the other hand, an efficient separation of the extract solution from the grains is a process which is not part of the laboratory extraction and will inevitably result in a certain amount of extract being lost in the draff at the end of the mash.

MILLING

An essential check on milling is to sieve the grist into its husk, 'grits' and flour fractions. This is usually done on site using the traditional sieving box, with two screens having mesh sizes of 1.4 and 0.2 mm. Samples can be taken from the bottom of some mills using a specially made sampling tray reaching from front to back. The utmost care must be taken to obtain samples representative of the whole grist, since the distribution of husk, grits and flour falling from the rolls is not uniform. Another type of mill employs sampling cups which can be used to obtain samples of malt which have passed through both sets of rolls or the top set only. In many distillery installations samples of grist can be obtained more satisfactorily from the shower of particles falling into the grist hopper. The sieving box must not be overfilled and the manner in which it is shaken is also important, different operators rarely achieving the same result. A skilled millman who is taking samples regularly should be able to detect changes in the grind due to different malts and adjust the mill accordingly.

In malt distilleries it is usual to aim at a grist containing approximately 18, 72 and 10% of husk, grits and flour respectively. Inspection of the fractions can often assist in deciding on an optimum grind for a particular malt and can indicate that the mill needs adjusting, servicing or the rolls reground. This would be the case, for example, where the grits fraction was found to contain significant quantities of shattered husk fragments, whilst leaving large pieces of starch in the husk fraction. From time to time samples of grist should be analysed in the laboratory, using a mechanical shaker as a check on results from the sieving box.

MASHING

As in grain distilleries, the conversion of polysaccharide to fermentable sugar is a two-stage operation. The first is a high-temperature reaction in the mashtun, where an accurate control of temperature, particularly at 'striking' when grist and liquor are mixed together, is essential. Fluctuations in temperature here can result in damage to enzymes required for the second stage, secondary conversion, which takes place in the fermenting vessel or wash back. Older mashing installations relying on gravity feed of grist and liquor to the mashing machine make it difficult to maintain the constant ratio of grist to liquor necessary to achieve a steady striking temperature. Modern plant with mechanical grist conveyors, and liquor flow made constant by pumping, has significant advantages in this respect. Automatic controls for levels and temperatures of mashing liquors may also be available, and a description of a fully automated mashing system, enabling one man to carry out both mashing and distilling at the same time has been described (Watson, 1983).

Cooling of the wort takes place as the mash is drained, and therefore quick drainage of the first water assists in the preservation of enzymes required for secondary conversion. The rate of drainage is strongly influenced by wort viscosity and fineness of milling and is always a compromise between the demands of extraction and fermentation. Lauter and semi-lauter mashtuns are becoming widely introduced in malt distilleries; the faster drainage from such plant can result in improved fermentation efficiencies as well as faster turn-round. Temperature at mashing in is monitored by means of a probe with its sensor positioned at the mashing spout and connected to a chart recorder. The final mashtun temperature is checked traditionally by means of a mercury in glass spear thermometer, which retains its reading on removal and is reset by shaking, or by the modern digital thermometer and probe.

DRAINING

As a check on draining efficiency a measure of the volume of extract recovered after each water is useful and can be carried out, under instruction, by any mashtun operator. The volume of worts recovered when the first water is drained is usually 80–85% of the volume of liquor added at mashing in. In some distilleries it is the practice to start adding the second water by spraying or 'sparging' before the first water is completely drained. Whatever method is used, the maximum amount of wort must be recovered at the end of the second water if a high extraction efficiency is to be achieved. In order to calculate the percentage draining efficiency at each water, it is necessary to know the volume of solution in the mashtun before draining starts. The volume of liquor added is readily found by measuring the contents of the hot liquor tank before and after adding, plus an allowance for the volume of any cooling water added during mashing. To this must be added the moisture in the malt and the volume taken up by the extract, which is approximately 0.61 litres per kg. An allowance should also be made for water of hydrolysis, which is about 4 litres per 100 kg starch hydrolysed.

It will be found for the first two waters that the specific gravity of the worts drained is greater than that calculated for the worts in the mashtun before drainage starts. This is because a concentration gradient exists in the mash, the worts remaining being weaker than those drained. If a record is kept of the recovery of worts any effects of changes in malt quality can be detected quickly and adjustments made either to the mill, or to the quantities of liquor added. Table 7.4 shows a mass balance for extract and water for a malt distillery three-water mash. The most important factor controlling the extraction efficiency in any water is the ratio of water (or liquor) added to the weight of malt or draff. The sugar concentration is a significant factor, especially in the first water. Using data from a number of mashes it is possible to derive equations to predict extract recoveries and to optimise mashtun performance.

DIAGNOSIS OF EXTRACTION PROBLEMS

To distinguish between poor conversion of starch and poor drainage, standard methods of spent grains analysis given in the *Recommended Methods* (1977) may be used. The method for

Table 7.4 Mass balance for a three-water mash at Cardhu distillery. 6.8 t malt with an extract of 75.3% and a moisture content of 4.9% were mashed.

	1st water	*2nd water*	*3rd water*
In mashtun before draining:			
Water in liquor added (litres)	31 647	8 300	28 100
Water in malt (litres)	333	8 317	7 864
Extract in liquor (kg)	413	0	0
Extract in malt (kg)	5 120	1 206	462
Extract in solution (%)	15.6	6.9	1.3
Volume of solution (litres)	35 377	17 577	36 468
In worts drained:			
Volume of worts (litres)	26 100	9 210	28 100
Extract concentration (%)	16.6	8.1	1.5
Water recovered (litres)	23 443	8 753	27 846
Extract recovered (kg)	4 327	744	413
% extraction	78.2	61.7	89.4
Left in mashtun:			
Water (litres)	8 317	7 864	8 118
Extract (kg)	1 206	462	49

Note: figures for water and extract left in mashtun include worts under mashtun false bottom as well as draff.

soluble extract gives an indication of losses from poor drainage, and is usually applied as routine to final draff samples. However, it is sometimes worth while to examine the soluble extract remaining after each water has been drained, when figures of approximately 13%, 4%, 1.5% and less than 1% 'sparge loss' would be expected for a four-water mash in a traditional mashtun. The method for total available extract includes unconverted starch as well as soluble extract. It is somewhat tedious and inaccurate because the figure for unconverted starch is found by determining the difference between the results for soluble and for total extract. Another test for unconverted starch is that using iodine (Schur *et al.*, 1981). Starch can be measured quantitatively using amyloglucosidase and glucose oxidase. However, the starch must be dissolved first and the method is not really suitable for a process control laboratory where large numbers of analyses have to be carried out at low cost. Instruments using immobilised preparations of these enzymes are now becoming available and may find application in laboratories carrying out spent grains analysis.

Fermentation

Accurate measurements of both the concentration of alcohol in wash and the bulk volume of wash are essential to measure fermentation efficiency, and in malt distilleries, where a certain amount of alcohol is left behind during distillation, to check on distillation losses. In distilleries without laboratories it is not easy to measure alcohol in wash directly, but indirect methods based on specific gravity determinations can be used. These measurements are employed by the distiller to adjust process conditions and maximise yields, and routine laboratory measurements are made to give a check on the accuracy of the indirect measurements.

MONITORING FERMENTATIONS IN THE DISTILLERY

Fermentations are monitored in all distilleries by following the decrease in specific gravity. Fully fermented all-malt wash has a specific gravity in the range 0.997 to 0.998. 'Final gravity', which is 1000 (specific gravity −1), is therefore in the range −2 to −3, or two to three degrees 'under' water. Final gravity is also the sum of residual gravity and spirit indication, usually about 8.0 and −10.5 respectively for malt wash. Final gravities for fully fermented grain wash vary according to the cereal in use, the cooking regime and whether backslopping is carried out. Some examples of those effects on final gravity are shown in Table 7.5.

The final gravity for a fermentation efficiency of, say, 85% is influenced little by the original gravity of the wort, varying by only 0.34° for 10° of original gravity. Provided its measurement is carried out carefully, it is a quick and accurate method. Specific gravity measurements are carried out every few hours using unfiltered samples and although these measurements are accurate only to one or two degrees, taken together with temperatures they provide an assurance (or otherwise) that fermentation is proceeding normally. Samples of fully fermented worts are filtered, with care being taken to avoid excessive filtration times and consequent losses of alcohol, and the specific gravity is then measured using a hydrometer graduated at intervals of 0.2°. A

Table 7.5 Effect of cereal type and cooking and mashing regimes on the final gravity of grain distillery wash.

Cereal	*Cooking*	*Mashing*	*Final gravity*
Maize	normal	normal	−2.5
Barley	normal	normal	+0.5
Wheat	normal	normal	−1.0
Maize	undercooked	normal	−3.5
Maize	normal	backslop used	−1.0

Note: normal conditions refer to batch cooking of unground cereal and batch conversion using green malt.

correction is made to adjust the reading to 20°C and by this means it is possible to obtain results agreeing closely with laboratory measurements using more accurate instruments.

Other techniques for monitoring fermentations include pH, % total acid and optical rotation; the changes in these parameters during the course of malt whisky fermentations, with acceptable and unacceptable levels of bacterial infections, are shown in Figs 7.1 and 7.2.

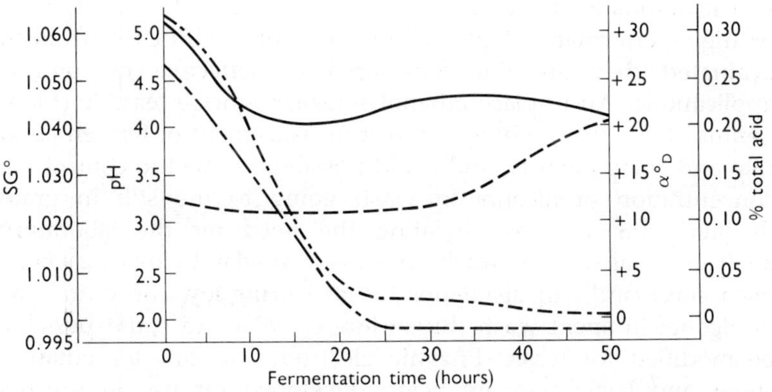

Fig. 7.1 Changes in SG, pH, acid content and α_D of wort with time in Scotch malt whisky fermentation – acceptable infection level (Dolan, 1976).
— — — = SG; —— = pH; — — = acid; - - - = α_D

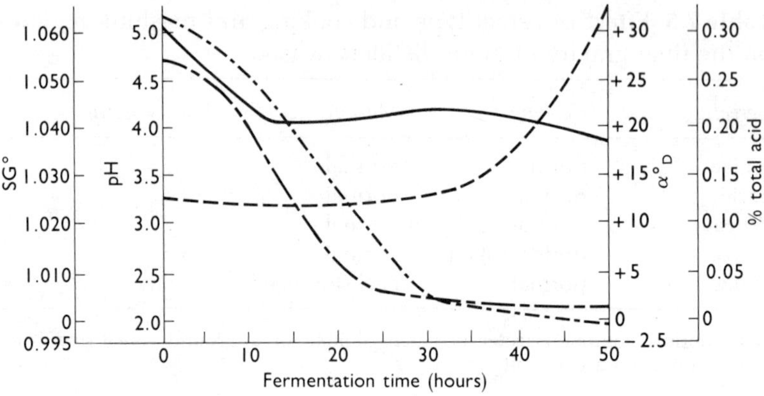

Fig. 7.2 Changes in SG, pH, acid content and α_D of wort with time in Scotch malt whisky fermentation – unacceptable infection level (Dolan, 1976).
– – – = SG; —— = pH; — – = acid; - - - = α_D

Whatever the method used, it is desirable that fermentation efficiency can be measured immediately after fermentation is completed. Even then, all off-line techniques fail, in that a fault at an earlier stage could have been repeated several times, particularly in grain distilleries. On-line measuring techniques might allow action to be taken earlier. Monitoring of pH is feasible and can provide early warning of abnormal fermentation or infection problems. Ethanol sensing would be useful, and whilst automatic sampling and analysis by gas chromatography or high performance liquid chromatography (HPLC) have been evaluated they are not considered cost-effective for on-line applications. An in-place ethanol sensor might be feasible (Mandenius *et al.*, 1984), giving a direct measurement of fermentation rate and completeness, and could possibly be used to control the concentration of alcohol in wash going to the still in grain distilleries as well as obviating the need for the laboratory analysis of alcohol in wash. A sensor similar to this has been used successfully in distilleries for measuring low concentrations of alcohol in spent wash (Bryan-Jones, 1985) and could possibly be modified for wash. Enzyme electrodes specific for ethanol, sugar and lactic acid are being developed for use in process control (Clarke *et al.*, 1984). On-line temperature measurement is used widely for monitoring fermentation rate. Such measurements also form the basis for the control of cooling in high-gravity grain distillery fermentations.

One factor rarely measured in distillery fermentations is yeast growth. Since up to 6% of carbohydrate available at the start of the process goes towards yeast growth and unwanted by-products of yeast growth such as glycerol, there may be scope for limiting growth and improving alcohol yields.

Ancillary controls are used by some distillers. Provision of air at the start and during fermentation is sometimes considered desirable, and circulation of carbon dioxide has been used. Control of fermentation temperature is relatively common practice in high-gravity fermentations. All of these procedures are best and most readily accomplished by means of automatic process control devices.

ATTENUATION CHARGE

The principle of 'attenuation charge', i.e. the relationship of alcohol concentration to apparent attenuation, has been in use ever since the imposition of Excise duty on Scotch whisky distillers. Apparent, or total attenuation, is found by adding the number of degrees under water to original gravity. The present relationship of attenuation charge, or alcohol allowance per degree attenuated, was agreed between the distillers and HM Customs & Excise in 1980 and is 0.00125 litres alcohol per litre wash per degree attenuated. This is an underestimate which allows the distiller some leeway before the imposition of Excise duty for unexplained losses. In the discussion on the prediction of alcohol yields from hot-water extract of malt it was shown that for a fermentation with a true attenuation (gravity lost) of 47°, 0.1615% alcohol is produced for each degree of gravity lost. In the example:

final gravity = spirit indication − residual gravity = 2.538 'under' giving an apparent attenuation of 55 + 2.538 = 57.538 therefore, for each degree of apparent attenuation there is production of

$$7.589/57.538 = 0.1319\% \text{ alcohol}$$

This factor (or its reciprocal, 7.58) is most useful in distilleries for calculating alcohol concentration in wash from measurements of apparent attenuation. The factor holds reasonably well over a wide range of degrees attenuated, as shown in Table 7.6, which also gives the relationships between true attenuation, apparent attenuation and per cent alcohol.

Table 7.6 Relationships between per cent alcohol, gravity lost and apparent attenuation for different values of gravity lost.

Gravity lost (degrees)	*Apparent attenuation (degrees)*	*Gravity lost/ apparent attenuation*	*% alcohol by volume*	*% alcohol/ gravity lost*	*% alcohol/ apparent attenuation*
42	51.46	0.8162	6.734	0.1603	0.1309
43	52.68	0.8162	6.908	0.1607	0.1311
44	53.90	0.8163	7.082	0.1610	0.1314
45	55.11	0.8165	7.249	0.1611	0.1315
46	56.33	0.8166	7.423	0.1614	0.1318
47	57.54	0.8168	7.589	0.1615	0.1319
48	58.75	0.8170	7.755	0.1616	0.1320
49	59.96	0.8172	7.922	0.1617	0.1321
50	61.17	0.8174	8.088	0.1618	0.1322
51	62.37	0.8177	8.246	0.1617	0.1322
52	63.58	0.8179	8.413	0.1618	0.1323

LABORATORY METHODS FOR MEASURING ALCOHOL IN WASH

Grain distillers have laboratory facilities on site and malt distilleries usually have access to the services of a central laboratory, at least for some checks on fermentation efficiency. Ethanol determination in wash is traditionally by the Revenue still method as used for original gravity determination (*Recommended Methods of Analaysis*, 1977). The method is essential for breaking down the overall efficiency of the distillery into its main stages – extraction, fermentation and distillation – and is therefore one of the most useful laboratory measurements. Gas chromatographic analysis using direct injection of wash samples (Hardy and McLeman, 1983) is much faster, but care must be taken to calibrate the method against results obtained using the distillation method. Other methods for the determination of alcohol in beer, also used in conjunction with original gravity, have been developed (Sartor and Sherry, 1983), using density and refraction (Kruger and Oliver-Daumen, 1980), near infra-red reflectance, and the 'Scaba' process (Piendl and Schneider, 1981). A further advantage can be gained where the output of automatic analysis equipment can be connected to a computer (Fig. 7.3), since then data acquisition, processing and storage, as well as analysis, can all be done automatically (Hardy and McLeman, 1983).

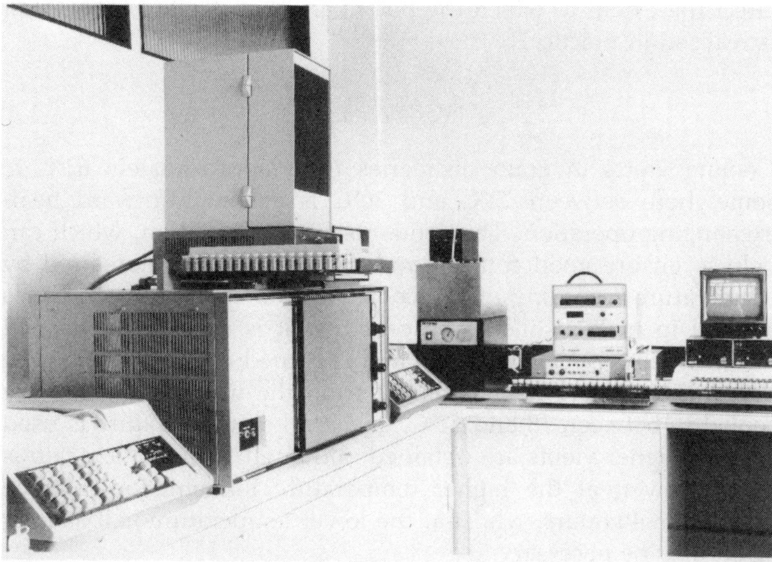

Fig. 7.3 Apparatus for automatic analysis of wash samples. From left to right: gas chromatograph, constant temperature bath for maintaining density-meter cell at 20 °C, digital density meter with autosampler, microcomputer (Hardy and McLeman, 1983).

The pycnometer or density bottle has been an essential feature of distilling laboratories for many years, but is now being superseded by the digital density meter employing an electronically excited U-tube maintained at 20°C. This equipment, which produces an output to five decimal places, is also capable of being connected to a computer and can be used for alcohol in distillates and for malt hot-water extracts, as well as distillery and laboratory wash final gravities (La Berge, 1979).

FACTORS CONTROLLING FERMENTATION EFFICIENCY

Distillery worts are not boiled, as they are in breweries, and this has the effect of allowing malt enzymes to carry on the process of starch degradation after worts have been transferred to the fermenter and yeast added. The effects which cooking and mashing operations can have on this secondary conversion, which may account for up to 15% and 25% of total yields in malt and grain fermentation respectively, have already been discussed. A number of process factors in the distillery may also affect the extent to which the potential fermentability of the malt is realised in practice.

Worts cooling

Cooling worts in grain distilleries from approximately 62°C to somewhere between 22°C and 30°C is a straightforward heat-exchanging operation. This lends itself to automation, which can help to ensure good control, and the operation is monitored by temperature recording of the cooled worts. The same process is applied in malt distilleries, except that it is still usually carried out manually, is much slower, and is carried out during the time that the worts are being drained from the mashtun. Worts are cooled to between 18 and 22°C, and if a low temperature is used slightly higher yields are obtained, although fermentation rate is much slower; at the higher temperature fermentation may be finished in 40 hours, whilst at the lower temperature half as long again may be necessary.

Yeasting

Yeast is added shortly after filling of the wash back commences but not until the tun room operator has made certain that the worts temperature is correct. Yeast is added by injection into the worts line beyond the heat exchanger and if the temperature probe is positioned beyond the yeast pipe a change of temperature in the worts can act as a useful event marker. The amount and batch reference number of the yeast should also be noted.

The amount of yeast added is critical and the optimum for malt is several times that for grain fermentations. Alcohol production and yeast growth occur together and the process is one of competition with bacteria for substrate. Addition of excess yeast may result in more substrate than necessary being used for yeast growth. Optimum seeding rates can only be found at present by distillery trials.

Some distilleries use pure culture distilling yeast on its own, others use it in a mixture with brewer's yeast, and there has always been controversy about the use of brewer's yeast in this way. If a mixture of the two is employed the optimum total yeasting rate for malt wort is about 0.42% w/v compared with 0.35% for distiller's yeast alone, but since brewer's yeast is cheaper this will result in similar yeast costs. Some companies have carried out sufficient research and trials to convince themselves that brewer's yeast has a positive effect on yields.

Bacteria and plant hygiene

Competition by bacteria with yeast for substrate reduces spirit yields and it is essential to keep populations below a certain level. Facilities should be available to allow bacteriological testing of samples from the distillery in order to monitor the state of hygiene and to identify the source of infections when abnormally high levels occur. Process control measures must be designed to minimise infection and this involves both plant design and operation.

Only anaerobic bacteria, predominantly lactobacilli, grow in distillery fermentations (Bryan-Jones, 1973; Makanjuola and Springham, 1984), so counts of total populations are of limited value. Even if lactobacilli are counted on their own only a few strains are harmful (Bryan-Jones, 1985).

By far the most dangerous source of infection is wash from a previous fermentation. Upwards of 100 million bacteria per ml are present in fully fermented wash. As a general rule, if more than 10 million bacteria per ml are present at 30 hours they are likely to have had a detrimental effect on yield (Dolan, 1976). After wash, the next most dangerous source is wort, which if allowed to cool below 65°C can be an excellent medium for bacterial growth. The design of pipework and positioning of valves in mashing and tun room installations can sometimes lead to the carry over of pockets of cool wort from one mash to another. Process control should as far as possible eliminate these defects and should allow efficient rinsing and sterilisation between mashes.

The fermenting vessel itself can be a serious source of infection, especially if traces of wash are left over from one fermentation to the next, and Dolan (1976) has given an indication of the losses of yield likely to be sustained in distilleries where cleanliness is not adequate. Cleanliness in stainless steel vessels is much easier to attain than in wooden wash backs, where the only really effective treatment is by steaming and then only if care is taken to ensure that all parts of the vessel surface are kept at least at 70°C for a minimum of 15 min (Bryan-Jones, 1985).

Cooking and mashing systems are generally kept at temperatures which ensure sterile conditions. In hot liquor tanks the temperature at the bottom and in the pipes leading out from the tank may not be sufficiently high. Worts lines, yeast lines, coolers and fermenters are used at temperatures which encourage bacterial growth and must be regularly cleaned. Hot water, steam and chemicals are all employed and whichever method is used, microbiological monitoring is essential. It may be accomplished using swabbing techniques or more conveniently and possibly more representatively by counting organisms in process samples such as worts, yeast cream and fermenting wash.

Dolan (1976) gives an account of the changes in specific gravity, pH, acid content and optical rotation that occur during fermentation, and the effects of bacteria, and therefore the usefulness of such measurements as indirect indices of infection (Figs 7.1 and 7.2). Procedures are now available for counting bacteria in relatively large numbers of samples on site without the facilities of a bacteriological laboratory (Bryan-Jones, 1973). Alternatively, samples can be taken by shift workmen and placed in vials containing a preservative mixture capable of holding numbers constant for several days for subsequent counting in the laboratory (Bryan-Jones, 1986). Such techniques are useful for tracing the source of infections, for example by sequential sampling of worts running into a wash back during the course of a mashing operation. This blanket coverage of the process means that samples can be taken routinely but analysed only if it is decided from yield efficiency results that it is relevant to plate out particular samples. It is conceivable that on-line measurement of, for example, pH or lactate using specific electrodes might soon give a method of monitoring bacteria in large distilleries where the production justifies the investment.

DIAGNOSIS OF FERMENTATION PROBLEMS

Because fermentation efficiency can be affected by so many variables, the job of accounting for and minimising the fluctuations that occur is one of the most difficult jobs in process control. It is helped significantly by having data from a number of distilleries using the same raw materials, so that comparisons can be made, for example, of the results from one batch of yeast or malt used at different distilleries. By this means it is usually possible to decide whether a fermentation problem is due to raw materials or to conditions at a particular distillery.

Analysis of wash for sugars remaining after fermentation is a most useful technique in the diagnosis of fermentation problems. Thin layer chromatography is a cheap but not a quantitative test for sugars (Hanson, 1975). HPLC is coming into use in the laboratory for the quantitative analysis of mono- and oligosaccharides (Slotema, 1978; Brenner, 1982; Honda, 1984), glucose and organic acids. The presence of fermentable sugars could be for any of the following reasons: too short a fermentation time, substandard yeast or incorrect pitching rates, high fermentation temperature, high bacterial infection, and in grain fermentations deficiency in nutritional quality of worts.

Glucose analysis of wash after pre-treatment by heating and/or addition of enzymes can be used to indicate residual starch and dextrins (Brown, 1984). Glucose may be measured conveniently by HPLC or by using an immobilised enzyme electrode. The nature of the residual carbohydrates in wash, together with other process measurements, is useful in diagnosing cooking, mashing and fermentation problems. Organic acid analysis can give an indication of plant infection. Variations in other metabolites, such as glycerol, may be monitored using HPLC and with appropriate detectors dextrins, fermentable sugars, organic acids, glycerol and ethanol may all be measured simultaneously. Gas chromatographic measurement of the major volatile congeners of wash gives useful indications of processing efficiency. In particular, higher alcohols in wash can give an indication of yeast quality, plant infection and deficiencies in mashing and cooking operations (Brown, 1984). Table 7.7 shows the diagnosis which typically may be consistent with measurements of the constituents of fermented wash.

VOLUME CHANGE DURING FERMENTATION

Most distillers assume that the volume of fermented wash is the same as that of the worts before fermentation. Unless it is considered that the worts volume measurement was not accurate, perhaps because of excessive frothing, only one determination will be made. Since there are profound changes in weight and specific gravity, with 0.957 kg of carbon dioxide being lost

Table 7.7 Diagnosis of grain fermentation problems based on composition of fermented wash.

Starch	*Dextrins*	*Fermentable sugar*	*Lactic acid*	*iso-Pentanol*	*Diagnosis*
high	low	low	normal	slightly high	undercooking
low	low	high	normal	normal	yeast fault
low	high	medium	normal	normal	mashing fault
low	high	medium	high	low	infection
medium	high	medium	low	low	overcooking

for every kg of ethanol produced, it seems fortuitous that the volume should not change. However, neglecting any other losses to the system, the following example shows that there is virtually no change in volume.

Taking the example used earlier for calculating predicted yields, a wort of SG 1.055 ferments to a final SG of 0.99746 with an alcohol concentration of 7.589 %vol.

Concentration of ethanol in alcoholic distillate of wash sample corresponding to 7.589 %vol = 6.058% by mass (*Laboratory Alcohol Table*, 1979).

Specific gravity of alcoholic distillate = 0.98946
therefore concentration of ethanol in wash

$$= 6.058 \text{ kg per } \frac{0.98946}{0.99746} \times 100 \text{ kg wash}$$

$$= 6.009 \text{ kg per } 100 \text{ kg}$$

Taking maltose as the substrate being fermented,

$$C_{12}H_{22}O_{11} + H_2O \longrightarrow 4C_2H_5OH + 4CO_2$$

for every 184 kg ethanol produced, 176 kg carbon dioxide are lost. Therefore, assuming no other losses, total weight loss during fermentation

$$= 176 \times \frac{6.009}{184} = 5.748 \text{ kg per } 100 \text{ kg wash}$$

Therefore, for every 1000 kg wash, 105.748 kg wort are fermented and since the specific gravities of wort and wash are 1.055 and 0.99746,

$$\frac{105.748}{1.055} \text{ litres wort gives } \frac{100}{0.99746} \text{ litres wash}$$

or 100 litres wort gives 100.02 litres wash

A more significant change in volume is the apparent increase due to thermal expansion during fermentation. However, even this only amounts to about 0.25% of the original bulk and may be ignored for all process control purposes. This makes it possible to use either wort or wash volume, as convenience dictates.

Distillation

MALT WHISKY DISTILLATION IN POT STILLS

The design of pot stills, and also the programme for making up charges and operating the stills, are regarded as essential for the maintenance of whisky character at individual distilleries. Until now there has been little quantitative information about the effects of still design and operation on whisky quality, and process control has largely been a matter of maintaining distillation efficiency by controlling cut points and end points and avoiding losses through leaks. Nevertheless, significant changes have been introduced where economic considerations have been considered paramount, such as the use of indirect heating by means of steam coils. These changes have also resulted in better control of the distillation process and the ability to increase throughput. Further modifications with the object of recovering heat are now being installed, or at least considered, by many companies.

Pot-still distillation end points and cut points

The concentration of ethanol in the vapour in equilibrium with a boiling solution of ethanol in water is related to its concentration in the liquid. During the later stages of pot-still distillation the concentration of alcohol in both vapour and liquid decreases exponentially and there comes a point beyond which the cost of recovery of spirit is greater than its value. This stage is reached when the alcohol concentration in the liquid is between 0.10% and 0.15%, but will be influenced by whatever measures are being used to recover heat and decrease fuel costs. All distilleries have some instruments for measuring alcohol in the distillate, but the ratio of alcohol in distillate to alcohol in liquid, which towards end of distillation is approximately 10 : 1 (Cornell and Montonna, 1933), is used to decide when to stop distilling.

Instruments for alcohol measurement in spirit safes usually consist of hydrometers calibrated in per cent alcohol at 20°C, and Celsius thermometers. Safes traditionally have been designed to allow readings to be taken without giving the operator access to samples of distillate. In order to allow use of the hydrometer at any temperature a correction table is provided which gives the alcohol reading at which to end distillation over the range of temperatures likely to be encountered. Provided that they are cleaned regularly and that they are displayed in such a way that readings are clearly visible, such instruments are quite satisfactory and regular checks on the alcohol content of discharge samples will soon show any inconsistencies. A more modern technique giving a digital display is the densitometer of the type employing an electronically excited U-tube, which has been described for use in the laboratory, but in this application is equipped to allow a constant flow of sample. Possible objections are cost, difficulty of keeping the instrument free of grease and solid particles, both of which critically affect the result, and the problem of working in a flame-proof environment.

Pot-still spirit safes have the facility of directing the distillate from the low wines (spirit) stills into either the feints or spirit receiver. The initial distillate, or 'foreshots', contains substances which are considered undesirable in spirit. These comprise very volatile compounds which are the first to be driven off when the liquid in the still comes to the boil, and also the least volatile substances remaining in the condenser or worm from the previous distillation and then flushed out with the strongly alcoholic first runnings. At present the only test, other than by timing, for the cut from foreshots to spirit is the 'water test' involving the mixing of equal volumes of distillate and water, again without direct access to spirit samples. The distillate is considered to be suitable for collection as spirit when no cloudiness is found.

The cut from spirit back to the feints receiver may be determined by the strength of spirit required to be collected. Since there is a gradual decline in alcohol concentration, the minimum strength for cutting from spirit will usually be lower by 5% or so than that required for filling. The strength of spirit to be collected is usually considered to have an influence on spirit composition and quality, although there is not a great deal of quantitative evidence to support this. At times, economic considerations have prevailed and it has been necessary to alter spirit strengths to meet changing filling requirements.

Losses in pot-still distillation

Comparison of the alcohol in wash with that recovered after distillation should give a reasonably accurate measurement of distillation losses. If the alcohol concentration in still discharge samples is being measured, it is possible to check whether the amount of alcohol being discarded in spent wash and spent lees is sufficient to account for these losses. Alcohol in spent wash and spent lees of 0.1% corresponds to an overall loss during distillation of about 1.1% of that in wash.

Assuming that alcohol concentrations have been measured accurately and that stills have not been discharged too soon, apparent distillation losses could be caused by incorrect measurement of worts or wash volume or by incorrect measurement of feints strength. If worts volume has been over-declared then mashtun extract will be over-declared also and evidence for this can be obtained by comparison with the extract of the malt mashed, taking into account the usual extraction efficiency achieved by the distillery. Measurement of feints strength is not always easy, as alcohol concentration may not be homogeneous throughout the bulk and stirring or rummaging of the liquid in the vessel can be difficult. Oil droplets floating in the feints can also affect hydrometer readings. Incorrect readings of feints strength will have the opposite effect the following week, as the distillery will be credited with more alcohol in the feints vessel than is actually present at the beginning of the week. Errors from this source will therefore cancel each other out after a few weeks' production.

If a genuine loss of alcohol has occurred a number of possibilities have to be considered, for most of which there are routine checks. These are: physical loss of wash due to operator error or leaking valves; evaporation from wash charger due to lids being left open; pin-holes in the body of the still (not always obvious if above the liquid level or behind lagging); leaks in condensers or worms; leaking flanges in steam coils; still discharge valves passing liquid when shut; and pin-holes in wash pre-heater plates.

GRAIN WHISKY DISTILLATION IN PATENT STILLS

Distillation can be considered simply as alcohol concentration and recovery from fermented wash. Wash at 6–12% volume ethanol is used to produce new make grain whisky of typically 94–94.5% volume (the legal upper limit is 94.8%). The new make whisky produced should have grain character and should contain essentially no iso-amyl alcohol. This is removed during distillation and is used to produce the by-product fusel oil. Stills used for the task vary widely. Coffey's Patent Still is well favoured but so too are multi-column stills which may be used for producing grain spirit or rectified spirit from fermented wash.

Efficient recovery of ethanol from wash by distillation is brought about by ensuring that no ethanol leaves the still in spent wash or fusel oil. The former may be assured by monitoring spent liquor for alcohol and may be done on-line, but crudely, by a floating bead indicator in spent liquor brought from the base of the analyser to the still head. Colorimetric reactions, typically using dichromate, may be carried out for testing for losses. Far more satisfactory and readily justified are on-line monitors. A device (the 'ambigometer') measuring the relative vapour pressure of spent wash is suitable for indicating loss of alcohol, but recent advances in gas sensor technology have allowed the development of a potentially cheap and useful ethanol probe (Bryan-Jones, 1985). The signal from such a device may be integrated to give a measure of alcohol lost over the duration of a distillation run. Fusel oil can be analysed satisfactorily off-line to ensure Customs & Excise limits for ethanol are not exceeded, either by density-based techniques or gas chromatography. Old-fashioned decanting of fusel oil to recover and return ethanol to the still has been replaced in some distilleries by fusel oil distillation columns which take a side-stream from the still and fractionate this to recover ethanol, propanol and iso-butanol and return them to the main still. Gas chromatography can be used off-line to monitor the efficiency of this operation.

At the end of a distillation period of one week the amount of alcohol recovered in receivers, corrected for that in feints at the start of the period, is used to calculate overall yield. As is the case in malt distilleries, comparison of this figure with the 'laboratory yield', i.e. alcohol in wash, will give an indication of any substantial losses in distillation.

Whisky quality

It is normally assumed by both distiller and customer that each distillery produces a whisky with a distinct, if not a unique, character and there is usually no question of attempting to alter the character of a whisky to suit different customers. The main purposes of quality control are to detect any off-odours or inconsistencies in quality or character that may occur from charge to charge, and which might result in customer complaints, whilst maintaining the distinct character of the spirit produced at each distillery. This can be a difficult problem when the distiller has to face changes in raw materials, production rate and even, when economic considerations are paramount, production methods.

Whereas yield variations of as little as 1% or 2% can be detected easily, techniques for the quantitative measurement of changes in spirit quality are much less precise. A great deal of information is available concerning the identification, quantitative analysis and influence of various process factors on the levels of individual volatile aroma substances (Berry, 1984). The substances dealt with are the volatile 'congeners', mainly alcohols, esters, fatty acids and phenols present at concentrations greater than 2 or 3 ppm that can be analysed by gas chromatography using either direct injection or with the minimum of sample preparation. Although gas chromatography is used routinely by some distillers it is an unfortunate fact that these congeners have at best a marginal effect on spirit quality. This can be demonstrated by preparing solutions of the substances at the concentrations reported in Scotch whisky. The aroma quality of such preparations is quite different from that of whisky and, more significantly, if the odour intensity of the solutions or of component substances is compared using threshold measurements with that of the whisky itself it is found that the greater part of the aroma of the latter is missing. The variations in character and quality of both malt and grain spirit are largely dependent on the presence of substances at concentrations of 0.1 ppm or less, and in grain whisky even below 0.1 ppb, some of which have odour intensities many orders of magnitude greater than those of the substances whose concentrations can be measured easily (Brown, 1974; Hardy and Perry, 1974; Masuda and Nishimura, 1980; Leppänen *et al.*, 1983). Techniques for the routine analysis of important minor constituents are being developed and will eventually become available for routine quality control work.

NOSING PANEL ASSESSMENTS IN THE CONTROL OF NEW MAKE WHISKY QUALITY

For the reasons stated above the distiller at present has to rely largely on organoleptic assessment, usually by means of nosing tests, for monitoring spirit quality. Procedures for carrying out nosing tests involve dilution with water to about 22% volume and placing samples in clear tulip-shaped glasses allowing plenty of room for swirling and for the build-up of aroma in the head-space. The time that the samples are allowed to stand before nosing affects the quality of the aroma because of the loss of substances whose volatility has been increased by dilution and because of changes in reaction systems where aroma volatiles are in equilibrium with ethanol and water.

Nosing panellists require the ability to discriminate reliably and reproducibly between different samples and to detect specific odours, especially off-odours, at low concentrations. Individuals differ greatly in sensitivity to each substance and to have any meaning an assessment of a spirit needs to be the consensus of a panel of at least four or five persons whose abilities are accurately known. Panellists can be trained, through the use of reference samples, to recognise and identify specific odours. In order to provide a basis for comparing remarks made by individual panellists, and also to gain understanding between customers and distillers, it is necessary to employ descriptive terms whose meanings are accepted by all concerned. Two quite complicated vocabularies have been developed for use with mature malt and blended whisky (Piggott and Jardine, 1979; Shortreed *et al.*, 1979). Many of the descriptions apply also to new malt and grain spirit but in addition there are some odours in new spirit rarely encountered in mature whisky.

THE CONTROL OF NEW MAKE GRAIN WHISKY QUALITY

It is recognised that for consistent spirit quality, good control of the operating conditions of the still is essential. Control of spirit off-take strength is of primary importance and is generally carried out automatically using density measuring devices. Quality control of the new make grain whisky relies mainly on organoleptic assessment. There remains little substitute for sampling and nosing assessment of spirit by distillery shift supervisors at regular intervals throughout the course of distillation runs, which may exceed 100 hours. Even though the composition of grain spirit is less complex than that of malt spirit no analytical technique on its own is capable of guaranteeing new make whisky quality. Consistency in composition is, however, an objective since it indicates stable distillation conditions, which are generally related to consistent quality. On-line gas chromatography of specific congeners has been used in some distilleries to give an immediate warning of changes in spirit consistency (Wilkin *et al.*, 1983) and could be used to provide control signals for the automatic adjustment of distilling conditions to ensure consistency of grain whisky composition.

Off-line gas chromatographic analyses are widely used for determining the composition of new make grain whisky. While nosing assessment at the still head by trained supervisors is an essential procedure for maintenance of spirit quality, it is essential that this is backed up by assessment by trained panellists as described above. Control of off-odours may be achieved by adjustment of distillation conditions but cooking, mashing and fermentation can also be important. Logging of still running conditions, especially wash strength and temperatures, and other distillery processing conditions is necessary for collation with quality assessments.

Examples of some off-odours recognised in new make grain whisky are 'feinty', 'acrolein', 'burnt', 'gassy', 'diacetyl', 'green' and 'acetal'. 'Acrolein' odours are generally associated with infection problems in wash from relatively long fermentations and other factors influencing its production have been reported (Mills *et al.*, 1954). Diacetyl, not unpleasant but occasionally causing an unacceptably sweet odour, is associated with yeast characteristics and certain fermentation conditions, including pH. 'Burnt' and 'gassy' odours are caused by very potent organic sulphur compounds. These are prevalent during the commissioning of new stills and it is known that copper has a crucial role to play in their formation and destruction (Brown, 1974). 'Feinty' odours are generally associated with poor control of distillation conditions. The chemicals responsible for most of these off-odours are known and may be measured by gas chromatographic and gas chromatographic-mass spectrometric techniques. On-line monitoring for some of the offending chemicals is not possible at present due to the extremely low concentrations present in new make grain whisky.

THE CONTROL OF NEW MAKE MALT WHISKY QUALITY

In spite of higher production costs, malt whisky is required for blending because of the flavour which it imparts when mixed with grain whisky. Malt whisky commands a premium when bottled and sold on its own, usually following extra years of maturation. Whereas blending will tend to even out any variations in the quality of the individual malts used, any such changes in character are perceived much more readily by the consumer of a single malt whisky.

Malt whisky distillers and blenders recognise a number of off-odours, which include those described for grain whisky. Because of the greater concentrations of aroma volatiles in the pot-distilled whisky the substances responsible for the off-odours have to be at considerably higher concentrations to become noticeable than they do in grain whisky. The routine analysis of malt whisky by gas chromatography is hampered by the presence of substances such as esters, alcohols and fatty acids, which occur in relatively large amounts and obscure the organoleptically significant components. Although the compounds of interest can be revealed by fractionation techniques and can be measured by gas chromatography-mass spectrometry, such procedures are not really suitable for routine quality control work. At the present time a considerable amount of effort is being put into the development of gas chromatographic methods employing derivatisation (Greenhoff and Wheeler, 1981) and specific detectors to make the analysis of substances such as sulphides and unsaturated carbonyl compounds possible as routine quality control procedures.

Certain whisky quality characteristics can be measured easily by gas chromatography or chemical analyses. A sour or sickly odour, due to *n*-butyric acid, is sometimes formed by bacterial action if mashing liquor is not kept at a high enough temperature, or if wash is allowed to stand too long in cast-iron vessels. Acrolein, although it is rarely present at a high enough level to be detected in the spirit by nosing, occasionally can be a problem in distillery filling stores due to its lachrymatory effects. Acrolein is formed during wash distillation and can be measured in spirit samples by a simple chemical analysis using tryptophan (Circle *et al.*, 1945). The development during fermentation of the precursor hydroxypropionaldehyde may be followed by the analysis of distillates of wash samples.

The peaty character of certain malt spirits, particularly those distilled in the islands of Islay and Skye, is related to the peatiness induced during kilning of the malt. Quantitatively, phenols are the most abundant class of substances of the volatiles in whisky derived from the peat smoke, and the analysis of whisky phenols using a method based on that of Mounib and Evans (1957) involving *p*-nitrosodimethyl aniline is used routinely. Care must be taken in interpreting these results. Only about 22% of the malt volatile phenol content finishes in the malt spirit; phenols, especially phenol itself, being water-soluble, remain in the residue after distillation. However, it should not be assumed that the recovery of peaty flavour is as low as this because other organoleptically more significant classes of compounds such as pyridines and thiazoles, also absorbed by the malt during peating, may be recovered with greater efficiency.

By-products

ANIMAL FEEDING STUFFS

Methods for the sampling and analysis of animal feeding stuffs are laid down by the Feeding Stuffs Regulations, 1988 (Statutory Instrument 1988 No. 396). These regulations include protein, oil, fibre, ash and moisture content, and trace element analysis may also have to be carried out and the results displayed on labels. In addition to these statutory requirements, monitoring of the process, which is basically one of drying and mixing, is necessary to ensure good process efficiency and fuel economy. If insufficient drying takes place, overheating and spoilage may occur during storage, whilst excessive drying wastes energy.

Rapid moisture measuring instruments are used, such as infra-red heating balances, capacitance and near-infra-red reflectance meters. For high throughput plants, on-line devices using the latter two techniques are employed. Grain distillery by-products plant nowadays may have to cope with material from a variety of different cereal sources. For certain cereals such as barley, attention to process conditions during cooking and mashing can alleviate problems in the subsequent evaporation of spent wash to produce syrup. Viscosity reduction is aided by ensuring optimum conditions for β-glucanase and other enzymes but addition of fungal enzymes to spent wash may be necessary.

The production of wet distillers' grains is either directly from the mashtun or, in the case of a grain distillery carrying out fermentations with all grains in, by screening, pressing or centrifuging spent wash. Wet grains deteriorate rapidly due to bacterial action and become unpalatable. An account of the chemical changes taking place is given by Miller (1969). Action must be taken to ensure prompt removal or consideration may have to be given to treatment with preservatives such as formic and propionic acids. The sale of pot ale syrup for feeding directly is increasing and as this product is priced according to solids content, this has to be determined for each load.

The level of copper permitted in feeding stuffs is a problem of particular significance to distillers. Copper is dissolved in wash during distillation. It may be affected by the acidity of wash after fermentation and the cleanliness of copper surfaces inside the still. In recent years regulations have been introduced to control also the levels of aflatoxins in feeding stuffs and very recently, draft regulations have been formulated to control pesticide residues. As with all regulations, analysis methods are specified and occasional monitoring has to be carried out.

CARBON DIOXIDE

Carbon dioxide is collected from fermenters at some grain distilleries, purified and compressed for sale as a by-product. Rigorous quality standards are required by most users of this by-product. Low residual moisture contents are required and the presence of nitrous gases in the product must be avoided. Monitoring of the product for these and some other parameters is essential and suitable laboratory methods are available. Monitoring the moisture of CO_2 gas after the drying stage and prior to compression may be considered prudent and can be accomplished by in-line electrolytic monitors. Nitrous gas problems arise from the presence of nitrate in worts. This in turn comes from the mashing water or cereal in use. Since catalytic hydrogenation facilities used to remove nitrous gases are expensive to operate it may prove to be cost-effective to control the use of the reactor from an in-line measurement of nitrous gases in raw carbon dioxide. In other instances regular off-line monitoring of nitrate concentration in mashing water may be sufficient.

FUSEL OIL

All grain distilleries produce fusel oil. This low volume by-product is easy to produce to the required quality standard. Off-line quality monitoring to ensure that ethanol content is below the upper limit allowed in the product by HM Customs & Excise is necessary. This may be done satisfactorily by methods based on a density measurement or is readily accomplished by gas chromatography if available.

Data acquisition and manipulation

Microcomputers in distilleries and laboratories can perform routine calculations and clerical duties that would otherwise have to be carried out manually. In contrast with research work that is aimed at a long-term improvement in understanding the processes involved in Scotch whisky production, process control laboratories are dealing with the particular batches of material that are currently in process. The speed of acquiring and disseminating data to those involved in process control is of primary importance and it is here that automatic systems for data acquisition, reduction and transmission can be employed profitably.

Even small malt distilleries without modern equipment for monitoring process variables have always been able manually to measure and record gravities and temperatures. Until now the job of collating these measurements with the analysis of raw materials and the effects of both on spirit quality and yields has been impossible because of the volume of work involved. Now that microcomputers are available at relatively low cost it has become possible to obtain a much clearer understanding of the complex relationships between raw materials, process variables, yields and quality. Ideally, all relevant information should be brought together for statistical analysis and interpretation using microcomputers either in the distillery laboratory or central production department. The techniques necessary are already available and several companies are actively engaged in this work. It is a costly operation and costs must be evaluated against the probability of such a system leading to savings in production cost.

The simplest and most useful statistical technique for the analysis of production data of the kind described is that of linear regression. This requires the setting up of a two-dimensional matrix where, for an adequate number of cases, there are complete sets of data for variables concerned with raw materials analysis, process and yield measurements. If computer files in the required format can be set up, software which is available commercially can be used for carrying out the statistical analysis. The correlation coefficients give an immediate indication of any pair of variables between which a significant correlation exists. Using stepwise and ordinary least squares multiple linear regression the complex relationships which exist between a number of variables may be examined.

These techniques are particularly useful if malting, as well as distillery, process variables are available since distillery results may well be influenced more by what has happened in the maltings than by any distillery process conditions, and the effects of variables such as barley analysis, steeping hours, moisture, temperatures and kilning conditions can then be examined. An example of an investigation into the effects of malting variables on fermentability is shown in Table 7.8. The value for a correlation coefficient to be significant at the 1% level of probability with 65 cases is 0.31, a negative value indicating an inverse correlation. It can be seen from this example that there are strong inverse relationships between fermentability and steeping hours, moisture contents during germination and hours on kiln. The reverse is true for the relationships between extract and the same process variables.

In grain distilleries the interactions between malting variables and data on cereal for cooking can be analysed to determine the optimum proportions and conditions to use. Feedback of in-

Table 7.8 Means, standard deviations and correlation coefficients for the relationships between malt extract, % fermentability, % fermentable extract and malting process variables ($n = 65$) (Hardy, 1985).

Variable	*Mean*	*Standard deviation*	*Correlation coefficients* Extract dry (%)	Fermentability (%)	Fermentable extract dry (%)
Steeping (hours):					
1st water	8.66	1.49	0.56	−0.63	−0.40
1st air-rest	9.05	1.01	0.56	−0.64	−0.40
2nd water	16.46	1.30	0.46	−0.54	−0.36
3rd water	14.95	2.24	0.35	−0.49	−0.37
Temperature (°C):					
End 1st water	57.1	0.595	−0.01	0.18	0.22
End 1st air-rest	60.3	1.05	0.30	−0.14	0.05
At cast	58.2	1.18	−0.29	0.37	0.26
% Moisture:					
Steep day 1	33.1	1.34	0.50	−0.50	−0.27
At cast	44.8	0.75	0.43	−0.52	−0.35
Germination day 2	45.2	0.62	0.50	−0.55	−0.33
Germination day 3	45.2	0.63	0.46	−0.46	−0.24
Kilning:					
Max. air on (°C)	74.4	3.32	−0.29	0.22	0.07
Max. air off (°C)	65.4	21.7	0.07	−0.20	−0.21
Hours on kiln	24.5	3.0	0.64	−0.72	−0.46
Finished malt:					
Moisture (%)	4.59	0.50	−0.13	0.27	0.25
Extract (dry, %)	80.9	0.71	1.00	−0.64	−0.08
Ferment-ability (%)	85.5	1.31	−0.64	1.00	0.82
Fermentable extract (dry, %)	69.2	0.82	−0.08	0.82	1.00

formation on starch conversion, fermentation efficiency and alcohol yield can be used to refine the accuracy of the computer model.

References

AHLUWALIA B, ELLIS E E 1984 A rapid and simple method for the determination of starch and β-glucan in barley and malt. *Journal of the Institute of Brewing* **90**: 254–9

ASSOCIATION OF OFFICIAL ANALYTICAL CHEMISTS 1980 *Methods of Analysis* 13th edn. AOAC, Washington, p 215

BAMFORTH C W 1985 Biochemical approaches to beer quality. *Journal of the Institute of Brewing* **91**: 154–60

BATHGATE G N 1983 The relationship between malt 'friability' and wort viscosity. *Journal of the Institute of Brewing* **88**: 324–8

BERRY D R 1984 The physiology and microbiology of Scotch whisky production. In Bushnell M E (ed) *Progress in Industrial Microbiology* **19**: 199–243

BERTOFT E, ANDTFOLK C, KULP S E 1984 Effects of pH, temperature and calcium ions on barley malt α-amylase iso-enzymes. *Journal of the Institute of Brewing* **90**: 298–302

BEUTLER H O 1978 Enzymatische Bestimmung von Starke in Lebensmittelm mit Hilfe der Hexokinase-Methode. *Starke* **30**: 309–12

BOURNE D T, WHEELER R E 1982 Laboratory prediction of brewhouse extract and performance. *Journal of the Institute of Brewing* **89**: 416–19

BRENNER K 1982 Advantages of HPLC as a separating procedure for use with raw materials, wort and beer. *Monatsscrift für Brauwissenschaft* **35**: 64–6

BRIGGS D E, HOUGH J S, STEVENS R, YOUNG T W 1971 *Malting and Brewing Science* 2nd edn. Chapman and Hall, London, vol 1, *Malt and Sweet Wort*, pp 194–221

BROWN H T 1914 Reports on the determination of the original gravity of beers by the distillation process. Part IV History of previous tables and a comparison of these with the 'mean brewery table', 1909–10. *Journal of the Institute of Brewing* **20**: 645–61

BROWN J H 1974 Unpublished results

BROWN J H 1984 Unpublished results

BRYAN-JONES G 1973 Lactic acid bacteria in distillery fermentations. In Carr J G, Cutting C V, Whiting G C (eds) *Lactic Acid Bacteria in Beverages and Food*. Academic Press, London, pp 165–75

BRYAN-JONES G 1985 Unpublished results

BRYAN-JONES G 1986 The preservation of distillery fermentation samples for bacteriological examination. *Journal of the Institute of Brewing* **92**: 253–4

CIRCLE S J, STONE L, BORUFF C S 1945 Acrolein determination by means of tryptophane. *Industrial and Engineering Chemistry. Analytical Edition* **17**: 259–62

CLARKE D J, BLAKE-COLEMAN B C, CALDER M R, CAN R J G, MOODY S C 1984 Sensors for bioreactor monitoring and control – a perspective. *Journal of Biotechnology* **1**: 135–8

CORNELL L W, MONTONNA R E 1933 Studies in distillation. II Liquid-vapour equilibria in the systems ethanol-water, methanol-water and acetic acid-water. *Industrial and Engineering Chemistry* **25**: 1331–5

DOLAN T C S 1976 Some aspects of the impact of brewing science on Scotch malt whisky production. *Journal of the Institute of Brewing* **82**: 177–81

DOLAN T C S 1979 Bacteria in whisky production. *The Brewer* 1979(2): 60–4

DOLAN T C S, DEWAR E T, GRAY J D 1981 Development and use of a method for determination of malt fermentability. *Journal of the Institute of Brewing* **87**: 352–5

GREENHOFF K, WHEELER R E 1981 Analysis of beer carbonyls at the part per billion level by combined liquid chromatography and high pressure liquid chromatography. *Journal of the Institute of Brewing* **86**: 35–41

HANSON S A 1975 Thin layer chromatographic method for identification of oligosaccharides in starch hydrolysates. *Journal of Chromatography* **105**: 388–90

HARDY P J 1985 Unpublished results

HARDY P J, MCLEMAN A D 1983 A gas chromatographic/density meter/computer system for the determination of original gravity. In Priest F G, Cambell I (eds) *Current Developments in Malting, Brewing and Distilling, Aviemore Conference 1982*. Institute of Brewing, London, pp 59–66

HARDY P J, PERRY D R 1974 Unpublished results

HASTIE S H 1925 The application of chemistry to pot still distillation. *Journal of the Institute of Brewing* **31**: 198–215

HENRY R J 1984 A rapid method for the determination of diastatic power. *Journal of the Institute of Brewing* **90**: 37–9

HONDA S 1984 High performance liquid chromatography of mono- and oligosaccharides. *Analytical Biochemistry* **140**: 1–47

JANATA J, BLACKBURN G F 1984 Immunochemical potentiometric sensors. *Annals of the New York Academy of Sciences* **48**: 286–92

KRUGER E, OLIVER-DAUMEN B 1980 Aktueller Stand der Konventionellen Bieranalytik. *Monatsscrift für Brauwissenschaft* **33**: 12–21

Laboratory Alcohol Table Density/Strength at 20°C for Laboratory Use (1979). HM Customs & Excise, London

LA BERGE D E 1979 Determination of specific gravity of malt extracts, worts and beer. *Journal of the American Society of Brewing Chemists* **37**: 105–6

LEE W J, PYLER R E 1982 Improved assay procedure for limit dextrinase in malt extracts. *Brewers Digest* 1982(7): 24–7

LEPPÄNEN O, RONKAINEN P, DENSLOW J, LAAKSO R, LINDEMANN A, NYKÄNEN I 1983 Polysulphides and thiophenes in whisky. In Piggott J R (ed) *Flavour of Distilled Beverages*. Ellis Horwood, Chichester, pp 206–14

LISLE D B, SAVAGE R I 1976 The thermal expansion of worts between 60°F and 20°C and a gravity lost table for use at 20°C. *Journal of the Institute of Brewing* **82**: 277–9

LONG D E 1982 Analysis of *N*-nitrosodimethylamine in malted barley. *Journal of the Institute of Brewing* **88**: 266–71

MACKENZIE K G, KENNY M C 1965 Non-volatile organic and pH changes during the fermentation of distillers worts. *Journal of the Institute of Brewing* **71**: 160–5

MAKANJUOLA D B, SPRINGHAM D G 1984 Identification of lactic acid bacteria isolated from different stages of malt whisky distillery fermentations. *Journal of the Institute of Brewing* **90**: 13–19

MANDENIUS C F, HEDMAN T, MATTIESON B 1984 An on-line sensor for monitoring of ethanol in beer. *Journal of the Institute of Brewing* **90**: 77–80

MARTIN H L, BAMFORTH C W 1983 Application of a radial diffusion assay for the measurement of β-glucanase in malt. *Journal of the Institute of Brewing* **89**: 34–7

MASUDA M, NISHIMURA K 1980 Occurrence and formation of damascenone, *trans*-2,6,6-trimethyl-1-crotonyl-cyclohexa-1,3-diene, in alcoholic beverages. *Journal of Food Science* **45**: 396–7

MATHEWSON P R, POMERANZ Y 1981 Rapid analysis of α-amylase in barley malts using a colorimetric assay. *Brewers Digest* 1981(10): 24–5

MILLER T B 1969 Evaluation of whisky distillery by-products 1. Chemical composition and losses during transport and storage of malt distiller's grains. *Journal of the Science of Food and Agriculture* **20**: 477–81

MILLS D E, BAUGH W D, CONNER H A 1954 Studies on the formation of acrolein in distillery mashes. *Applied Microbiology* **2**: 9–13

MOUNIB M S, EVANS J V 1957 Methods for the determination of phenols and sulphide. *Analyst* **82**: 518–23

MURTAGH J E 1973 *Microbial enzymes in grain alcohol production*. PhD thesis, University of North Wales.

NETTLETON J A 1913 *The Manufacture of Whisky and Plain Spirit*. Cornwall and Sons, Aberdeen

PALMER G H 1980 Malting and mashing. In Rainbow C, Float G E S (eds) *An Introduction to Brewing Science and Technology*, Part I. Institute of Brewing, London, pp 10–27

PIENDL A, SCHNEIDER K 1981 Comparative determination of original gravity, alcohol and extract content of beer by means of 'Scaba' and distillation. *Brauwissenschaft* **34**: 81–5

Piggott J R, Jardine S P 1979 Descriptive sensory analysis of whisky flavour. *Journal of the Institute of Brewing* **85**: 82–5

Recommended Methods of Analysis 1977. Institute of Brewing, London

Ronkainen P, Leppänen O, Harju K 1978 The reuse of stillage water in the mashing of grain as a means of a energy conservation. *Journal of the Institute of Brewing* **84**: 115–17

Rutherford A G 1981 Mashing for malt whisky production. *The Brewer* **67**: 82–6

Sartor D J, Sherry N B 1983 Automated alcohol and original gravity analyser: a new aid to brewing. *Technical Quarterly, Master Brewers Association of the Americas* **20**: 160–3

Savage R I 1971 Metrication in the brewery laboratory I. A 'Gravity Lost' table for 20°C. *Journal of the Institute of Brewing* **77**: 261–3

Schur F, Anderegg P, Pfenniger H 1978 Characterisation of materials inhibiting filtration, Part 2. Photometric iodine test. *Brauerei-Rundschau* **89**: 129–32

Schur F, Anderegg P, Pfenniger H 1981 Evaluation of malt using the iodine value of laboratory spent grains. *Brauerei-Rundschau* **92**: 49–53

Shortreed G W, Rickards P, Swan J S, Burtles S M 1979 The flavour terminology of Scotch whisky. *Brewers Guardian* 1979(11): 55–62

Slotema F P 1978 Application of chromatography in the brewery. *Brewers' Digest* 1978(8): 48–52

Statutory Instrument 1979 No. 1146, Customs & Excise, *The Alcoholic Liquors (Amendment of Units and Methods of Measurement) Regulations 1979*

Statutory Instrument 1988 No. 396, Agriculture, *The Feeding Stuffs Regulations 1988*

Statutory Instrument 1982 No. 6111, Customs & Excise, *The Spirits Regulations 1982*

Thorpe E, Brown H T 1914 Reports on the determination of the original gravity of beers by the distillation process. Part I Memorandum and tables. *Journal of the Institute of Brewing* **20**: 569–93

Watson J G 1983 Microprocessor-controlled mashing in a malt whisky distillery. In Priest F G, Campbell I (eds) *Current Developments in Malting, Brewing and Distilling, Aviemore Conference 1982*. Institute of Brewing, London, pp 239–44

Wilkin G D 1983 Raw materials – mashing and extract recovery. In Priest F G, Campbell I (eds) *Current Developments in Malting, Brewing and Distilling, Aviemore Conference 1982*. Institute of Brewing, London, pp 35–44

Wilkin G D, Webber M A, Lafferty E A 1983 Appraisal of industrial continuous still products. In Piggott J R (ed) *Flavour of Distilled*

Beverages. Ellis Horwood, Chichester, pp 154–65
UNGER E D, WILLKIE H F, BLANKMEYER H C 1944 The development and design of a continuous cooking and mashing system for cereal grain. *Transactions of the American Institute of Chemical Engineers* **40**: 421–3

CHAPTER 8

Maturation and maturation chemistry

K NISHIMURA and R MATSUYAMA

Introduction

Freshly distilled whiskies usually have pungent, unpleasant odours and a sharp taste. Thus storage in oak casks for some years is needed to modify their aroma to a pleasant and agreeable one. Maturation, which may be defined as the improvement of sensory quality during storage, is the important final phase in whisky manufacturing to produce a product of good quality.

Spirits such as whisky, brandy and rum are traditionally stored in oak casks for maturation. The following points are known about the maturation of spirits in oak casks:

1. Storage varies from a few years to more than 10 years.
2. Flavour quality differences between unmatured and matured spirits are generally very significant.
3. Both volume and strength are lost owing to evaporation of ethanol and water through the porous wood of oak casks.
4. Quality varies with the individual cask because rate of maturation and the quality of the spirits are affected by alcoholic strength and type of whisky, and the size, nature, history and prior treatment of the casks, as well as the temperature and humidity of the environment.
5. The mechanisms of maturation have not been completely elucidated and no reliable chemical or physical index is available for indicating the progress of maturation.

The storage periods of whiskies are subject to legal requirements in some countries. Scotch malt whisky, Scotch grain whisky and Irish whiskey are stored for at least three years and bourbon whiskey and American grain whiskey are stored for at least one year (Lyons and Rose, 1977). Canadian whisky is stored for at least three years (Rannie, 1976).

Oak cask

OAK SPECIES

The principal oak species used for making casks can be divided into two groups, the white oak group grown in North America and the European oak group mainly consisting of *Quercus robur* and *Q. sessilis* (Singleton, 1974).

The white oak group contains varieties of very similar oak species, such as *Q. alba, Q. prinus, Q. bicolor, Q. muehlenbergii, Q. stellata, Q. macrocarpa, Q. lyrata* and *Q. durandii. Quercus alba* is considered to be the best and accounts for about 45% of the group. Among the European oaks, *Q. robur (Q. pedunculata)*, whose common names are French oak, English oak and summer oak, is the predominant species for making casks, with *Q. sessilis*, which is considered a variety of *Q. robur*, ranking second. *Quercus sessilis* is said to be the same as *Q. petraea* and *Q. sessiliflora* and its common names include dry oak and winter oak (Singleton, 1974).

TYPES OF CASK

The casks for ageing whisky at distilleries around the world are classified mainly into four types from their sizes and shapes, as follows:

	(litres)
American barrels	180
Hogsheads	250
Butts	480–520
Puncheons	480–520

American barrels, which are made of white oak and used to store bourbon whiskey, are usually charred on the inside surface with a burner flame before filling with whisky. As American law requires that bourbon casks be used only once, some of them are reused for storage of other types of whiskies such as Scotch, Canadian, Irish and Japanese whiskies. In many cases, these casks are broken down in the United States and shipped to Scotland, Japan and other countries and reassembled, supplementing the damaged staves.

Hogsheads are an example of such casks reassembled from the staves of used bourbon casks; the charred areas may be removed before reassembly. Sherry butts are casks previously used to store Spanish sherry. For more than a hundred years Scotch whisky has been stored in used sherry casks. Recently, owing to the shortage of used sherry casks, many Scotch distilleries have begun using American oak casks which have previously been used for maturing bourbon whiskey. Puncheons hold the same volume as butts, but puncheons are short and round whereas butts are long and narrow.

These casks are usually used over and over again, with the damaged staves being repaired or replaced. This indicates that there are a great many varieties of cask stored in the warehouse of a distillery, considering the diversity of their history.

Mechanism of maturation

The mechanism of maturation in oak casks is basically common to all distilled spirits without regard to their raw materials. Changes in the compounds in wood casks during maturation of spirits can be divided into the following seven types:

1. Direct extraction of wood components.
2. Decomposition of the macromolecules forming the framework of wood, such as lignin, cellulose and hemicellulose, followed by elution into the spirits.
3. Reactions of wood components with components of the unaged distillates.
4. Reactions involving only the extracted wood materials.
5. Reactions involving only the distillate components.
6. Evaporation of the low-boiling compounds through the wood of the cask.
7. Formation of stable molecular clusters of ethanol and water.

OAK WOOD EXTRACTS

The heart wood of American white oak (*Q. alba*) is composed of cellulose (49–52%), lignin (31–33%), hemicellulose (22%) and a fraction extractable by hot water or ethyl ether (7–11%) (Ritter and Fleck, 1923). European oak has a similar composition (Singleton, 1974). The extractable fraction contains volatile oils, volatile and non-volatile acids, sugars, steroids, tannic substances, pigments and inorganic compounds.

Volatile oils

Over a hundred volatile components have been identified from oak wood shavings by Nishimura *et al.* (1983): 35 aliphatic compounds (7 hydrocarbons, 24 acids, and 4 other compounds), 54 aromatic compounds (28 hydrocarbons, 17 phenols, 2 alcohols, 3 acids and 4 other compounds), 3 furan compounds and 26 terpene compounds. Seven additional fatty acids were found by Nykänen (1984).

Isomeric 3-methyl-4-octanolides, which were named *Quercus* lactone-a and -b or oak lactones, are the major components of the volatile fraction of oak wood, together with 4-nonanolide and eugenol (Masuda *et al.*, 1971). These lactones were first erroneously assigned and corrected later as follows (Masuda and Nishimura, 1981b):

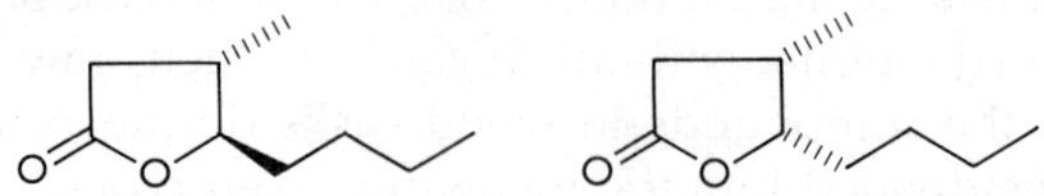

trans-3-methyl-4-octanolide (*Quercus* lactone-b)   *cis*-3-methyl-4-octanolide (*Quercus* lactone-a)

They are distributed in North American white oak, European oak such as Limousin, Gascony and Troncais oaks and some deciduous Japanese oak wood (Masuda and Nishimura, 1971; Otsuka *et al.*, 1974). These lactones were identified in distilled spirits stored in oak wood casks (Suomalainen and Nykänen, 1970; Nishimura and Masuda, 1971; Guymon and Crowell, 1972; De Smedt and Liddle, 1978), contributing to the matured flavour with low threshold values (Salo *et al.*, 1972; Otsuka *et al.*, 1974). The amounts of oak lactones in some commercial distilled liquors are shown in Table 8.1.

Tannic substances

Tannins are water-soluble polyphenols commonly present in plants and can be divided into two groups: derivatives of

Table 8.1 The amounts of *Quercus* lactone-a and -b in distilled liquors.

			Quercus lactone-a (cis-lactone) (ppm)	*Quercus lactone-b (trans-lactone) (ppm)*
Brandy	(cognac)	* * *	0.14	0.17
		Napoleon	0.16	0.36
		extra	0.22	0.43
Whisky	(Scotch)	ordinary	0.26	0.70
		medium	0.31	0.85
		high	0.75	1.42
	(bourbon)		0.39	3.84
	(Canadian)		0.07	0.95
	(Irish)		0.21	0.58
Rum	(Jamaica)		0.05	1.21

Otsuka *et al*. (1974), reproduced with permission from *Agricultural and Biological Chemistry* **37**: 487.

flavanols, the so-called condensed tannins, and hydrolysable tannins which yield gallic acid and ellagic acid by hydrolysis.

The polyphenolic substances extracted from white oak heart wood contain gallic acid, ellagic acid, gallotannins and ellagitannins (Chen, 1970). Seventeen American oak wood samples contained an average of 6.4% of 55% aqueous alcohol solubles, with a total phenols concentration (calculated as gallic acid) of 365 mg g^{-1} of solid extract, while seven European oak samples contained an average of 10.4% of extractables with 561 mg g^{-1} of total phenols. The ratios of non-flavanoids to total phenols in both oaks had the same value of 87.7% (Singleton, 1974).

Liebman and Rosenblatt (1943), Liebman and Scherl (1949), Baldwin and Andreasen (1974) and Reazin *et al*. (1976) have demonstrated that a rapid increase in tannin concentration occurs during the first six months of whisky maturation, followed by a decrease in the rate and then an indefinite linear increase during further maturation, as shown in Fig. 8.1.

Otsuka *et al*. (1965) analysed the total phenol contents of commercial whiskies, calculated as tannic acid equivalents per 100 ml, and found that Scotch whiskies contained 7.6–18.1 mg, Japanese 7.4–18.2 mg, Canadian 15.0–27.0 mg and American 41.4–59.8 mg, and cognac 30.6–74.8 mg. Of the tannic substances, they identified gallic acid in whiskies and gallic acid and

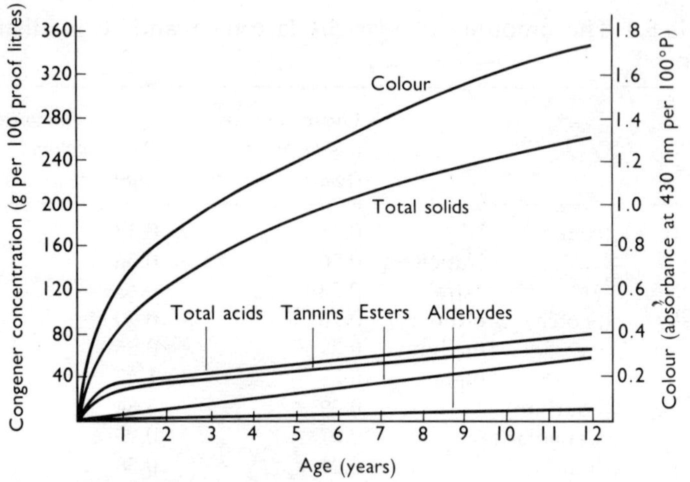

Fig. 8.1 Congener changes during maturation (Reazin, 1981; reproduced with permission from *American Journal of Enology and Viticulture* **32**: 286).

ellagic acid in cognacs. Lehtonen (1983) quantified the gallic acid content, together with other lignin degradation compounds, in various distilled spirits and found that whiskies contain less gallic acid than cognacs.

The florescent compound scopoletin (6-methoxy-7-hydroxy coumarin) was first detected in bourbon whiskey and neutral spirits aged in white oak casks (Baldwin *et al.*, 1967). Otsuka and Zenibayashi (1974) found it in white oak wood. They measured its concentration in whiskies of various ages and found that the amount of scopoletin increased with ageing.

Sugars and glycerol

Arabinose, xylose, glucose, rhamnose, fructose and galactose were found in aqueous ethanol extracts of oak wood (Lichev and Panaiotov, 1958; Dzhanpoladyan *et al.*, 1969). Together with the above sugars, deoxyinositol, fucose, mannose, inositol, protoquercitol and glycerol were identified in various kinds of whiskies (Otsuka *et al.*, 1963; Duncan and Philp, 1966; Martin *et al.*, 1965; Martin and Eib, 1968; Nykänen, 1984).

Black and Andreasen (1974) and Reazin (1981, 1983) studied the development of arabinose, glucose, xylose, fructose and galactose in whisky during 12 years' maturation and found that arabinose, glucose, xylose and galactose were produced at a faster rate during the first year of maturation than in the following years, while the fructose and glycerol formation rate increased in the later stages of maturation, indicating differences in their precursors in oak wood (Fig. 8.2). Recently Nykänen (1984) reported that glucose, proto-quercitol and arabinose were the most abundant sugar components in one-year-old whisky distillates, followed by xylose, mannose, galactose and rhamnose in descending order.

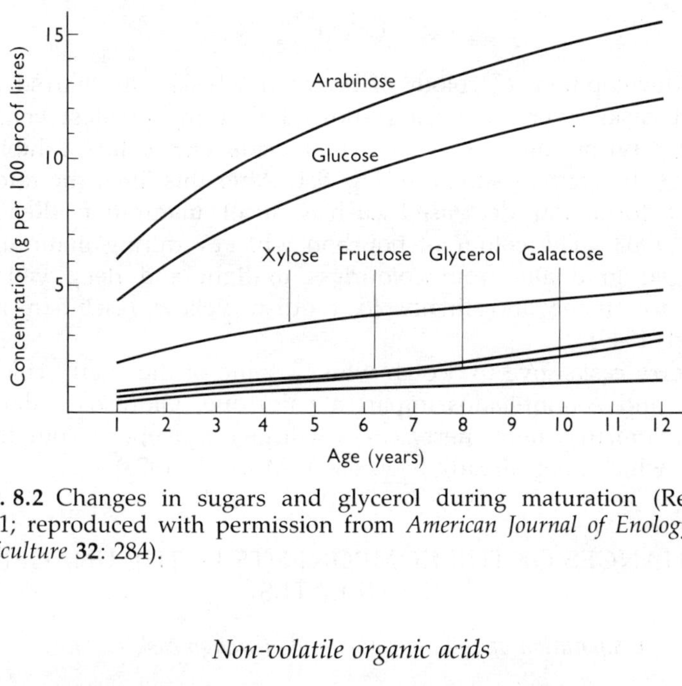

Fig. 8.2 Changes in sugars and glycerol during maturation (Reazin, 1981; reproduced with permission from *American Journal of Enology and Viticulture* **32**: 284).

Non-volatile organic acids

The aqueous ethanol extract of oak chips contains oxalic acid, fumaric acid, succinic acid, methyl succinic acid, mesaconic acid, adipic acid, methoxysuccinic acid, pimeric acid, furan dicarboxylic acid, suberic acid, phthalic acid, azelaic acid, sebacic acid and trimethyl benzene tricarboxylate (Nykänen, 1984). Although these dicarboxylic acids have not been reported found in matured spirits, compounds such as diethyl fumarate, succinate and azelate have been identified in matured whisky, obviously originating from wood, as they are subject to esterification in spirits (Nykänen, 1984).

Steroids

Chen (1970) analysed the steroid fraction from the heart wood of *Quercus alba* and found it to contain 85% β-sitosterol, 7% stigmasterol, 3% campesterol and a trace of dihydro-β-sitosterol. Braus *et al.* (1957) and Black and Andreasen (1973) isolated β-sitosterol and β-sitosterol-D-glucoside as the major components from whisky flocculent precipitate. In addition, they found β-sitosterol as well as a small amount of stigmasterol and campesterol in extracts of the residue from whisky filter pads.

Colour

The development of colour of bourbon whiskey in charred oak wood casks over four years showed that the greatest change occurs during the first year. Total acids and solids exhibited similar changes, as shown in Fig. 8.1. After this time, the rate of colour formation decreased each year of maturation (Reazin, 1981, 1983). The colour of bourbon whiskey during maturation changed gradually from colourless to light and deep yellow, then to amber and finally to reddish yellow (Liebman and Scherl, 1949).

Sherry casks give to Scotch whisky some of the sherry colour. Finos and Amontillados impart a pale tone, Olorosos a darker tone. Similarly, new sherry casks will give a deeper colour than those which have already been used (Murphy, 1978).

CHANGES OF THE COMPONENTS IN THE UNAGED DISTILLATES

Evaporation of volatile compounds through cask surface

Ethanol and water are lost due to diffusion through the cask staves and cask head and subsequent evaporation into the atmosphere. The alcohol content of spirits in a cask increases during ageing under low to moderate relative humidity but decreases under high humidity (Ohnishi *et al.*, 1977). During ageing in the unheated humid warehouses of Scotland, the proof goes down. In the higher altitude, warmer and dryer storage conditions of the United States, the proof goes up during storage. The loss during storage in Scotland is approximately 8% in three years (Valaer, 1940).

The evaporation ratios of the small molecular weight compounds other than water from the casks during one year of storage were measured using a model ageing system, and the losses were up to 32% of acetaldehyde, 12.7% of ethanol, 10.6% of *n*-propyl alcohol, 10.5% of ethyl acetate, 5.2% of iso-amyl alcohol, 1.3% of ethyl caproate and 1.0% of acetic acid. This indicates that the losses of a compound through the cask do not always depend on its boiling point (Yoshizawa *et al.*, 1981).

Chemical changes of components in unaged distillates

Studies on bourbon whiskey by Liebman and Rosenblatt (1943), Liebman and Scherl (1949), Baldwin and Andreasen (1974) and Reazin (1981, 1983) have shown that the volatile aldehyde, ester and total acid contents increase during maturation. Aldehydes and esters increase roughly linearly throughout the maturation period, while the acid content increases most quickly during the first few years.

Aldehydes in whisky are mainly acetaldehyde, which is already present in the distillate as a fermentation product. Reazin *et al.* (1976) showed that the concentration of acetaldehyde in bourbon whiskey increased more than four times during maturation for four years and, using a radioactivity assay, it was found that most of the increased acetaldehyde was derived from ethanol. Nishimura *et al.* (1983) demonstrated that oak wood extract is indispensable for oxidation of ethanol to acetaldehyde.

Acetic acid is the major component of the total acids in matured spirits (Ohnishi *et al.*, 1977). Using a radioactivity technique, Reazin *et al.* (1976) showed that acetic acid was produced from ethanol via acetaldehyde during maturation. They estimated that part of the acetic acid increase during maturation was due to some wood components containing the acetic acid moiety, on the basis of data showing that specific radioactivity of the increased acetic acid was lower than that of ethanol. Nishimura *et al.* (1983) confirmed that a considerable part of the acetic acid increase is derived from a precursor present in oak wood. The precursor is insoluble in distilled spirits but seems to gradually decompose to yield acetic acid under acidic conditions during maturation. Caprylic, capric and palmitic acids increased two to three times during four-year maturation.

The increase in total esters is mainly due to the increase in ethyl acetate, which is produced by a shift of the equilibrium owing to the increase in acetic acid. Ethyl acetate increased to more than four times that of the new distillate during four years of maturation (Reazin, 1981, 1983). Ethyl esters of caproic through myristic acids remained unchanged or showed small decreases in bourbon whiskey. Ethyl caprate, palmitate and linoleate all decreased significantly. 2-Phenethyl acetate also decreased (Reazin, 1981).

Some of the compounds present in the new distillates can be consumed in chemical reactions. Acrolein, a lachrymator with a pungent odour, disappears over two to three years of ageing to yield 1,1,3-triethoxypropane (Kahn *et al.*, 1969). Some sulphur compounds also decrease during maturation. Masuda and Nishimura (1981a) analysed sulphur compounds in malt whiskies of various ages and found that dimethyl sulphide disappeared within one year, and 3-(methylthio)-propyl acetate, dihydro-2-methyl-3(2H) thiophene and ethyl 3-(methylthio)-propanoate disappeared within three years, as shown in Table 8.2. Dimethyl disulphide and dimethyl trisulphide decreased gradually during maturation. Leppänen *et al.* (1983) suggested that the ratio of dimethyl sulphide to dimethyl trisulphide may be used as a fair measure of the maturity of a sample because the rate of decrease of dimethyl trisulphide is slower than that of dimethyl disulphide.

Oak wood extract is indispensable for decreasing some sulphur compounds such as dimethyl sulphide, dimethyl disulphide, methionyl acetate and ethyl methionate (Nishimura *et al.*, 1983).

CHANGES IN LIGNIN-RELATED COMPOUNDS DURING MATURATION

Among the compounds derived from oak casks, lignin-related compounds are one of the most significant groups and are closely associated with the development of matured whisky flavour. Russian chemists were the first to note the importance of lignin-related compounds on wood ageing, as early as 1960. They identified vanillin, syringaldehyde, coniferaldehyde and *p*-hydroxybenzaldehyde as the lignin degradation products in brandy (Sisakyan and Egorov, 1953; Egorov and Borisova, 1957; Lichev and Panaiotov, 1958). These compounds were also found in malt whisky by Yamada and Fukui (1963), in bourbon

Table 8.2 Concentrations (μg per litre at 100% alcohol) of sulphur compounds in malt whiskies of various ages.

Compounds	*Years** 0	*1*	*2*	*3–5*	*6–10*
Dimethyl disulfide	359	345	272	149	91
Dimethyl trisulfide	12	16	15	19	14
3-(Methylthio)-propanal	7	t†	0	0	0
Dihydro-2-methyl-3(2H)-thiophenone	77	53	3	0	0
Ethyl 3-(methylthio) propanoate	21	11	t†	0	0
3-(Methylthio)-propyl acetate	274	139	3	0	0
2-Thiophenecarboxaldehyde	66	83	49	60	54
3-(Methylthio)-propanol	8	0	0	0	0
5-Methyl-2-thiophene-carboxaldehyde	15	15	10	16	16
Benzothiophene	39	27	27	31	38
Benzothiazole	18	16	14	15	32

*Malt whisky age. †Trace.
Masuda and Nishimura (1981a), reproduced with permission from *Journal of Food Science*, **47**: 104.

whiskey by Martin *et al.* (1965) and in brandy by Guymon and Crowell (1968), in rum and brandy by Puech *et al.* (1977) and Puech (1981). In addition to these compounds, Otsuka *et al.* (1965) and Puech (1981) identified phenolic acids such as vanillic acid, syringic acid and ferulic acid in whisky and brandy respectively. More recently, Lehtonen (1983) quantitatively analysed those lignin-related compounds, together with gallic acid, in various spirits including whiskies of different origin; he suggested the possibility of distinguishing different classes of spirits by comparing their ratios.

Mechanisms for aromatic aldehyde formation from lignin have been proposed by Baldwin *et al.* (1967), Puech *et al.* (1977),

Reazin (1981) and Nishimura *et al.* (1983). Nishimura *et al.* verified the existence of the following four pathways for the origin of lignin-related compounds in matured spirits using simulation studies:

1. Degradation of lignin to aromatics with toasting or charring of the oak cask.
2. Extraction of monomeric compounds present in the free state and of native lignin by spirits.
3. Formation of aromatics by ethanolysis of lignin.
4. Further conversion of compounds existing in the spirits.

Nishimura *et al.* (1983) found that direct extraction of oak wood chips toasted or charred on the surface yielded vanillin, propiovanillone, syringaldehyde, acetosyringone, coniferaldehyde, vanillic acid and sinapaldehyde, whilst untoasted oak chips yielded none or very small amounts of these compounds. This indicated that these aromatics from lignin result from the toasting or charring of oak wood.

When uncharred oak chips were stored in aqueous alcohol for six months, the solution contained some of those compounds, although not as much as a solution with charred chips, as shown in Table 8.3. This implies that ethanolysis of lignin in oak wood occurs to yield those aromatics.

Puech (1981) analysed the content of vanillin, syringaldehyde, coniferaldehyde and sinapaldehyde in the outer and inner faces

Table 8.3 Lignin-related compounds from uncharred and charred oak wood chips stored in 60% EtOH solutions (ppm).

	Uncharred	*Charred*
Vanillin	0.14	6.25
Acetovanillone	0.02	0.37
Syringaldehyde	0.27	12.40
Acetosyringone	—	0.66
Coniferaldehyde	trace	8.40
Vanillic acid	0.35	2.63
Sinapaldehyde	0.04	5.40

Note: Amounts of oak chips used were 1.0% (wt/v) to 60% EtOH.
Nishimura *et al.* (1983), reproduced with permission from Piggott J R (ed) *Flavour of Distilled Beverages: Origin and Development*. Ellis Horwood, Chichester, p 247.

of staves that had contained armagnac for 20 years and found that the inner face was about 16 times richer in those compounds than the outer face. He concluded that a fraction of oak wood lignin undergoes intense oxidation when in contact with spirits and oxygen, leading to the formation of aromatic aldehydes which are then solubilised in the spirits.

Baldwin *et al.* (1967) isolated a kind of soluble lignin, ethanol lignin, which was shown to be formed with some influence from ethanol in the distillate by Reazin (1981). Ethanol lignin is an intermediate to the aromatic aldehydes from lignin. Nishimura *et al.* (1983) prepared Braun's native lignins from ethanol extracts of uncharred and charred oak wood chips and showed that both samples yielded aromatic aldehydes by ethanolysis. Lignin-related aromatic aldehydes and ketones were subjected to further oxidation or esterification when stored in aqueous ethanol solution in glass bottles in which the head spaces had been replaced by oxygen gas (Nishimura *et al.*, 1983). The conversion of some lignin-related aromatics is shown in Fig. 8.3.

A proposed scheme (Nishimura *et al.*, 1983) of changes of lignin and its related compounds occurring in oak wood and in

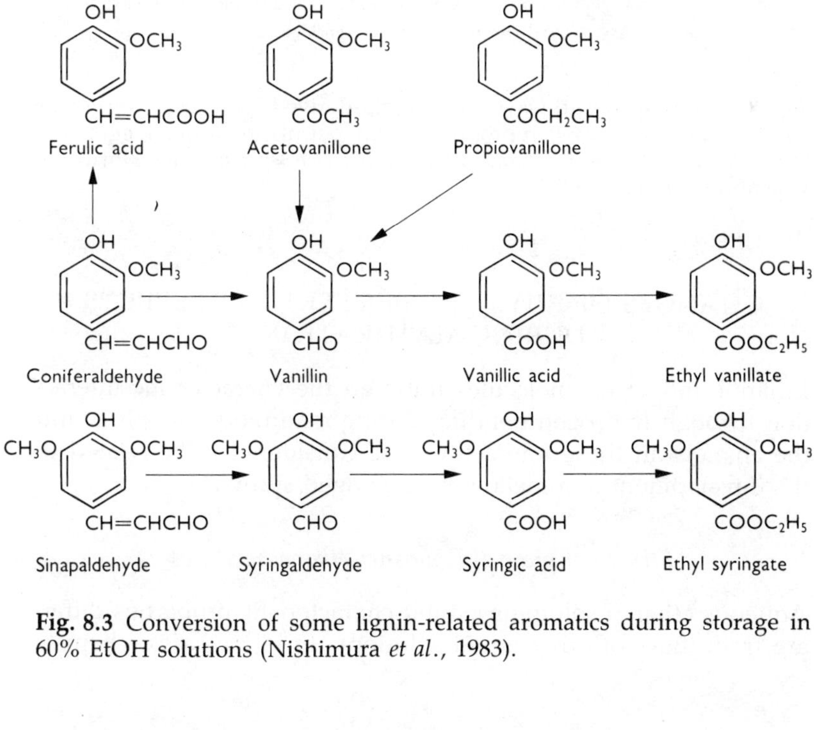

Fig. 8.3 Conversion of some lignin-related aromatics during storage in 60% EtOH solutions (Nishimura *et al.*, 1983).

spirits during maturation is shown in Fig. 8.4. Lignin-related compounds in matured spirits such as bourbon whiskey stored mainly in charred oak barrels are mainly derived through the pathway [A]→[B]→[D] and pathway [C] is relatively less important, while pathway [C]→[D] is the main route for Scotch whisky and cognac.

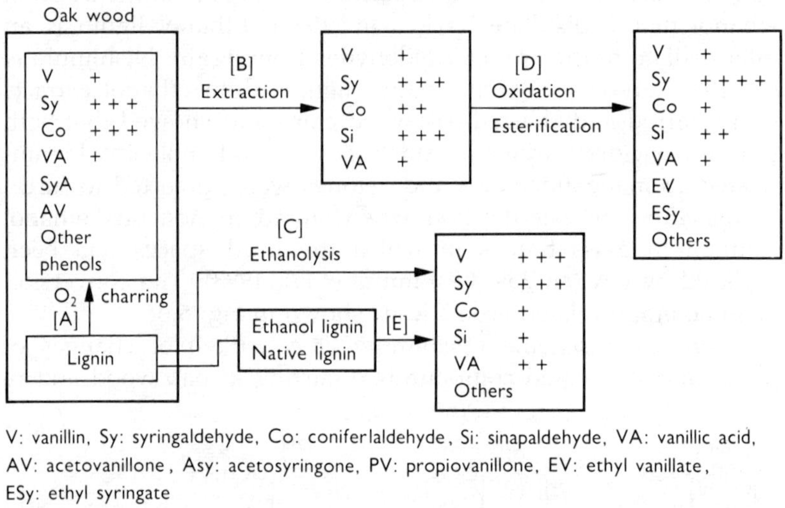

Fig. 8.4 Proposed pathways for lignin-derived congener formation (Nishimura *et al.*, 1983; reproduced with permission from Piggott J R (ed) *Flavour of Distilled Beverages: Origin and Development*, Ellis Horwood, Chichester, p 249).

CHANGES OF PHYSICAL PROPERTIES IN SPIRITS DURING MATURATION

Ethanol and water molecules undergo the characteristic interaction through hydrogen bonding during maturation of spirits and the changes in these interactions are considered to be related to the development of a mellow taste in aged spirits.

Structural properties of water-ethanol mixtures

Aqueous ethanol solutions exhibit characteristic properties differing from those of other liquid solutions. In many cases, changes in the physical constants of the aqueous ethanol solution show bell-shaped curves with maximum or minimum points as a function of the ethanol concentration.

When ethanol and water are mixed, the maximum contraction in volume is exhibited at 60% at 17°C, as shown in Fig. 8.5 (Akahoshi, 1963). Heat evolution upon mixing of ethanol and water in various ratios gives a minimum at 40% at 25°C (Boyne and Williamson, 1967) and the viscosity of aqueous ethanol solutions has a maximum at 40% (National Bureau of Standards, 1918).

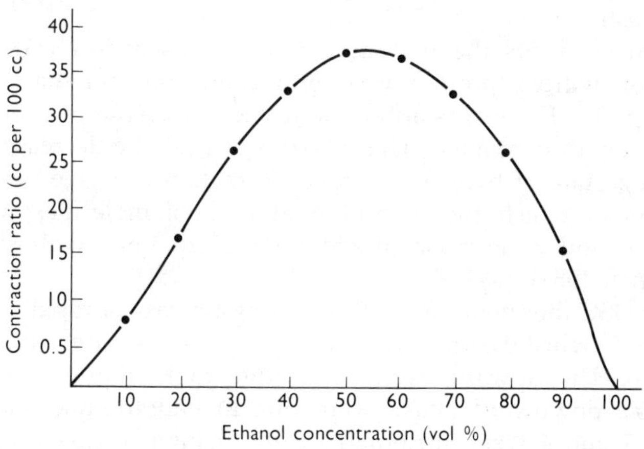

Fig. 8.5 Contraction in volume in mixing ethanol and water at various proportions (Akahoshi, 1963; reproduced with permission from *Nippon Nogeikagaku Kaishi* **37**: 435).

These phenomena are some of the features of aqueous solutions and are considered to be attributable to the interaction between ethanol and water molecules through hydrogen bonding and the formation of stable clusters.

Changes in structural properties in spirits during maturation

The increase in the interaction between ethanol and water molecules and the progress in the formation of stable clusters of ethanol and water during storage of spirits in oak casks can be found by various physico-chemical measurements such as laser-raman spectroscopy (Nakamura, 1952b), light scattering measurements (Nakamura, 1952a), proton magnetic resonance spectroscopy (Akahoshi and Ohkuma, 1984), dielectric measurements (Akahoshi, 1963) and differential scanning calorimetry (DSC) (Koga and Yoshizumi, 1977, 1979; Akahoshi and Ohkuma, 1985).

From dielectric measurements of whisky aged 10 years and unaged aqueous ethanol solutions at various frequencies, the dielectric constants of the matured spirits were lower than those of unmatured aqueous ethanol, and differences between the samples were larger with increasing frequency. This indicated that the molecules of water and ethanol in matured spirits could not easily be oriented due to cluster formation during maturation (Akahoshi, 1963).

In aged spirits the nuclear magnetic resonance (NMR) spectrum of hydroxy protons was significantly broader than that of new spirits. There was a half band width increase in proportion to the length of storage. These results indicated a decrease in the exchange rate of hydroxy protons, formation of stable molecular clusters in which the migration of ethanol molecules was restricted, and a decrease in monomeric ethanol (Akahoshi and Ohkuma, 1984).

The DSC thermograms of the melting process of rapidly frozen matured whisky and water-ethanol mixture are shown in Fig. 8.6. The upward peaks were due to exothermic reactions and the downward peaks were due to endothermic reactions. Peaks 1 and 4 were attributed to the melting of free water and ethanol molecules, while peak 3 was due to the eutectic of alcohol and water molecules. Comparing the DSC thermogram of matured whisky with that of the water-ethanol mixture, the areas of peaks 1 and 4 of matured whisky were smaller than those of the unmatured one, and the area of peak 3 was larger. These results indicated the increasing interaction between alcohol and water molecules with maturation (Koga and Yoshizumi, 1977).

A non-volatile extract of oak wood was observed to promote the interaction between alcohol and water molecules in spirits. The DSC thermogram in Fig. 8.7 shows that the interactions between alcohol and water observed in matured whisky were destroyed by distillation, but appeared again on the addition of the dillation residue to the distillate (Nishimura *et al.*, 1983).

Thermograms of the melting of rapidly frozen samples prepared by adding non-volatile components isolated from matured

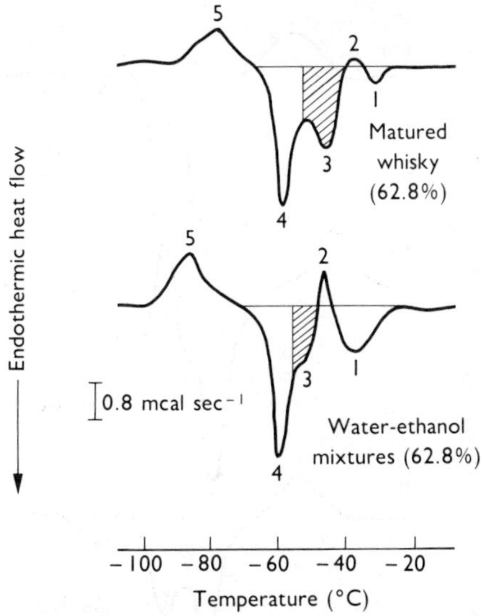

Fig. 8.6 DSC thermograms of the melting of rapidly frozen whisky and water-ethanol mixture of the ethanol concentration of 62.8% (Koga and Yoshizumi, 1977; reproduced with permission from *Journal of Food Science* **42**: 1216).

whisky in various amounts to unmatured whisky, as shown in Fig. 8.8, indicated clearly that interaction between ethanol and water increased with an increased concentration of the non-volatile components (Koga, 1981). Among the non-volatile components of matured whisky, inorganic salts and a fraction of molecular weight above 1000 containing wood components such as lignin and tannins, were found to contribute to the formation of stable clusters of water and ethanol molecules.

Maturation conditions related to the constituents and quality of the spirits

ALCOHOLIC PERCENTAGE OF STORED SPIRITS

Alcoholic strength of whiskies at the point at which barrelling takes place varies for different types of whisky: Scotch malt whisky is at about 11° overproof British (63%), Scotch grain whisky is up to 20° overproof British (68%), Irish whiskey is at

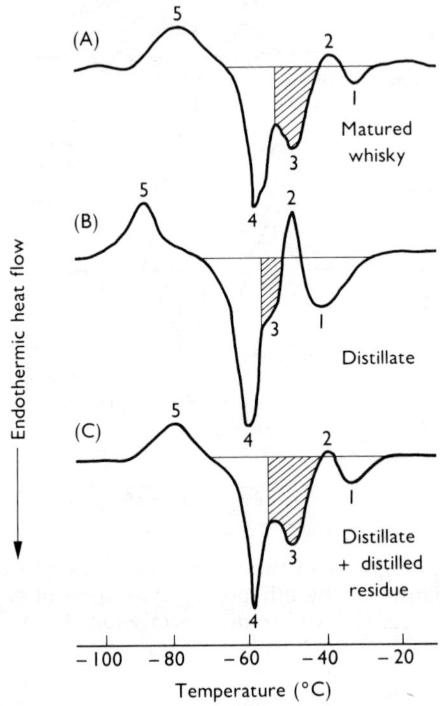

Fig. 8.7 DSC thermograms of the melting of rapidly frozen matured whisky, distillate of matured whisky and restored whisky (Nishimura *et al.*, 1983; reproduced with permission from Piggott J R (ed) *Flavour of Distilled Beverages: Origin and Development*, Ellis Horwood, Chichester, p 252).

25° overproof British (71%), bourbon whiskey is at 125° American proof (63%), American grain whiskey is at or above 190° American proof (95%) (Lyons and Rose, 1977). Canadian grain whisky is filled into casks for ageing at about 70% (Rannie, 1976) and Japanese malt whisky is at 60 to 70%.

The amount of the extracts liberated from oak wood by ethanol-water mixtures during maturation is variable, depending upon the alcohol concentration. When oak chips were dipped in aqueous ethanol solutions of 20, 40, 60, 80 and 100% ethanol concentration for five months, both the amount of the total extract and the total phenols attained a maximum at 60% ethanol concentration, as shown in Table 8.4 (Nykänen, 1984).

The relationships between alcoholic percentages at which the distillate enters the casks and the amounts of the components of

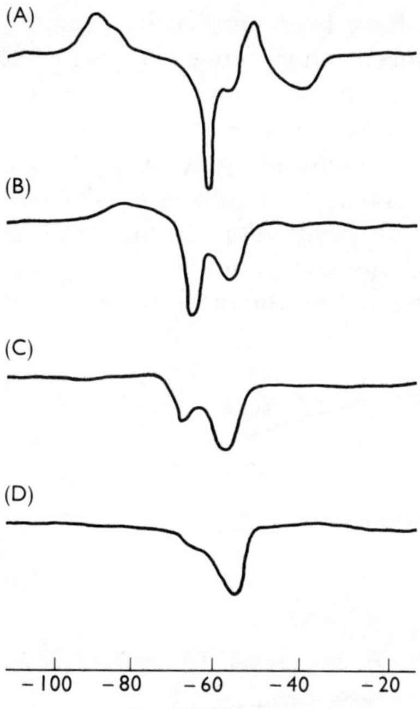

(A) 0 ppm, (B) 3000 ppm, (C) 7000 ppm
(D) 10 000 ppm

Fig. 8.8 The DSC thermograms of the melting of rapidly frozen aqueous ethanol solutions contained non-volatile fractions of matured whisky in various amounts (Koga, 1981; reproduced with the permission of Dr K Koga).

Table 8.4 The total extract and total phenols liberated from oak chips by alcohol-water mixtures during five months.

Alcohol content (% by volume)	*Total extract (%)**	*Total phenols (mg per litre)†*
20	6.31	62.5
40	7.44	74.8
60	7.68	78.8
80	6.53	66.3
100	3.70	38.5

*Calculated on the amount of dry chips.
†Determined as gallic acid.
Nykänen (1984), reproduced with permission from Nykänen L, Lehtonen P (eds) *Flavour Research of Alcoholic Beverages: Instrumental and Sensory Analysis*. Foundation for Biotechnical and Industrial Fermentation Research, Helsinki, p 143.

bourbon whiskey have been studied by Reazin (1983). When the stored alcohol concentration ranges from 109° to 155° American proof, the production of colour, volatile acids and tannins decreases as the proof increases, as shown in Fig. 8.9. Extraction of sugars such as arabinose, glucose and xylose is depressed with increasing alcohol percentages, as shown in Table 8.5. On the other hand, the production of the components originating solely from distillate components, such as esters and fusel alcohols, is not affected by the alcohol concentration.

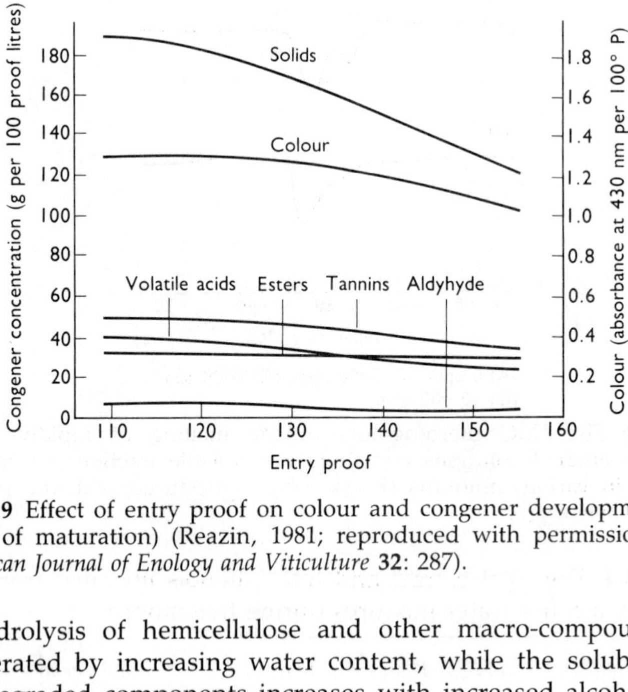

Fig. 8.9 Effect of entry proof on colour and congener development (six years of maturation) (Reazin, 1981; reproduced with permission from *American Journal of Enology and Viticulture* **32**: 287).

Hydrolysis of hemicellulose and other macro-compounds is accelerated by increasing water content, while the solubility of the degraded components increases with increased alcohol content. The two studies mentioned above showed that an ethanol content of around 60% is the most effective concentration for taking the extracts from oak wood into spirits.

Boruff and Rittschof (1959) studied the flavour development of bourbon whiskey stored at 118°, 127° and 154° American proof, as well as its chemical analysis. They found that flavour after eight years was normal in the whisky stored at 118° proof, slightly less mature at 127° proof and different at 154° proof because of a spicy green oak taste of the 154° proof sample compared with the 110° proof sample.

Table 8.5 Effect of proof on sugar and glycerol in spirits aged four years.

	Concentration (g per 100 proof litres)	
	109°P	*141°P*
Arabinose	1.19	0.62
Xylose	0.24	0.14
Glucose	0.85	0.43
Galactose	0.05	0.03
Fructose	0.17	0.10
Glycerol	0.36	0.19

Reazin (1981), reproduced with permission from *American Journal of Enology and Viticulture* **32**: 287.

SIZE AND NATURE OF THE CASKS

It is well known that the size of cask influences the speed of maturation. The concentrations of the components resulting from an extractive process tend to be higher in smaller casks, where the ratio of surface area to volume of whisky is greater (Schoeneman *et al.*, 1971). For instance, there is 68 cm^2 per litre (48 $in.^2$ per gal) of wood contact with whisky in a 500 litres (110 gal) butt, but 85 cm^2 per litre (60 $in.^2$ per gal) of wood contact in a 250 litre (55 gal) hogshead (Cooper, 1978).

The decrease of some compounds during maturation also seemed to be influenced by the cask size. Masuda and Nishimura (1981a) found that the decrease of dimethyl disulphide in malt whisky during maturation occurred more rapidly in hogshead casks than in a puncheon. This result is consistent with the observation that oak extract takes part in the decrease of sulphur compounds, including dimethyl disulphide as mentioned above (Nishimura *et al.*, 1983).

Spirits stored in the smaller size of cask undergo more loss by evaporation. The Customs & Excise in Scotland formally recognises this natural and inevitable loss. They allow an overall 2% loss per year plus a further 2% for whisky in butts of 500 litres (110 gal) or a further 3% for whisky in hogsheads of 250–300 litres (55–66 gal) (Murphy, 1978).

As mentioned above, the inside surface of the casks used for bourbon whiskey maturation is usually charred strongly to form a charcoal layer and many other whiskies are stored in casks which have been toasted inside to some extent to activate them and to remove the raw woody aroma and astringent taste. The products yielded by toasting or charring are dependent on the treatment temperatures. The amounts of compounds derived from lignin degradation produced by toasted oak wood chips at 100, 150 and 200°C and charred chips are shown in Table 8.6 (Nishimura *et al.*, 1983).

Table 8.6 Aromatic aldehydes produced by toasting or charring oak chips (ppm).

Products	*Toasted temperature* 100°C	150°C	200°C	*Charred*
Vanillin	1.1	3.8	13.5	2.8
Propiovanillone	0.6	1.1	1.4	0.9
Syringaldehyde	0.1	3.8	32.0	9.2
Acetosyringone	—	0.025	1.5	0.6
Coniferaldehyde	trace	4.3	24.0	4.8
Vanillic acid	—	1.8	6.1	1.1
Sinapaldehyde	trace	6.5	60.0	9.0

Note: Amounts of oak chips used were 2.0% w/v in 60% ethanol solution.
Nishimura *et al*. (1983), reproduced with permission from Piggott J R (ed) *Flavour of Distilled Beverages: Origin and Development*, Ellis Horwood, Chichester, p 245.

It was shown that these products increased remarkably with higher toasting temperatures, while untoasted chips contained hardly any aromatic compounds. The lesser amounts of aromatics in charred chips compared to toasted chips at 200°C may have been due to their loss by carbonation of the aromatics. Reazin (1981) studied the effect of barrel charring as well as barrel reuse on aromatic aldehyde formation and found that the production of aromatic aldehydes in a new charred cask after 1.5 years exceeded that found in uncharred or reused casks, as shown in Table 8.7.

In addition to the lignin-related phenolic compounds, some volatile compounds such as furfural, 2-methyl furfural, guaiacol and 4-methyl guaiacol increased with toasting of oak wood chips (Nishimura *et al.*, 1983). 2-Hydroxy-3-methyl-2-cyclopentenone and maltol were isolated from the toasted oak chips at 220°C. Both of these compounds were found in bourbon whiskey, giving it the characteristic sweet and burnt aroma (Nishimura *et al.*, 1983).

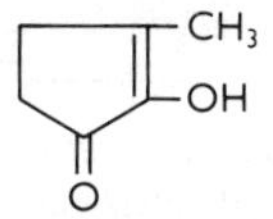

2-hydroxy-3-methyl-2-cyclopentenone

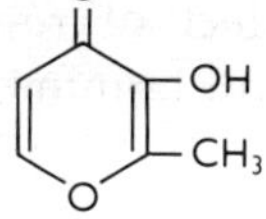

maltol

Table 8.7 Aromatic aldehyde content of aged spirits.

	Barrel type		
	New uncharred	*Used*	*New charred*
Age (years)	6	6	1.5
Aromatic aldehydes (g per 100 proof litres)			
Coniferaldehyde	0.03	0.03	0.12
Sinapaldehyde	0.02	0.04	0.13
Syringaldehyde	0.12	0.14	0.45
Vanillin	0.12	0.06	0.07
Total	0.29	0.27	0.77

Reazin (1981), reproduced with permission from *American Journal of Enology and Viticulture* **32:** 288.

CASK HISTORY

The history of a cask also affects the production of the components of spirits during maturation. The amounts of volatile acids, esters, tannins, colour, total acids, fixed acids and furfural produced during maturation decrease rapidly as the number of cask refillings increases (Reazin, 1981). To rejuvenate a used cask the effect of re-charring was studied. It led to an increase in the components over that found in the uncharred used cask, but the amount did not equal that in whisky aged in a new charred cask, as shown in Table 8.8 (Reazin, 1981). These data imply that re-charring can replenish some precursors of the components depleted by previous cask use, but not to the levels found in a new cask.

Scotch malt whisky, as well as Scotch grain whisky, may be aged in casks previously used many times and the development of secondary constituents is exceedingly slow: less in 12 years in such casks than in a year in new charred white oak casks or 2 years in new plain white oak casks (Valaer, 1940).

Table 8.8 Effect of re-charring used barrels on colour, solids, fixed acids and tannins.

	Concentrations at 2 years age			
	New	*Re-charred*		*New*
	1st fill	*After one fill*	*After two fills*	*2nd fill*
Colour (Klett units)	98	42	31.6	27
Congeners (g per 100 proof litres)				
Solids	97	24.6	18.6	16
Fixed acids	5.1	1.6	1.2	0.8
Tannins	32	12.6	10.3	8

Reazin (1981), reproduced with permission from *American Journal of Enology and Viticulture* **32**: 288.

ENVIRONMENTAL CONDITIONS (US)

Warehouse temperature influences the ratio of the various components which characterise whisky maturation. Reazin (1981, 1983) determined the amounts of the constituents in bourbon

Table 8.9 Effect of warehouse temperature on congener development.

	Tier level		
	Bottom	*Middle*	*Top*
Average temperature (°F)	64.5	66.6	73.4
Proof (°P)	101.5	103.9	105.5
Congeners (g per 100 proof litres):			
Acetaldehyde	1.9	2.2	2.8
Acids: fixed	6.0	6.3	8.9
volatile	43.0	46.7	52.1
Ethyl acetate	24.0	27.0	31.0
Furfural	1.0	1.2	0.9
Oak lactone	0.3	0.3	0.3
Solids	131.0	138.0	173.0
Sugars: Arabinose	7.3	7.7	8.9
Fructose	0.6	0.6	0.8
Glucose	4.5	4.8	5.8
Tannins	39.0	40.0	47.0

Reazin (1981), reproduced with permission from *American Journal of Enology and Viticulture* **32**: 288.

whiskey aged for four years on tiers 1, 6 and 12 of a 12-tier iron-clad warehouse, as shown in Table 8.9. The yearly average temperatures on these tiers were 18.1, 19.2 and 23°C. Acetaldehyde and fixed acids showed an increase of about 48% between the top and bottom tiers. Other compounds such as ethyl acetate, solids, sugars and tannins showed increases of 21–38%. No changes in furfural and oak lactones were detected.

Puerto Rican rum stored for 18 weeks in warehouses in which temperatures were controlled at 26.7°C and 46.1°C showed increases of 10% in acetaldehyde, 250% in ethyl acetate, 100% in tannins and 48% in total acids at the higher temperature (Torres *et al.*, 1979).

Reazin (1981) pointed out that, although the various reactions occurring during maturation can be increased with temperature, there does exist an optimum temperature for the production of the desired product quality. This is because the organoleptic impact differs for each of the components.

References

Akahoshi R 1963 Studies on aged spirits and their dielectricity. Part I. The dielectric coefficient of aged spirit and ethanol aqueous solution which contain some impurities. *Nippon Nogeikagaku Kaishi* **37**: 433–8

Akahoshi R, Ohkuma H 1984 NMR spectra of hydroxy protons in aged spirits. *Nippon Nogeikagaku Kaishi* **58**: 357–65

Akahoshi R, Ohkuma H 1985 Thermal properties of ethanol aqueous solutions and aged spirits at low temperature. *Nippon Nogeikagaku Kaishi* **59**: 1–9

Baldwin S, Andreasen A A 1974 Congener development in Bourbon whiskey matured at various proofs for twelve years. *Journal of the Association of Official Analytical Chemists* **57**: 940–50

Baldwin S, Black R A, Andreasen A A, Adams S L 1967 Aromatic congener formation in maturation of alcoholic distillates. *Journal of Agricultural and Food Chemistry* **15**: 381–5

Black R A, Andreasen A A 1973 Steroids in aged whisky. *Journal of the Association of Official Analytical Chemists* **56**: 1357–61

Black R A, Andreasen A A 1974 Gas-liquid chromatographic determination of monosaccharides and glycerol in aged distilled spirits. *Journal of the Association of Official Analytical Chemists* **57**: 111–17

Boruff C S, Rittschof L A 1959 Effects of barreling proof on the ageing of American whiskeys. *Journal of Agricultural and Food Chemistry* **7**: 630–3

BOYNE J A, WILLIAMSON A G 1967 Enthalpies of mixing of ethanol and water at 25°C. *Journal of Chemical and Engineering Data* **12**: 318

BRAUS H A, ECK J W, MUELLER W M, MILLER F D 1957 Isolation and identification of a sterol glucoside from whisky. *Journal of Agricultural and Food Chemistry* **5**: 458–9

CHEN C L 1970 Constituents of *Quercus alba*. *Phytochemistry* **9**: 1149

COOPER D 1978 *Guide to the Whiskies of Scotland*. Pitman House, London

DE SMEDT P, LIDDLE P A P 1978 Identification of 1,1-diethoxy-propan-2-one in spirits aged in wood. *American Journal of Enology and Viticulture* **29**: 286–8

DUNCAN R E B, PHILP J M 1966 Methods for the analysis of Scotch whisky. *Journal of the Science of Food and Agriculture* **17**: 208–14

DZHANPOLADYAN L M, MNDZOYAN E L, SAAKYAN I G, AKHNAGARYAN F A 1969 Change of oak wood carbohydrates in an alcoholic medium. *Vinodelie i Vinogradarstvo SSSR* **29**: 13–16; C A 72, 88898 (1970)

EGOROV I A, BORISOVA N B 1957 The aromatic aldehydes of Cognac spirits. *Biokhim Vinodeliya Sbornik* **5**: 27–37; C A 51, 18458g (1957)

GUYMON J F, CROWELL E A 1968 Separation of vanillin, syringaldehyde, and other aromatic compounds in the extracts of French and American oak woods by brandy and aqueous alcohol solutions. *Qualitas Plantarum et Materiae Vegetabiles* **16**: 320–33

GUYMON J F, CROWELL E A 1972 GC-Separated brandy components derived from French and American oaks. *American Journal of Enology and Viticulture* **23**: 114–20

KAHN J H, SHIPLEY P A, LAROE E G, CONNER H A 1969 Whisky composition: Identification of additional components by gas chromatography-mass spectrometry. *Journal of Food Science* **34**: 587–91

KOGA K 1981 Unpublished data

KOGA K, YOSHIZUMI H 1977 Differential scanning calorimetry (DSC) studies on the structures of water-ethanol mixtures and aged whiskey. *Journal of Food Science* **42**: 1213–17

KOGA K, YOSHIZUMI H 1979 Differential scanning calorimetry (DSC) studies on the freezing processes of water-ethanol mixtures and distilled spirits. *Journal of Food Science* **44**: 1386–9

LEHTONEN M 1983 High performance liquid chromatographic determination of nonvolatile phenolic compounds in matured distilled alcoholic beverages. *Journal of the Association of Official Analytical Chemists* **66**: 71–8

LEPPÄNEN O, RONKAINEN J, DENSLOW J, LAAKSO R, LINDEMAN A, NYKÄNEN I 1983 Polysulphides and thiophenes in whisky. In Piggott J R (ed) *Flavour of Distilled Beverages: Origin and Development*. Ellis Horwood, Chichester

LICHEV V I, PANAIOTOV I M 1958 Reducing sugars in Cognac spirits and plum brandy. *Bulgarska Akademiya na Naukite Izvestiya na Instituta za Khimiya* **6**: 167–79; C A 54, 817e (1960)

LICHEV V I, PANAIOTOV I M 1959 Identification of aldehydes in the spirit of Cognac. *Vinodelie i Vinogradarstvo SSSR* **18**: 10–15; C A 52, 19010 (1958)

Liebman A J, Rosenblatt M 1943 Changes in whisky while maturing. *Industrial and Engineering Chemistry* **35**: 994–1002

Liebman A J, Scherl B 1949 Changes in whisky while maturing. *Industrial and Engineering Chemistry* **41**: 534–43

Lyons T P, Rose A H 1977 Whisky. In Rose A H (ed) *Alcoholic Beverages*. Academic Press, London

Martin G E, Eib N K 1968 Determination of sugars in distilled spirits by GLC. *Journal of the Association of Official Analytical Chemists* **51**: 925–7

Martin G E, Schmit J A, Schoeneman R L 1965 Various components of Bourbon whisky by thin layer and gas-liquid chromatography. *Journal of the Association of Official Analytical Chemists* **48**: 962–64

Masuda M, Nishimura K 1971 Branched nonalactones from some *Quercus* species. *Phytochemistry* **10**: 1401–2

Masuda M, Nishimura K 1981a Changes in volatile sulphur compounds of whisky during ageing. *Journal of Food Science* **47**: 101–5

Masuda M, Nishimura K 1981b Absolute configurations of *Quercus* lactones, (3S,4R)- and (3S,4S)-3-methyl-4-octanolide, from oak wood and chiroptical properties of mono-cyclic-γ-lactones. *Chemistry Letter*: 1333–6

Murphy B 1978 *The World Book of Whisky*. William Collins and Company, Glasgow and London

Nakamura K 1952a Depolarization of the light scattered by aqueous solutions of ethyl alcohol. *Science Reports of the Tohoku Imperial University, Sendai, Japan. Series 1* **35**: 200–5

Nakamura K 1952b Raman spectrum of aqueous solutions of ethyl alcohol. *Science Reports of the Tohoku Imperial University, Sendai, Japan. Series 1* **35**: 195–9

National Bureau of Standards 1918. *National Bureau of Standards Bulletin* **14**: 59

Nishimura K, Masuda M 1971 Minor constituents of whisky fusel oils. I. Basic, phenolic and lactonic compounds. *Journal of Food Science* **36**: 819–22

Nishimura K, Ohnishi M, Masuda M, Koga K, Matsuyama R 1983 Reactions of wood components during maturation. In Piggott J R (ed) *Flavour of Distilled Beverages: Origin and Development*. Ellis Horwood, Chichester

Nykänen L 1984 Aroma compounds liberated from oak chips and wooden casks by alcohol. In Nykänen L, Lehtonen P (eds) *Flavour Research of Alcoholic Beverages: Instrumental and Sensory Analysis*. Foundation for Biotechnical and Industrial Fermentation Research, Helsinki

Ohnishi M, Guymon J F, Crowell E A 1977 Changes in some volatile constituents of brandy during aging. *American Journal of Enology and Viticulture* **28**: 152–8

Otsuka K, Imai S, Morinaga K 1965 Studies on the mechanism of aging of distilled liquors. Part II. Distribution of phenolic compounds in aged distilled liquors. *Agricultural and Biological Chemistry* **29**: 27–31

Otsuka K, Morinaga K, Imai S 1963 Sugars in aged distilled spirits and oak wood. *Nippon Jozo Kyokai Zasshi* **59**: 448–9

Otsuka K, Zenibayashi Y 1974 On the determination of scopoletin in aged distilled liquors. *Agricultural and Biological Chemistry* **38**: 1079–80

Otsuka K, Zenibayashi Y, Itoh M, Totsuka A 1974 Presence and significance of two diastereomers of 4-methyl-γ-octalactone in aged distilled liquors. *Agricultural and Biological Chemistry* **38**: 485–90

Puech J-L 1981 Extraction and evolution of lignin products in Armagnac matured in oak. *American Journal of Enology and Viticulture* **32**: 111–14

Puech J-L, Jouret C, Deibner L, Alibert G 1977 Substances phenoliques des eaux de vie d'Armagnac et de rhum. II. Produits de la degradation de le lignine: Les aldehydes et les acides aromatiques. *Industries Alimentaries et Agricoles* **94**: 483–93

Rannie W F 1976 *Canadian Whisky: the product and the industry* Rannie, Lincoln, Ontario

Reazin G H 1981 Chemical mechanisms of whisky maturation. *American Journal of Enology and Viticulture* **32**: 283–9

Reazin G H 1983 Chemical analysis of whisky maturation. In Piggott J R (ed) *Flavour of Distilled Beverages: Origin and Development*. Ellis Horwood, Chichester

Reazin G H, Baldwin S, Scales H S, Washington H W, Andreasen A A 1976 Determination of the congeners produced from ethanol during whisky maturation. *Journal of the Association of Official Analytical Chemists* **59**: 770–6

Ritter G T, Fleck L C 1923 Chemistry of wood. IV. The results of analysis of heart wood and sapwood of some American woods. *Industrial and Engineering Chemistry* **15**: 1055–6

Salo P, Nykänen L, Suomalainen H 1972 Odour thresholds and relative intensities of volatile aroma components in an artificial beverage imitating whisky. *Journal of Food Science* **37**: 394–8

Shoeneman R L, Dyer R H, Earl E M 1971 Analytical profile of straight Bourbon whiskey. *Journal of the Association of Official Analytical Chemists* **54**: 1247–61

Singleton V L 1974 Some aspects of the wooden container as a factor in wine maturation. In Webb A D (ed) *Chemistry of Winemaking*. American Chemical Society, Washington DC

Sisakyan N M, Egorov I A 1953 The chemical mechanism of the maturing of brandy spirits. *Biokhim Vinodeliya Sbornik* **4**: 121–41; C A 49, 563 (1955)

Suomalainen H, Nykänen L 1970 Investigation on the aroma of alcoholic beverages. *Naringsmiddelindustrien* **23**: 15–30

Torres C S, Aguiar J L, Batiz H, Hernandez I 1979 Aging of rum in a hot chamber. *Journal of Agriculture of the University of Puerto Rico* **63**: 64–77

VALAER P 1940 Scotch whisky. *Industrial and Engineering Chemistry* **32**: 935–43

YAMADA K, FUKUI Y 1963 Studies on whisky flavour. Part II. Studies on high-boiling components. *Nippon Nogeikagaku Kaishi* **37**: 53–7

YOSHIZAWA K, MOMOSE H, HASUO T 1981 Substances evaporated through barrel of whisky. *Nippon Nogeikagaku Kaishi* **55**: 1063–8

CHAPTER 9

Cask quality and warehouse conditions

J M PHILP

Introduction

Maturation is the final step in the whisky production process prior to blending and bottling, and traditionally is surrounded by more mystique than any other stage of the production process. The new make whisky is put into casks as a raw, immature distillate and emerges, as if by magic, as a round, smooth, flavoured product a number of years later. Until recently, little was known of the maturation mechanisms involved in the production of these desirable flavours, and even less about the influence of cask type, cask history and warehouse conditions. A review by Sharp (1983) emphasised the lack of published information on maturation conditions relative to flavour production. Studies on the influence of American warehouse conditions on the maturation of American whiskey were last published by Valaer and Frazier (1936), Liebmann and Rosenblatt (1943), and Liebmann and Scherl (1949), using the classical volumetric and colorimetric analytical techniques available at that time. Similar studies have never been published for whiskies matured under the climatic conditions prevailing in Scottish warehouses. Analytical techniques have also changed dramatically over the past 40 years, principally due to the introduction of gas and liquid chromatography. Consequently, mature Scotch whiskies are now known to contain in excess of 400 components, many of which have important flavour characteristics. This review will attempt to evaluate the role of the cask and the role of warehouse climatic conditions in influencing the changes in composition which occur during maturation and in the development of the mature whisky flavour characteristics.

Whisky flavour development

The development of the flavour of a fully mature whisky can be traced through a number of quite discrete and independent steps, the cumulative result being the perceived flavour from each cask of whisky after a given number of years in the warehouse.

Initially, there is, of course, the basic organoleptic character of the immature distillate. In Scotch whisky terms this is either new make malt spirit with a high congener content or new make grain spirit with a relatively low congener content. In either case the principal constituents of the congener fraction are higher alcohols and fatty acid esters. Superimposed on this grain or malt spirit organoleptic base are individual distillery characteristics and charge to charge variations within each distillery. Many of the congeners in this category are present in extremely small amounts but are extremely potent contributors to flavour, such as long-chain, unsaturated aliphatic carbonyls (Nishimura and Masuda, 1985), sulphur-containing compounds (Leppänen *et al.*, 1983) and nitrogen-containing compounds (Viro, 1984).

Finally, the cask and warehouse conditions exert an influence on the total organoleptic character of the distillate. The role of the cask may be organoleptically subtractive, in that it may remove by adsorption, evaporation, or interaction some or all of the 'undesirable' congeners in the distillate. It may also perform an organoleptically additive role by the straight addition of flavour congeners to the distillate. In addition, it may perform a much more complex indirect role, in that it may produce a wood extract which will interact oxidatively with the distillate to produce desirable flavour tones in the mature whisky (Wildenradt and Singleton, 1974). The relative extent to which each of these additive, subtractive or interactive processes takes place will be governed principally by the previous history of the cask, and to some extent by the warehousing conditions.

Descriptive sensory analyses of whisky flavour have been carried out by Woolley (1973), Piggott and Jardine (1979), Swan *et al.* (1981) and Canaway (1983). Correlations between these descriptive flavours and analyses by Jounela-Eriksson (1978), Swan *et al.* (1981), Viro (1984), and Nishimura and Masuda (1985) have shown that the principal flavour-related congeners in mature Scotch whisky are esters, carbonyls, sulphur components, lactones, phenols and nitrogenous bases.

Cask quality

The quality of a whisky cask may be assessed in two quite clear and distinct ways. On the one hand, it may be classified as a good container. The wood is sound, joints are tight, there is little or no leakage and little attention is required from the cooper. On the other hand, it may be classified as having good organoleptic qualities relative to the production of a particular flavour in mature whisky. The two properties do not, however, necessarily go together. The traditional attitude has been largely to emphasise the economics of the cask as a cost-effective container and to assume that the desired flavour could be obtained by judicious blending. In the absence of any clear understanding of the production and control of whisky flavour there was clearly no alternative. However, in recent years, advances in analytical techniques relative to maturation mechanisms and flavour chemistry have led to a position where a greater degree of control can be exercised in the production and treatment of whisky casks and therefore in the control of whisky flavour.

In the maturation of Scotch whisky, casks are used, repaired, and reused indefinitely until the cask is no longer a sound and viable container or until it has lost all its 'virtue'. Virtue is a rather vague term but in essence a cask with virtue will give a mature whisky while one without will not. The very nature of this process means that the vast majority of the casks in a whisky warehouse at any one time are refill casks. That is, they have been refilled an unknown number of times, they have had various degrees of repair work carried out on them, they produce variable levels of wood extract relative to time, and consequently they have various degrees of virtue relative to the production of flavour in the maturing whisky. Normal practice is to average out these variables in the make-up of the blended whisky by blending various nominal wood types and ages.

Apart from nominal refill casks, the Scotch whisky industry also uses casks, of various sizes, types and previous history, which are on their first fill to either Scotch grain or malt whisky. Four principal categories can be readily discerned (Fig. 9.1), and these in turn may be associated with three geographical regions of origin (UK, USA and Spain).

In the first category are 558 litre (123 imp. gal) American oak puncheons, made and charred in the UK mainly between

Puncheon	IFWT	ASB	Butt
558 litres AO 1959–72	254 litres AO 1960–78	191 litres AO 1965–70	500 litres AO/SO
ex UK	ex UK/USA	ex USA	ex SPAIN
New wood charred/WT ↓ Grain ↓ Refill M/G	Bourbon (ASB shooks) ↓ IFWT ↓ Grain ↓ Refill M/G	Grain ↓ IFWT ↓ M/G	1950–68 HE (SO) 1965–74 Gomez (AO) treated 1974–85 Wine seasoning (AO/SO) Malt refill WT/M

M = malt spirit
G = grain spirit
HE = home-emptied
AO = American oak
SO = Spanish oak
ASB = American standard barrel
IFWT = first fill WT dumphogshead
WT = wine-treated

Fig. 9.1 Examples of some of the cask types used in the Scotch whisky industry between 1950–85.

1959–74, used for grain spirit on their first fill, and then subsequently filled with either grain or malt spirits. The second and third categories (IFWT and ASB) are interrelated in that the staves in both cases had one fill (at least four years) with bourbon prior to the first filling with Scotch whisky. The American standard barrels (ASB) were received as standing barrels and used untreated. They would normally be increased in capacity to the larger dumphogshead size on refill by introducing additional staves and new American oak ends. An alternative method was to purchase the ex-bourbon staves only as cask bundles or shooks. These were then made up to 254 litres (56 imp. gal) dumphogsheads in a similar manner by the addition of new American oak ends. After wine treatment these casks would be designated first fill wine-treated (IFWT). In both the ASB and IFWT categories a first fill with grain spirit would be followed in all subsequent fills by either malt or grain spirit. In the fourth principal cask category (and the third by geographical origin) are all the casks which have been purchased, either directly or indirectly, from the Spanish sherry industry. These casks have been mainly nominal 500 litre (110 imp. gal) butts together with relatively small numbers of sherry hogsheads and puncheons. Most of the casks received from Spain between 1950 and 1985 may be classified in subsections as either home-emptied, wine-seasoned, or ammonia-steam-treated, and each category is discussed under the section on wood treatment.

All first fill casks used in the Scotch whisky industry can therefore be classified in terms of physical parameters, wood type and wood treatment. Some examples of the composition of a grain spirit and a Lowland malt spirit after eight years' maturation in various cask types are shown in Table 9.1.

PHYSICAL PARAMETERS

Little can be said regarding any correlations between the physical parameters of cask construction and the production of flavour in the whisky during maturation. In general, there appear to be more stringent quality control regulations on the supply of American white oak than on the supply of Spanish or European oak to the Scotch whisky industry. For example, heading for new ends would normally be supplied from America as quarter sawn, prime quality and flat dressed, and must comply with the regulations laid down by the Associated Cooperage Industries of America, Inc. (1958) which specify that 'staves, cut from straight-grained bolts, must be sawn so that a straight line, drawn at right angles across the thickness at the ends of a stave, must pass through not less than three lines of grain at any one place, but in no case shall the angle of the grain be more than 45 degrees'. Similar regulations do not appear to exist for Spanish cooperages and this apparent lack of quality control may have contributed significantly towards a preference, both within the cooperages and the Scotch whisky industry as a whole, for tight American oak cooperage. In either case, any physical leakage or excessive evaporation due to poor cooperage would not only be economically undesirable, but would also lead to atypical ethanol concentration changes relative to time, and hence to anomalous flavours.

The principal physical dimensions of five cask categories are shown in Table 9.2. Relative to flavour development the most important is the surface to volume ratio, which ranges from 100 cm^2 per litre in the ASB to 67.5 cm^2 per litre in a puncheon. Traditionally, given casks of the same wood type and wood history, the smaller casks with a higher surface to volume ratio

Table 9.1 Eight-year-old Lowland malt and grain spirits filled at 68.5% v/v ethanol.

Cask type	*Distillery type*	*pH*	*Colour* (units)*	*Solids (mg per 100 ml)*	*Acids (mg per litre)*	*Esters (mg per litre)*	*Aldehydes (mg per litre)*	*Tannins (mg per litre)*	*Sugars (mg per litre)*	*Galloyl esters (mg per litre)*
WTDH	Lowland malt	4.5	8	99	224	1030	151	155	555	29
RDH	Lowland malt	4.6	7	30	215	775	156	145	160	25
SB	Lowland malt	4.7	16	282	485	2680	227	520	2540	195
RB	Lowland malt	4.4	14	77	259	1025	216	355	445	130
RP	Grain	4.9	7	27	117	370	60	140	125	37
IFP	Grain	4.7	11	93	195	870	93	295	510	65

*% absorbance (430 nm, 2 mm)

WTDH wine-treated dumphogshead
SB sherry butt
RP refill puncheon
RDH refill dumphogshead
RB refill butt
IFP first fill puncheon

Table 9.2 Physical dimensions of the five main cask categories.

Cask type	*Volume (litres)*	*Height (m)*	*Maximum diameter (m)*	*Internal surface (m^2)*	*Surface : Volume ratio (cm^2 per litre)*
Butt	500	1.27	0.90	3.55	71
Dump-hogshead	254	0.86	0.72	2.25	88.5
ASB	191	0.88	0.63	1.90	100
Puncheon	558	1.13	0.96	3.78	67.5
Dump puncheon	463	0.99	0.95	3.34	72

would be expected to produce a higher extract and mature the whisky in a shorter period of time.

WOOD TYPE (BOTANICAL)

The traditional wood used for Scotch whisky cooperage is American white oak, which is a cooperage trade classification used in the United States for at least 10 botanical species, the principal being *Quercus alba* (Alexander, 1972; Singleton, 1974). Therefore all oak casks either purchased in the United States, or manufactured in Scotland from timber imported from the United States, will fall into the same general botanical category.

Sherry producers in Spain, on the other hand, adopt a selective procedure in that they use American oak for fino and amontillado sherries, and Spanish oak for oloroso sherry (Rickards, 1983). Imported 'sherry' casks from Spain could therefore be both a mixture of American white oak botanical species (mainly *Quercus alba*), and Spanish oak botanical species (*Quercus sessilis* and *Quercus robur*).

The selective use of cooperage timber for specific beverages and the corresponding influence on extract and flavour is well established. For example, armagnac is matured in Gascony oak casks, and cognac in Limousin oak. Rous and Anderson (1983) established regression curves for the phenols produced in white wine aged over a six-month period in French and American oak barrels. They concluded that 'there were strong sensory differences between French oak and American oak barrels, at least through to the third fill'. Similar maturation studies by Guymon and Crowell (1970, 1972) on six-year-old brandies, aged in either new American oak or new French Limousin oak, showed an increase in total tannin, colour and extract in the brandies matured in French Limousin oak, and an organoleptic preference for the associated aroma and taste. Of particular interest was the presence of the 'oak or woody lactone', β-methyl–γ-octalactone, in brandy aged in American oak and its virtual absence in brandies aged in French oak.

Similar associations should therefore exist between wood botanical species, and Scotch whisky analysis and flavour. Maclean (1977) found that the concentrations of the *cis*- and *trans*-isomers of the oak lactone (odour thresholds 0.5 and 0.07 mg per litre 23% v/v ethanol respectively) in both oak extracts and in mature whiskies were in the ratio of 1 : 15. American oak contained about 20 times as much of both the *cis*- and *trans*-isomers as did Spanish oak. Nine-year-old Scotch whisky from wine-treated dumphogsheads (American oak) contained 2.36 mg *trans* woody lactone per litre of whisky at cask strength. Similar analysis on blended whiskies ranged from 0.23–1.72 mg per litre 40% v/v spirit.

Other analyses of various congeners, and groups of congeners, relative to the botanical species of the wood are shown in Table 9.3. The influence of the botanical species of oak on the whisky analyses, although modified slightly by age and degree of wine-seasoning, is undoubtedly a major factor.

WOOD TREATMENT

Modification of the surface chemistry in the cask may be carried out in a number of ways in order to achieve the desired extraction, interaction and adsorption characteristics. Processes such as hydrolysis, ethanolysis and anaerobic pyrolysis may be accepted as part of the 'natural' production process (and possibly little understood), or, as has happened in recent years, the wood surface can be deliberately altered to produce a specific result.

Sherry treatment

The terminology 'sherry cask' or 'home-emptied sherry cask' should really be retained only for a 500 litre (30 arrobas) shipping or export butt which was manufactured in Spain (probably from

Table 9.3 Examples of analyses on whiskies from a selection of treated American and Spanish oak casks.

Wood treatment	*Age (years)*	*pH*	*Colour units**	*Total tannin (mg per litre)*	*Gallic acid (mg per litre)*	*Ellagic acid (mg per litre)*	*Total sugars (mg per litre)*
American oak							
Ammonia steam	6	5.7	7	195	4.7	7.8	500
ANFL18	5	4.5	15	215	11.0	19.8	915
ANOL10	5	5.0	8	150	5.0	16.8	770
AOFL7	8	4.5	17	220	8.8	18.6	630
ANOH10	5	4.5	9	165	6.2	15.4	655
ASFL	5	4.0	12	225	12.7	13.3	630
Spanish oak							
Shipping butt	6	4.3	30	600	19.2	24.0	2250
SOFH24	5	4.3	46	630	27.3	40.9	1985
SXFL3/4	8	4.3	51	720	36.8	55.4	1640
SXFL20	8	3.6	52	815	26.8	36.2	2230
SXFH18	5	4.0	46	650	24.0	34.2	1220
SWFL36	5	4.0	32	520	17.5	25.8	855

*% absorbance (430 nm, 2 mm)

Cask codes

1st letter	A	American oak
	S	Spanish oak
2nd letter	N	Treated with ammonia
	S	Sulphur dioxide treated (SO_2)
	O	Not treated with SO_2, ammonia or water
	X	Unspecified treatment
	W	Water washed only
3rd letter	F	Used for fermentation of must
	O	Not used for fermentation of must
4th letter	L	Low-strength fino (15%)
	H	High-strength shipping sherry (18%)
	D	Dulce or Pedro Ximenez
Followed by a number		Total number of months the cask has held fortified wine (not must)

Spanish oak), well-seasoned by use in the grape fermentation process, and then used to ship various types of fully matured sherries to the United Kingdom. Total contact time with the mature sherry could be of the order of 6–9 months. These traditional sherry butts were used in the Scotch whisky industry until about 1965. Thereafter, most of the sherry was shipped by bulk tankers and alternative measures were introduced both in Spain and in the United Kingdom to simulate the seasoning and sherry treatment processes. For example, ammonia and steam treatment (Jeffs, 1970), and various forms of wine-seasoning have been used. Examples of mature whisky analyses from some of these casks are also shown in Table 9.3.

Steam and ammonia treatment

This was a treatment developed by Señor Gomez of M M Gomez, SA, in Spain and these casks were often referred to as 'Gomez treated' casks. In an attempt to simulate the seasoning achieved by fermentation these American oak casks were treated under pressure with steam and ammonia (1 litre 0.88 SG ammonium hydroxide with steam at 55 kPa (8 psig) for 60 min), which effectively stripped the tannins from the internal surface of the cask. Subsequent use in Scotch whisky maturation produced a very light-coloured, low-extract and slow-maturing whisky of a relatively high pH value (see Table 9.3).

Wine-seasoned wood

This was an attempt to simulate both cask seasoning and the hydrolytic effect of sherry contact. These casks would be manufactured in Spain, from either American or Spanish oak, specifically for the Scotch whisky industry. After use for grape fermentation and variable periods in contact with maturing sherry (Table 9.3), these casks would be shipped empty to the United Kingdom for use in the Scotch whisky industry.

Wine treatment

An apparatus for the testing and seasoning of casks was first introduced by W P Lowrie in 1888 using pressurised steam. This procedure apparently evolved into the present-day wine treatment process, which is to use a very sweet, dark sherry (Paxarette) under pressure. Paxarette is a brown sherry blended with solera olorosos and raya wines of all ages, sweetened with Pedro Ximenez and moscatel. It usually contains vino de color

(Jeffs, 1970) and is consequently extremely rich and sweet in flavour, and also dark in colour. Typical analyses on Paxarette sherry relative to other well-known types of sherry are shown in Table 9.4.

Wine treatment has been carried out in a variety of ways but a typical current cooperage procedure in the Scotch whisky industry is to add 500 ml of Paxarette per hogshead, or 1 litre of Paxarette per butt, pressurise at 48 kPa (7 psig) for 10 min and then disgorge any unabsorbed Paxarette. After bulking, the Paxarette is then recycled. Prior to the cask being refilled at the distillery it will be checked and any surplus Paxarette will again be drained off. Examples of analyses of Scotch whiskies matured in wine-treated casks are presented in Table 9.1.

Most of the changes in analytical composition created by wine treatment of the cask (for example, the increase in total sugars, total esters, solids and colour) appear to be of very little significance relative to flavour and the true 'sherry shipping cask' flavour (as defined above) has never been reproduced by this treatment. Some caution should be exercised in both the *in situ* sampling of wine-treated casks and also in the interpretation of these whisky analyses. It has been demonstrated (Philp, 1966) that the high-gravity sugars and pigments from the Paxarette sherry form a layer on the bottom of the cask which may take up to two years to disperse uniformly in the spirit. During this period the analytical pattern may become distorted, and sampling technique therefore becomes important.

Thermal degradation

In the construction and subsequent treatment of oak casks in the United States, Spain and Scotland, thermal degradation of the wood has traditionally taken place to various degrees, either as firing during the raising of the cask, or by mild or heavy charring after raising. Firing or mild charring applied mainly to the casks from Spain and heavy charring to both the imported bourbon staves from the United States and the new American oak puncheons manufactured in Scotland between 1959–74.

Relative to whisky flavour development the benefits from thermal degradation of the wood are threefold:

(a) A layer of active carbon is produced which can remove 'undesirable' flavour elements during maturation.

Table 9.4 Comparative analyses on Paxarette sherry and other types of sherry.

*Sherry types**	*pH*	*Colour units†*	*Acids (g per litre)*	*Esters (g per litre)*	*Aldehydes (g per litre)*	*Phenols (g per litre)*	*Tannins (g per litre)*	*Sugars (g per litre)*
Fino	3.3	10.1	3.27	3.47	0.30	0.02	0.65	0.63
Bone dry	3.3	13.1	3.48	3.88	0.30	0.02	0.58	1.74
Amontillado	3.2	20.5	4.00	24.00	0.25	0.02	0.56	29.00
Rich golden	3.4	58.8	4.00	65.00	0.37	0.03	0.58	70.00
Oloroso	3.6	66.9	3.20	78.00	0.37	0.02	0.83	115.00
Rich brown	3.5	97.8	4.20	117.00	0.80	0.03	1.34	136.00
Paxarette	3.7	>99.0	22.00	278.00	4.90	0.05	4.70	375.00

*All at nominal 16% v/v.
†% absorbance (430 nm, 2 mm).

(b) The oak lignin is anaerobically degraded in the layer immediately under the charcoal and flavour components such as vanillin are released and subsequently extracted into the whisky.
(c) The total wood extract, colour and phenols in the maturing whisky are increased, and by a series of oxidative interactive reactions (Wildenradt and Singleton, 1974) flavour components can be produced.

Analyses on maturing whiskies relative to these subtractive, additive and interactive flavour mechanisms are shown in Tables 9.5 and 9.6. The subtractive flavour production mechanism is illustrated both relative to maturation time and cask treatment (Table 9.5). Similarly, the additive and interactive flavour production mechanisms are illustrated in Table 9.6 relative to the treatment of nominal refill dumphogsheads. The additive and interactive congeners derived from the wood also increased relative to maturation time in every case.

In summary, therefore, it is apparent from the whisky analyses presented in Tables 9.3, 9.5 and 9.6 that the selection of the oak botanical species and the thermal degradation of the oak are both significant aspects of cask quality control relative to the production of flavour in Scotch whisky. Wine seasoning demonstrated a certain degree of tannin hydrolysis in the American oak, which increased with sherry contact time. The tannin in the Spanish oak, on the other hand, was apparently stripped by excessive contact with sherry, which therefore reduced the tannin available to the whisky. In both cases, the influence on the whisky analysis was relatively small compared with the effect of change in botanical species. Other treatments, such as by steam and ammonia, had little effect on the rate or quantity of cask extractives, and appear to contribute little towards whisky flavour development.

Warehouse conditions

Until 25 years ago Scotch whisky was traditionally matured in stone-built, single- or multi-storey warehouses, which were usually located beside the distillery. The single-storey warehouses had cinder floors and the multi-storey floors were of wooden construction. The casks were stored in 'stows', usually

Table 9.5 Analyses over a 12-month period of grain spirit in treated second fill and nominal refill dumphogsheads.

Cask history	*No.*	*Cask treatment*	*μg per litre (mean analysis)*					
			Dimethyl disulphide, after months:			*Dimethyl trisulphide, after months*		
			0	*6*	*12*	*0*	*6*	*12*
Refill	10	None	134	125	118	3.4	2.9	2.7
	10	Scraped	134	126	132	3.4	3.1	3.0
	10	Charred	134	77	41	3.4	2.1	1.4
	10	Scraped and charred	134	82	45	3.4	2.8	1.9
Second fill	10	None	134	118	119	3.4	2.9	2.5
	10	Scraped and charred	134	69	36	3.4	2.4	1.4

Table 9.6 Analyses from 12-month-old grain spirits from both treated second fill dumphogsheads and treated refill dumphogsheads.

			mg per litre (mean analysis)					
Casks		*Cask treatment*	***Additive***			***Interactive***		
Type	*No.*		*Syringic acid*	*Vanillin*	*Syringal-dehyde*	*Total tannin*	*Gallic acid*	*Ellagic acid*
Refill	10	None	0.3	0.2	0.5	47	0.7	1.6
	10	Scraped	0.2	0.2	0.4	32	0.8	0.9
	10	Charred	0.3	0.3	0.8	55	0.8	1.3
	10	Scraped and charred	0.5	0.5	2.3	118	1.0	3.5
Second fill	10	None	0.3	0.3	0.6	44	1.1	2.4
	10	Scraped and charred	0.5	0.6	2.5	98	1.2	3.5

two or three high, which meant that the casks were sitting on top of one another with wooden runners in between each layer of casks. Very damp conditions were often created on the ground floor of sloping sites and these were reputed to produce the best quality whiskies, especially if matured in sherry butts.

Expansion in whisky production meant that warehouse accommodation at the distilleries was insufficient, and so large, centralised units were built. The basic construction was brick walls with insulated aluminium roofs. Steel racking with wooden runners was installed to allow casks to be stored up to 10 high, depending on size. These warehouses were usually constructed in adjoining units of three, with each section containing as many as 11 500 butts or 23 000 hogsheads.

NORMAL CLIMATIC CONDITIONS

Liebmann and Rosenblatt (1943) and Liebmann and Scherl (1949) studied changes in acidity, esters, aldehydes, furfural, fusel oil, solids, colour, tannins and pH in American whiskey and developed hyperbolic regression curves for these analytical parameters relative to maturation time. Valaer and Frazier (1936) carried out similar studies, using the same analytical parameters, and attempted to correlate these analyses with warehouse construction (brick, concrete, wood or metal) and climatic conditions within the warehouse (hot-air heating, steam heating and no heating). Although the cask is serving primarily as a container, it is also acting as a semi-permeable membrane which permits the passage of alcohol and water vapours from the interior of the cask to the outside. This phenomenon is an integral part of maturation, it is relevant to the differences between warehouse conditions particularly as commonly found in the USA and in Scotland, and it helps ultimately to extend the differences between American whiskies and Scotch. Climatic conditions in the United States (relatively hot and dry) encourage a preferential loss of water vapour relative to ethanol and hence the ethanolic strength of the spirit increases during maturation. In contrast, under the relatively cool and humid climatic conditions found in Scotland the concentration of ethanol diminishes relative to time. These changes in ethanol/water ratio affect the polarity of the spirit as a solvent and reaction medium, and therefore exert different influences on the composition of the wood extract and the production of flavour components.

Duncan and Philp (1966b) studied the climatic conditions in four types of Scottish warehouse and in 10 locations within these warehouses (Table 9.7). Over a period of three years, the mean temperatures in the different warehouses were similar and ranged from 7.4°C to 9.7°C. The annual temperature fluctuation ranged from 9°C in the most stable site to 20°C in the least stable site. Daily temperature fluctuations of up to 15°C were recorded. The mean humidity in the various sites ranged from 70–80% throughout the year. Sites closest to the warehouse roofs were found to be least stable in temperature and also the driest, whereas those near to the ground were most stable in temperature and the wettest. Single-storey sites, on the other hand, tended to be both unstable in temperatures and relatively wet.

Table 9.7 Scottish warehouse locations used for climatic studies.

Warehouse type		*Site location*	
		Centre section:	
1	Modern, three-section, racked warehouses; brick-built; aluminium insulated roof.	1	bottom rack
		2	middle rack
		3	top rack (6th)
		Outside section:	
		4	bottom rack
		5	middle rack
		6	top rack (6th)
2	Old multi-storey, slate-roofed, stone-walled, ash (ground) and wooden floors.	7	ground floor
		8	top floor
3	Single-storey, centre section, brick, ash floor, insulated aluminium roof.	9	bottom stow
4	Single-storey, stone-walled, slate roof, ash floor.	10	bottom stow

In a subsequent maturation trial, grain whisky, three Highland malt whiskies, an Islay whisky and a Lowland malt whisky were stored in five different cask types on each of the 10 site locations. Analytical data, produced over a four-year maturation period, using the methods of Duncan and Philp (1966a), showed that the distillery was the dominant factor in determining the concentration of volatile congeners, and that few of the volatile components measured changed significantly during maturation. Changes in acetal and ethyl acetate concentrations were significant organoleptically and were related to cask type. No effect due to warehouse site location was observed in any of the volatile components measured. Cask type, on the other hand, was the dominant factor in determining the amounts of non-volatile components present in the matured whisky. Colour, solids, acids, esters and sugars were all found to be very much influenced by wine treatment of the cask, whereas low pH and high total acids were associated with sherry butts. The warehouse site and location had some influence on the volume and strength of the whisky in the cask but this was not sufficient to affect any of the other measured components. An organoleptic assessment of the four-year-old spirits by an expert nosing panel was unable to detect any influence of the warehouse site in which the casks were stored on the character of the matured whisky. It was concluded, therefore, that the rate of maturation was least affected by the warehouse climatic conditions, and was governed principally by the nature of the cask (especially sherry casks relative to refill casks) and the distillery from which the distillate was derived.

CONTROLLED CLIMATIC CONDITIONS

A further warehouse trial by Duncan and Philp (1966b) used 117 casks, of three different types, filled with the same grain whisky, and 234 casks, of three different types, filled with one Highland and one Lowland malt whisky. All the casks were housed in a series of 13 controlled chambers, which could be maintained at constant temperature and humidity, or programmed as required (Table 9.8).

After three years' maturation under controlled climatic conditions the overall indications were, as found in the previous experiment under normal warehouse conditions, that only the non-volatile components derived from the wood were significantly influenced by temperature, cask type and to a lesser extent by humidity.

Table 9.8 Temperature and humidity conditions set in the various controlled chambers.

Constant conditions

Storage unit no.	*Room no.*	*Temperature (°C)*	*Humidity (%RH)*
1	4A	17	73
2	1A	14	61
3	1B	14	85
4	2A	9	56
5	2B	9	90
6	3A	3	61
7	3B	3	85
8	4B	1	73
9	6A	9	73
10	6B	9	73

Varying conditions

Storage unit no.	*Daily temperature cycle (°C)*	*Duration (weeks per annum)*	*Daily humidity cycle (%RH)*	*Duration (weeks per annum)*
11	1-2-3	10	95-85-75	10
	4-6-8	10	83-73-63	10
	11-13-14	7	83-73-63	7
	15-17-19	8	68-58-48	8
	11-13-14	7	83-73-63	7
	4-6-8	10	83-73-63	10
12	9	permanent	95-85-75	10
			83-73-63	17
			65-58-48	8
			83-73-63	17
13	1-2-3	10	73	permanent
	4-6-8	10		
	11-13-14	7		
	15-17-19	8		
	11-13-14	7		
	4-6-8	10		

The volatile components measured (Duncan and Philp, 1966a) were largely unaffected by climatic conditions, with the exception of acetal and iso-valeraldehyde, which both showed changes in concentration apparently related to the storage conditions.

The non-volatile components and component groups (pH, colour, solids, total acids, total esters, total aldehydes, total phenols, total tannins and total sugars) generally showed significant cask and warehouse effects and a correlation with the temperature of storage. Humidity, on the other hand had little effect on the concentration of these wood-related components. Obviously, humidity must first induce a change in spirit strength by preferential evaporation before it can affect the concentration of any individual component in the whisky. Since, however, the range of spirit strengths observed after three years' maturation varied by only 1.74% v/v, it is not surprising that at this point there was no correlation between the concentration of the congeners measured and the humidity in the controlled chambers. Interactions of effects were not common.

These general observations were consistent with the hypothesis that changes in volatile component contents are largely due to chemical reactions and that even over a period of years a difference of 15°C in storage temperature can have only a very limited effect. Changes in non-volatile or cask-extractive components, in contrast, are largely physical and depend upon solution or extraction of the material from the cask timber. These will be much more temperature-dependent operations. They would also be more likely to be influenced by the strength of the whisky in the cask, which, in turn, would be influenced by the storage conditions.

After three years' maturation under controlled climatic conditions of variable temperature and humidity (Table 9.8) it was observed that the strength was influenced by whether the climatic conditions were variable or constant but only to the extent of 0.87% v/v ethanol. No other component quantity measured was consistently influenced by the variable warehouse conditions used in this part of the experiment. It was concluded, therefore, that seasonal variations in warehouse humidity and temperature were less significant relative to ethanol loss and flavour development than the annual mean temperature and the annual mean relative humidity.

Similar analytical patterns were detected after six and eight years' maturation of the same casks under the same controlled climatic conditions (Philp *et al.*, 1975). Humidity influenced the volume and strength of the whisky and appeared to have little influence on the other measured components of the whiskies. Temperature influenced not only the volume and strength of the whisky but also the pH, ethanol content, colour, total solids, total acids, total aldehydes, total phenols, total tannins and total sugars present in the mature whisky.

Multiple regression analyses (time, temperature and humidity) were carried out on all the analytical data collected over the eight-year maturation period. Initially, three linear regression equations were employed:

$$\log Y = c + m \log t \tag{1}$$
$$\log Y = c + mt \tag{2}$$
$$\log Y = c + mt + nt^2 \tag{3}$$

where Y represents the mean amount of component present in a group of samples at time t, and c, m and n are constants.

In addition, the following equation was also developed:

$$\log Y = a + bT + cH + dT^2 + eH^2 + fTH \tag{4}$$

where T and H represent temperature and relative humidity, respectively, a, b, c, d, e and f are constants, and Y is the mean concentration of each component during maturation. It was found that the term in T was always significant, whereas of the others, T^2, H and TH terms were only occasionally significant.

The final stage in the equation development was to introduce a logarithmic function which would account for the varying rate of concentration with time in the first one to two years (Liebmann and Rosenblatt, 1943; Reazin, 1983), before the rate of increase becomes constant. These considerations led to the formulation of a combined multiple regression equation:

$$Y = a + b\log_e M + cMd^{T} + fMH \tag{5}$$

where Y represents the mean concentration of the component, M represents the length of storage time (in months), and T and H represent the temperature and relative humidity of storage, respectively, with a, b, c, d and f as constants. Further terms in M^2, T^2 and TH were originally incorporated but were not found to make a significant contribution. Using this equation, multiple regression analyses were carried out on each analytical parameter, and coefficients for the regression equation were calculated which gave the best fit to the overall analytical results and also the analyses from each of the three cask types independently (refill butts, wine-treated dumphogsheads and refill dumphogsheads) (see Table 9.9).

Also included in this table is a figure of index of determination

Table 9.9 Regression equation constants for 13 measured parameters from three cask types.

	Cask	*a*	*b*	*c*	*d*	*f*	*ID*
Ullage	O	6.95E+1	0.00E+0	−9.00E−2	1.04E+0	1.07E−3	9.69E+1
	RB	1.10E+2	0.00E+0	−1.24E−1	1.05E+0	7.23E−4	9.44E+1
	WTDH	5.50E+1	0.00E+0	−7.42E−2	1.04E+0	8.63E−4	9.15E+1
	RDH	5.56E+1	0.00E+0	−7.84E−2	1.04E+0	1.34E−3	9.61E+1
Strength	O	2.03E+1	−4.33E−1	5.50E−3	1.19E+0	−2.64E−3	8.16E+1
	RB	1.99E+1	−3.02E−1	2.60E−3	1.23E+0	−1.81E−3	8.04E+1
	WDTH	2.05E+1	−4.84E−1	5.50E−3	1.19E+0	−3.08E−3	8.06E+1
	RDH	2.05E+1	−5.11E−1	8.30E×3	1.17E+0	−3.01E−3	8.29E+1
% content	O	9.93E+1	0.00E+0	−1.27E−1	1.05E+0	1.65E−3	9.75E+1
	RB	9.95E+1	0.00E+0	−1.14E−1	1.04E+0	9.30E−4	9.38E+1
	WTDH	9.93E+1	0.00E+0	−1.34E−1	1.05E+0	2.05E−3	9.78E+1
	RDH	9.97E+1	−2.57E−1	−1.29E−1	1.05E+0	1.98E−3	9.85E+1
% ethanol content	O	9.92E+1	−3.08E−1	−1.23E−1	1.03E+0	0.00E+0	9.57E+1
	RB	9.93E+1	−3.37E−1	−1.07E−1	1.04E+0	0.00E+0	9.22E+1
	WTDH	9.85E+1	0.00E+0	−1.38E−1	1.04E+0	0.00E+0	9.61E+1
	RDH	9.91E+1	−2.71E−1	−1.38E−1	1.03E+0	0.00E+0	9.70E+1

O: overall; RB: refill butt; WTDH: wine-treated dumphogsheads; RDH: refill dumphosgheads.

Table 9.9 (*cont.*)

	Cask	*a*	*b*	*c*	*d*	*f*	*ID*
pH	O	4.97E+0	0.00E+0	−2.50E−3	1.05E+0	0.00E+0	5.47E+1
	RB	4.97E+0	0.00E+0	−3.10E−3	1.05E+0	0.00E+0	3.49E+1
	WTDH	4.92E+0	0.00E+0	−1.80E−3	1.06E+0	0.00E+0	5.10E+1
	RDH	5.04E+0	0.00E+0	−2.60E−3	1.05E+0	0.00E+0	6.18E+1
Colour	O	4.71E+0	3.61E+0	1.83E−1	1.08E+0	0.00E+0	9.29E+1
	RB	6.77E+0	8.29E+0	2.23E−1	1.07E+0	0.00E+0	7.88E+1
	WTDH	9.09E+0	2.05E+0	1.63E−1	1.09E+0	0.00E+0	9.12E+1
	RDH	2.30E+0	2.26E+0	1.64E−1	1.08E+0	0.00E+0	9.44E+1
Total solids	O	7.73E+0	3.19E+0	1.60E−1	1.08E+0	0.00E+0	9.35E+1
	RB	6.21E+0	5.64E+0	4.20E−1	1.05E+0	0.00E+0	8.77E+1
	WTDH	3.42E+1	0.00E+0	7.90E−2	1.13E+0	0.00E+0	5.21E+1
	RDH	4.67E+0	1.14E+0	1.00E−1	1.09E+0	0.00E+0	8.81E+1
Total acids	O	8.34E+1	2.09E+1	6.86E−1	1.07E+0	0.00E+0	9.52E+1
	RB	9.07E+1	2.53E+1	1.02E+0	1.06E+0	0.00E+0	9.14E+1
	WTDH	8.64E+1	1.81E+1	5.96E−1	1.09E+0	0.00E+0	9.56E+1
	RDH	8.35E+1	1.31E+1	8.37E−1	1.07E+0	0.00E+0	9.68E+1
Total esters	O	6.53E+2	−2.43E+1	2.99E+0	1.03E+0	0.00E+0	8.84E+1
	RB	5.90E+2	0.00E+0	3.93E+0	1.03E+0	0.00E+0	8.39E+1
	WTDH	7.85E+2	−2.81E+1	2.89E+0	1.04E+0	0.00E+0	6.25E+1
	RDH	5.62E+2	−2.40E+1	1.99E+0	1.04E+0	0.00E+0	7.09E+1

O: overall; RB: refill butt; WTDH: wine-treated dumphogsheads; RDH: refill dumphogsheads.

Table 9.9 (*cont.*)

	Cask	*a*	*b*	*c*	*d*	*f*	*ID*
Total aldehydes	O	8.57E+1	1.01E+1	4.85E−1	1.05E+0	0.00E+0	9.19E+1
	RB	9.50E+1	6.47E+0	9.33E−1	1.04E+0	0.00E+0	8.57E+1
	WTDH	8.58E+1	1.20E+1	3.32E−1	1.07E+0	0.00E+0	8.57E+1
	RDH	8.20E+1	1.19E+1	2.51E−1	1.06E+0	0.00E+0	8.70E+1
Total phenols	O	1.68E+0	4.81E−1	5.75E−2	1.04E+0	0.00E+0	9.82E+1
	RB	1.88E+0	1.27E+0	4.52E−2	1.04E+0	0.00E+0	8.98E+1
	WTDH	1.68E+0	2.86E−1	5.82E−2	1.05E+0	0.00E+0	9.72E+1
	RDH	2.05E+0	0.00E+0	6.70E−2	1.04E+0	0.00E+0	9.78E+1
Total tannins	O	1.67E+1	−3.32E+0	1.70E+0	1.04E+0	0.00E+0	9.70E+1
	RB	2.67E+1	0.00E+0	3.38E+0	1.03E+0	0.00E+0	8.75E+1
	WTDH	1.53E+1	−3.26E+0	1.29E+0	1.04E+0	0.00E+0	9.65E+1
	RDH	1.33E+1	−4.36E+0	1.29E+0	1.04E+0	0.00E+0	9.61E+1
Total sugars	O	7.57E+1	2.39E+1	1.15E+0	1.07E+0	0.00E+0	8.88E+1
	RB	5.06E+1	4.31E+1	2.41E+0	1.04E+0	0.00E+0	7.98E+1
	WTDH	3.44E+2	0.00E+0	1.33E+0	1.08E+0	0.00E+0	4.63E+1
	RDH	3.57E+1	6.19E+0	7.24E−1	1.06E+0	0.00E+0	9.39E+1

O: overall; RB: refill butt; WTDH: wine-treated dumphogsheads; RDH: refill dumphogsheads.

(ID), which can vary from 0 for no fit to 100% for a perfect fit. In the present instance, with 90 results involved in setting up each equation, an index of approximately 12% indicates a significance level of 1%. It can be seen that all the ID values presented in Table 9.9 are very highly significant. Typical examples of hyperbolic regression curves for colour, total acids, total aldehydes and total tannin for whiskies from three cask types and three storage temperatures are presented in Figs. 9.2 and 9.3. Additional regression curves (Fig. 9.4) show the influence of various temperature and humidity conditions on the ullage and strength of the spirit over a six-year period. In each case the overall mean

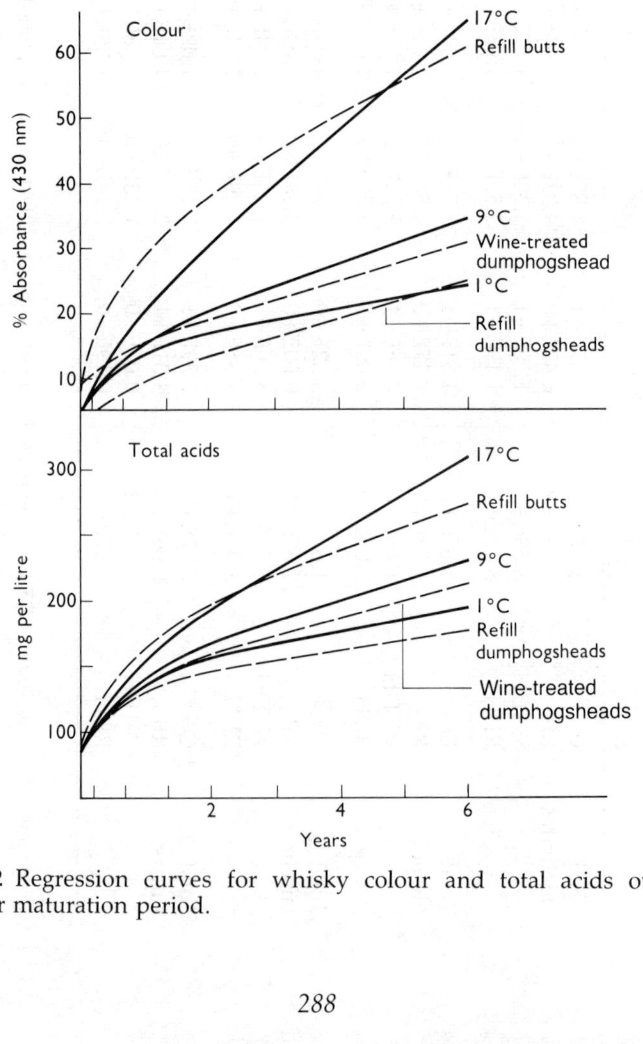

Fig. 9.2 Regression curves for whisky colour and total acids over a six-year maturation period.

change of each variable is expressed relative to time for each of the cask types, the storage temperature and the temperature humidity combinations, irrespective of any other parameters.

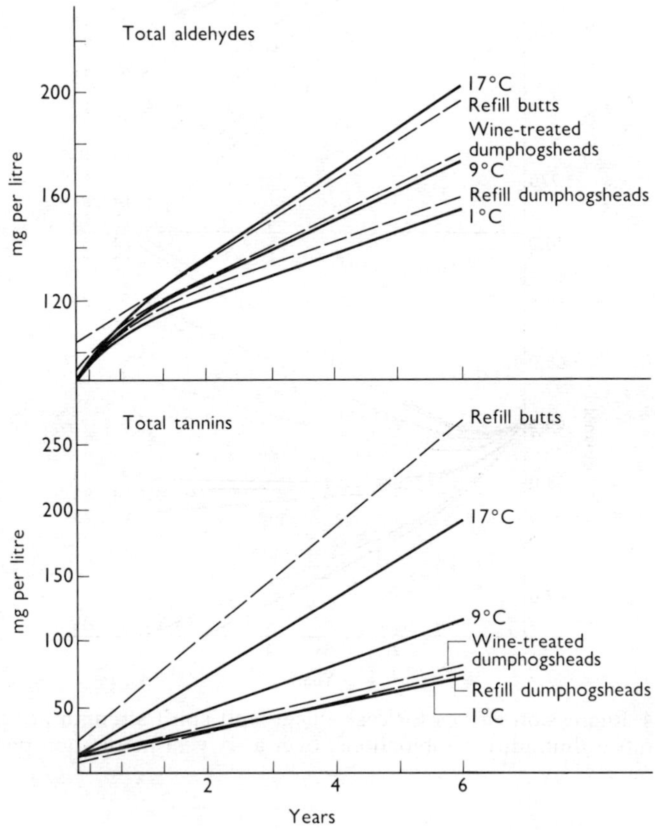

Fig. 9.3 Regression curves for total aldehydes and total tannins in whisky over a six-year maturation period.

It was concluded that the changes in spirit composition which took place during eight years' maturation under conditions of controlled temperature and humidity could be summarised as:

1. Malt and grain spirits showed similar analytical patterns relative to time if allowance was made for the amounts of congeners present in the original new make spirit.

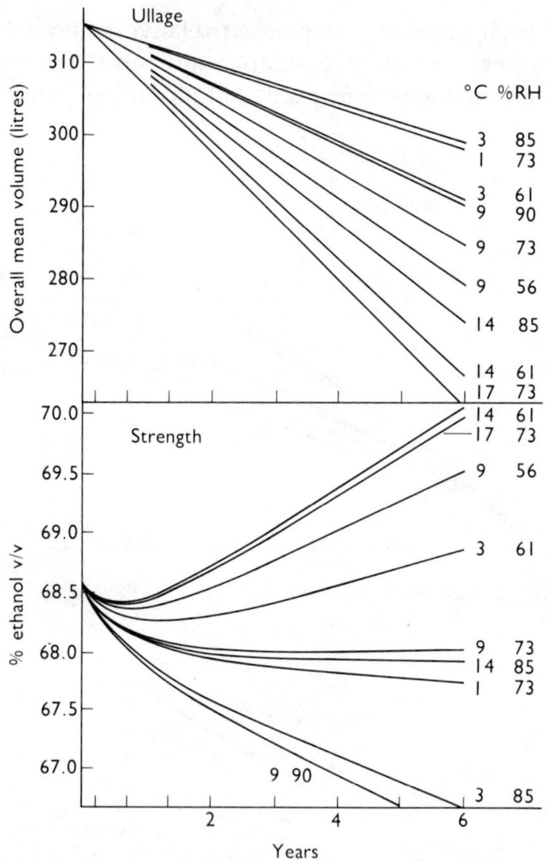

Fig. 9.4 Regression curves for cask ullage and spirit strength at various temperature/humidity combinations, over a six-year maturation period.

2. The concentration of wood extractives in both the malt and grain spirits was dependent on the temperature in the warehouse chambers and on the storage time.
3. Humidity of storage only affected significantly the volume and the strength of the spirit in the cask. On average, the strength diverged by 5.8% v/v for both malt and grain spirits stored for eight years at 56% and 90% relative humidity.
4. Ethanol loss was dependent on the storage temperature. After eight years maturation approximately 11% was lost at 3°C, 15% at 9°C and 22% at 17°C.

FLAVOUR DEVELOPMENT RELATIVE TO WAREHOUSE CONDITIONS

Nine eight-year-old samples from each of the malt and grain distilleries, representing three cask types and three sets of storage conditions, were evaluated to determine to what extent the various conditions of temperature and humidity had influenced flavour (Woolley, 1975). The flavours of the 27 whiskies were assessed by nose and taste by a panel of 63 members using a questionnaire containing 18 word-pair descriptions. Analyses of variance were carried out on the flavour scores for each word-pair scale for the nine whiskies in each distillery matured in the three types of cask under three sets of conditions. No consistency was seen regarding the effect of temperature and humidity, nor of cask type, on the flavour of the Highland and Lowland malt spirits as measured by six principal flavour attributes: activity, pleasantness, heaviness, cleanliness, sweetness and texture. However, for the grain spirits it was concluded that the whiskies matured at 9°C and 73% relative humidity were the most active and least pleasant. Furthermore, the grain spirits matured at the higher temperatures were less clean, but sweeter, whilst the smoother and more pleasant whiskies were matured at lower temperatures.

It is widely held that Scotch whiskies mature to the greatest perfection in cool, damp warehouses. In the study described this was not evident although it had been expected that systematic odour differences, related to the ambient conditions, would have occurred. The probable explanation is that these spirits had been put in bond at 68.5% v/v, a filling strength being commonly considered at the time for both malt and grain spirits. Even after eight years the lowest strength recorded was 64.8% v/v. In the context of Scotch whisky higher strength will give a slower rate of maturation and flavour development. Thus the distiller can exert some control over maturation rate by his choice of filling strength. In general the most favoured strength is 63.4% v/v.

Summary and conclusions

Published literature on the influence of cask quality and warehouse conditions on the development of whisky flavour has been reviewed. Additional material based on current cooperage and warehousing practice within the Distillers Company PLC, and also on cask experiments carried out at Glenochil Research Station, are also included. These studies have shown that cask history plays an important part in the development of Scotch whisky flavour. In particular, the botanical species of oak used and thermal degradation by charring were shown to be of special importance. Warehouse conditions, on the whole, played a less significant role, although it was established that the mean warehouse temperature was important.

The interpretation of any analytical results relative to whisky flavour and the control of whisky flavour by cask quality and warehouse environment, clearly assumes an acceptable definition of whisky flavour in analytical terms, and that the analyses used in the maturation experiment were relevant to any one or a number of whisky flavours. The analyses applied by Valaer and Frazier (1936), Liebmann and Rosenblatt (1943), Liebmann and Scherl (1949) and Duncan and Philp (1966a) were the best available at the time, but in retrospect it could be claimed that many of these analyses were of no great significance relative to whisky flavour. Over the past 20 years our knowledge of flavour chemistry, and maturation mechanisms relative to the production of whisky flavour, has become much more refined, and some of these warehouse experiments should be repeated with analyses for components in other analytical categories, such as lignin degradation products (Reazin, 1983), unsaturated aliphatic carbonyls (Nishimura and Masuda, 1985), sulphur-containing compounds (Leppänen, 1983) and nitrogen-containing compounds (Viro, 1984). Only then will all the maturation mechanisms and physical parameters related to the production of Scotch whisky flavour be fully understood and controlled.

Acknowledgements

The author is grateful to the Directors of the Distillers Company PLC for permission to publish this review. I also gratefully acknowledge contributions from my colleagues and former colleagues (Dr J L Beveridge, Mr A L Cunningham, Mr J Drummond, Dr I Maclean and Dr D Woolley) at Glenochil Research Station, Menstrie, Clackmannanshire, Scotland.

References

ALEXANDER G C 1972 *Characteristics of oak timbers and their implications in cask manufacture*. PhD thesis, University College of North Wales (Bangor)

ASSOCIATED COOPERAGE INDUSTRIES OF AMERICA, INC. 1958 *Grade rules and specifications*

CANAWAY P R 1983 Sensory aspects of whisky maturation. In Piggott J R (ed) *Flavour of Distilled Beverages*. Ellis Horwood, Chichester

DUNCAN R E B, PHILP J M 1966a Methods for the analysis of Scotch whisky. *Journal of the Science of Food and Agriculture* **17**: 208–14

DUNCAN R E B, PHILP J M 1966b Studies of the maturation of whisky. I. Design of trials carried out under normal warehouse conditions and under conditions of controlled temperature and humidity. Internal Report No. 302, Glenochil Research Station, Menstrie

GUYMON J F, CROWELL E A 1970 Brandy aging (some comparisons of American and French oak cooperage). *Wines and Vines* **51**(1): 23–5

GUYMON J F, CROWELL E A 1972 GC-separated brandy components derived from French and American oaks. *American Journal of Enology and Viticulture* **23**(3): 114–20

JEFFS J 1970 *Sherry*. Faber & Faber, London

JOUNELA-ERIKSSON P 1978 The aroma composition of distilled beverages and the perceived aroma of whisky. In Charalambous G (ed) *Flavour of Foods and Beverages*. Academic Press, New York

LEPPÄNEN O, RONKAINEN P, DENSLOW J, LAAKSO R, LINDEMAN A, NYKÄNEN I 1983 Polysulphides and thiophenes in whisky. In Piggott J R (ed), *Flavour of Distilled Beverages*. Ellis Horwood, Chichester

LIEBMANN A J, ROSENBLATT M 1943 Changes in whisky while maturing. *Industrial and Engineering Chemistry* **35**(9): 994–1002

LIEBMANN A J, SCHERL B 1949 Changes in whisky while maturing. *Industrial and Engineering Chemistry* **41**(3): 534–43

LOWRIE W P 1888 Patent (Lond) No. 3769

MACLEAN I 1977 Personal communication

NISHIMURA K, MASUDA M 1985 Identification of some flavour characteristic compounds in alcoholic beverages. In Nykänen L, Lehtonen P (eds) *Flavour Research of Alcoholic Beverages*. Foundation for Biotechnical and Industrial Fermentation Research, Helsinki

PHILIP J M 1966 Preliminary observations on the colour of maturing whisky. *Internal Report No. 300*, Glenochil Research Station, Menstrie

PHILP J M, CUNNINGHAM A L, DRUMMOND J, WOOLLEY D 1975 Unpublished internal report

PIGGOTT J R, JARDINE S P 1979 Descriptive sensory analysis of whisky flavour. *Journal of the Institute of Brewing* **85**: 82–5

REAZIN G H 1983 Chemical analysis of whisky maturation. In Piggott J R (ed) *Flavour of Distilled Beverages*. Ellis Horwood, Chichester

RICKARDS P 1983 Scotch whisky cooperage. In Priest F G, Campbell I (eds) *Current Developments in Malting, Brewing and Distilling, Aviemore Conference 1982*. Institute of Brewing, London: 199–203

ROUS C, ANDERSON B 1983 Phenolic extraction curves for white wine aged in French and American oak barrels. *American Journal of Enology and Viticulture* **34**(4): 211–15

SHARP R 1983 Analytical techniques used in the study of whisky maturation. In Priest F G, Campbell I (eds) *Current Developments in Malting, Brewing and Distilling, Aviemore Conference 1982*. Institute of Brewing, London, 143–156

SINGLETON V L 1974 Some aspects of the wooden container as a factor in wine maturation. In Gould R F (ed) *Chemistry of Winemaking*. American Chemical Society Series No. 137

SWAN J S, HOWIE D, BURTLES S M, WILLIAMS A A, LEWIS M J 1981 Sensory and instrumental studies of Scotch whisky flavour. In Charalambous G (ed) *The Quality of Foods and Beverages*. Academic Press, New York, 201–224

VALAER P, FRAZIER W H 1936 Changes in whisky stored for four years. *Industrial and Engineering Chemistry* **23**(1): 92–105

VIRO M 1984 Heterocyclic nitrogen compounds in whisky and beer. *Chromatographia* **19**: 448–51

WILDENRADT H L, SINGLETON V L 1974 The production of aldehydes as a result of oxidation of polyphenolic compounds and its relation to wine aging. *American Journal of Enology and Viticulture* **25**(2): 119–26

WILLIAMS L I 1983 Supply of wood for cooperage with specific reference to Scotch whisky. In Priest F G, Campbell I (eds) *Current Developments in Malting, Brewing and Distilling, Aviemore Conference 1982*. Institute of Brewing, London, 193–198

WOOLLEY D 1973 Personal communication

WOOLLEY D 1975 Personal communication

CHAPTER 10

Blending and bottling

—— M BOOTH, W SHAW and L MORHALO ——

Blending

THE WHISKY BLEND

The blending side of whisky production is perhaps the one facet of the manufacturing process that, above all others, provides the distiller with an opportunity to make his product unique. To openly discuss one's blending practices would be an invitation to others to learn why a particular blend may enjoy more success than another. It could be said, therefore, that the 'whisky blend' is a distiller's 'hallmark', and has for good reason remained shrouded in secrecy. Whisky blending is an elusive topic and will probably continue to be so for years to come.

There are certain blending practices that fall within the category of general knowledge, which will be presented here. It must be emphasized that blending is such a personal and private practice that there will be distillers whose methods may not always fall within the boundaries we will draw around the general subject. For the sake of expediency, the discussions will be limited to the practice of whisky blending in five of the major whisky producing countries of the world: Scotland, Japan, Canada, Ireland and the United States.

BLENDED WHISKY AND THE BLENDER

Blended whisky is as the name suggests – a mixture. The blender is an individual or a small group of individuals, charged with the task of ensuring that a company's blended whiskies are consistent from one production batch to the next. He may also be assigned on occasion to create a totally new whisky blend, if there is a need for such a product in the market.

The mixture of ingredients that constitutes a blend is determined in part by the legal definition of whisky as written by the country of origin. The blender must work within the broad guidelines of the legal definition; given this, he is then at liberty to exercise his talents in the selection of components for his blend. In the creation of new blends, the blender may have an opportunity to display some creative talents. It will be his knowledge of available components, his experience in knowing how they work with one another in combination, and his skill in selecting the participants and their properties, that will spell success.

New blend creations, although important at the time, are not the major responsibility of a blender. Successful manufacturers of whisky have become so because of certain successful blends that have been established and renowned for their quality and consistency for many years. When these flagship brands are the issue, the blender's prime responsibility is to use his talents to ensure that these whiskies remain consistent despite minor or even significant variations in the components with which he has to work. His knowledge of blending components and his experience in their use are often called upon to guarantee blend consistency.

THE MAJOR WHISKY BLENDERS

Scotland

Blended Scotch whisky is what it is today because of Scotland itself – the land and the people. Only those who have attempted to duplicate the product, elsewhere in the world, can fully appreciate the accuracy of this statement. It is truly a product unique to Scotland.

Scotch whisky is a distinctive product of Scotland, manufactured in Scotland in compliance with the laws of the United Kingdom regulating the manufacture of Scotch whisky for consumption in the United Kingdom, and containing no distilled spirits less than three years old. Scotch whisky is commonly a blend of several whiskies, each of which satisfies the legal definition. The minimum age of three years is usually greatly exceeded by the blenders, as most whiskies are not sufficiently mature by this time to be at their best.

The Scottish blender has a large number of individual whiskies from which to select the components in his blend. There are two distinctly different whisky types used in the blend. Grain whiskies are produced primarily from maize. The maize starch is converted to fermentable sugar by malted barley utilised in fairly high proportions in the mash bill. The distillate is produced by continuous distillation using column stills. Grain whiskies are distilled at a high proof in order to create a light-bodied whisky distillate that is ideally suited after maturation to serve as the base whisky in a blend. The base whisky provides background to the blend and can tone down some harsher notes that might otherwise be present. There are at the time of writing 14 grain whisky distilleries in Scotland, the majority of which are located in the south, in close proximity to the major blending and bottling facilities.

The second type of whisky available to the blender is malt, produced entirely from malted barley and distilled in a double distillation process at a lower proof, utilising small, copper pot stills. These whiskies are used in various combinations and proportions and are the basis for the blender's unique flavouring component of his blend. There are at the time of writing more than one hundred malt whisky distilleries located throughout Scotland, with each distillery producing its own distinctive product, which may vary only slightly from year to year. The reasons for there being differences between the various malt whiskies have been debated for years. Although some individuals will expound with great confidence that it is the source of the water, the malt, the design of the stills or the microflora peculiar to a given distillery, the truth is that these are all theories and have not been substantiated.

The fact that each malt whisky is distinctive makes it difficult to categorise these whiskies into groups. However, a system of classifying malt whiskies by geographical area has been adopted. Blenders utilise this system when discussing or working with the malts. The malts have been grouped into Lowland, Cambelltown, Islay and Highland malts. The Highland malts, being the major category, have in turn been divided into three subcategories, based on their blending attributes. These are first, second and third class whiskies. Blenders are reluctant to generalise about the characteristics of the malts from each of the four regions. However, there are some broad comments that can be made about the malts within each region that will hold true. The

Islay malts are full-flavoured and tend to have smoky or peaty qualities. They are powerful and rich whiskies and their presence in a blend can have a significant impact on the mix. Islay malts are important blending whiskies, but because of their strength of flavour only small quantities are used in most blends.

Lowland malts, by comparison with the malts of the other regions, are rather difficult to describe due to the lack of distinctive characteristics. They are, however, important to the blender because they can be used in reasonably large quantities without imparting too much of a character-building influence on the blend. They are malts that give weight to a blend. Although they are not generally outstanding as 'singles', they are an essential component of a good blended whisky.

The Campbelltown malts are now only two in number, so I will avoid generalising about this minor category. The Highland malts are made in the Highlands of Scotland, which is the region north of the Highland line, stretching from Dundee in the east to Greenock on the west coat. In this group will be found most of the great malt whiskies that have gained solid reputations for their strength as singles and their blending value. The category harbours whiskies of a wide variety, with some commanding more demand than others. Because of the variety available in this large category, it is difficult to generalise about the group as a whole. It would be necessary to look at each whisky independently, attaching to them such commonly used adjectives in the trade as full, heavy, fragrant, mellow, fruity, robust, smoky, peaty, light, subtle, flowery etc. We will therefore simply say that the Highland malts offer the blender an exciting array of quality and diverse whiskies, and if selected and combined in the right proportions with other whiskies, they will give to his blend the personality to rise above the description of average.

A blended Scotch whisky will typically contain more grain whisky than malt. There are, however, no hard and fast rules as to the percentages of each used. There is a common misconception that blended whiskies having a higher percentage of malt whisky are superior to those with a lower percentage of malt. This is not necessarily true. A blend comprised of 50% malts selected poorly, matured in inferior wood and proportioned without skill, would certainly be an inferior blend. On the other hand, a blend having 75% grain whiskies and 25% malts can be of exceptional quality, given the right choice of malts proportioned skilfully and combined with fully matured grain whiskies. There are, however, some limitations on what a blender can accomplish without a reasonable proportion of quality malts in his blend. A blend of 95% grain and 5% malt would certainly not come up to the expectations of a good blend.

A good blend is only as successful as it is consistent. The blender knows the importance of consistent aroma, taste and overall quality to the viability of a key brand. In order to be assured of maintaining consistency, the blender will select several different malts to be included in the malt whisky portion of the blend. Again, there are no set rules as to the numbers used – some blenders might use as few as 20 in a blend and others might use as many as 50. The advantage in using a large number of individual malts in a blend is apparent if any given malt becomes unavailable. The impact of a loss is likely to be less extreme if there are 40 malts present rather than 20. There are, however, certain powerful malts that are influential enough in a blend to be obvious in their absence. It is in situations such as this that the blender must know the malt whiskies sufficiently well to substitute the missing component with a replacement that is suitably compatible with the rest. In the interest of blend consistency, he must know the whiskies that are interchangeable and where they can be secured in the required quantities.

Shortages of certain whiskies can occur for a number of reasons. Inaccurate forecasting of consumption is the most likely; losses due to unforseeable catastrophes such as fire are always a possibility. Deficient whisky stocks can often be remedied in emergency situations through the exchange of matured stocks with competitors, or by purchasing them from independent brokers on the open market. If neither option is possible, the blender is called upon to do some skilful juggling of the malt complement of a blend by utilising what he perceives to be suitable substitutes for the absentee musicians in his orchestra.

When blending formulations have been decided on, the company must rely on reasonably accurate sales forecasts in order to be assured of having sufficient stocks of properly aged whiskies on hand. Inaccurate forecasts can either result in whisky surpluses or deficiencies, neither of which is desirable. A certain number of the whiskies required will be pulled from the company's own stock of matured whiskies. The balance of whiskies in the blend will be secured from competing distilleries by either purchasing or exchanging them as fillings (new whiskies). The fillings are matured, as a rule, in the producer's warehouses and delivered to the user when they have reached the maturity level desired by the blender.

It is essential that large blending companies have adequate whisky stocks available for their blends when the market demands them. The company's own malt distilleries are operated today at a rate that will meet the perceived demand for the matured whisky. The perceived demand is the quantity required for their own blends as well as the forecast demand for their malt whisky by other blending houses. Distillery production forecasting for anticipated demands several years hence is a difficult task, to say the least.

New whiskies destined for a company's blend, whether their own production or whiskies purchased from other distilleries, are evaluated organoleptically in the blender's sample room. Blenders have their own particular styles when it comes to evaluating samples. They seem to take a great deal of pride in their chosen ritual of sample evaluation and rarely deviate from their acquired method. The sample is poured into a tulip-shaped snifter and water is added. The proportions of water and sample are personal preference. Some will look at a sample reduced to 30% alcohol while others prefer 20%. Some blenders will look at a sample at both dilutions. The usual practice for evaluating new or matured whiskies is to simply swirl the contents of the snifter and smell the volatile components that emanate from the sample. Another technique is to place the palm of the hand over the mouth of the snifter and vigorously shake snifter and contents. After this performance, the product adhering to the palm of the hand is rubbed against the other palm. The palms of the hands are subsequently smelled to detect residual whisky aroma. This latter technique is generally reserved for the evaluation of new whisky distillates rather than matured whiskies.

Prior to the commencement of a commercial blend, samples are taken from casks of all whiskies to be used in the blend. When the component whiskies are approved, the mixing of the whiskies can begin. The actual process of mixing the whiskies is a simple procedure. However, when large batches are involved, an error in judgement by the blender for a particular component whisky can be costly, if the finished blend is not as it should be.

Approved whiskies are delivered to the blending house and are drained from the casks, in the correct proportions, into passivated stainless steel troughs. The troughs convey the whiskies to a blending vat where they are thoroughly mixed with mechanical agitators and compressed air. The mix is then sampled and evaluated by the blender. At this point, the blender will suggest the addition of minimal bits and pieces of selected whiskies to round out the blend if the original mix is lacking certain properties when compared with a reference standard for the brand. When the blend is finally correct in the opinion of the blender, deproofing water is added to the blend to reduce the strength from that of a barrelled whisky to that required for bottling. Bottling strength is generally 40% ethanol but may vary slightly for export markets.

The quality of water used for blending must be impeccable. If it is not, a blender's efforts and the reputation of a brand can easily be destroyed. In Scotland, the natural water supplies are generally very soft and may be used for blending purposes without pre-treatment. However, some areas of the country do find it necessary to use demineralised water. Where mineral salts are present in large enough concentrations to react with whisky to form precipitates, cation and anion exchange resins are employed to remove them. The water may also be passed through a carbon bed as a supplemental treatment to demineralisation in order to remove any traces of organic odours that might be present. A blender's water supply is judiciously monitored by quality control personnel, and only after their approval is it utilised for deproofing.

A small quantity of food grade caramel is added as a final step, to ensure product colour consistency. Whisky caramel is organoleptically inert at the levels used and does not impart any aroma or taste to the blend. The final product is filtered prior to bottling. This subject will be addressed later for whiskies in general.

Japan

Blended Japanese whisky bears a resemblance to blended Scotch whisky in that it is generally a mixture of lighter distillates produced continuously in column stills, and heavier malt whiskies produced in copper pot stills.

Japanese whisky blends must, by law, contain a portion of grain whisky. The required portions are specified in the legal definitions for the three classes of blended whisky in Japan.

Premium whisky is designated 'Special Class' and contains not less than 30% grain whisky. 'First Class' whisky contains less than 27% grain whisky. The final category, 'Second Class', contains a lower grain whisky complement than 'First Class' but must contain not less than 10% grain whisky.

Grain whisky, by Japanese interpretation, is spirit produced from any type of grain, including malted barley. Blended whiskies containing higher grain content are taxed more heavily than blends with a lesser amount of grain whisky. 'Special Class' whiskies are therefore the most heavily taxed category. Japanese blenders may be encouraged to blend within the confines of a particular whisky class definition in order to gain a tax advantage. Japanese blended whisky quite often contains a proportion of imported malt whiskies that are combined with their domestic whiskies in the grain whisky complement of the blend. Since there are only a few malt whisky distilleries in Japan, and the exchanging of whiskies between competitors is not practised, it is not surprising that individual whiskies are imported to give the blender more flexibility in formulating his blends.

The benefits of having a varied selection of whiskies available to the blender were discussed earlier. The Japanese blender certainly appreciates the advantages of having a wide selection of whiskies available to him and would enjoy this luxury. However, in a country isolated geographically and with only a few malt distilleries, this is not easily accomplished. In an effort to increase the number of domestic malts available for blending, some Japanese distillers have attempted to produce different malt whisky types at a given distillery. Different strains of yeast have been employed to produce a desired characteristic in certain distillates. Mashing conditions, grain type, fermentation conditions and distillation methods may be modified slightly to create subtle differences between new whiskies produced at the same facility.

It is quite obvious that the challenges facing the Japanese blender differ greatly from those facing the Scottish blender. He does not have the same resources, so consequently his product will be different. He is also governed by laws that define whisky differently. One should therefore not expect Japanese whisky to taste like Scotch whisky. Japanese blended whisky is a unique product that has evolved as a reflection of the economy of the country, the legal definitions of the product and the tastes and desires of the Japanese consumer. They are excellent whiskies with their own identity. They should be accepted and appreciated as such.

The actual practice of whisky blending in Japan is similar to techniques used in Scotland, with the exception that Japanese blenders rely to a large extent on whisky distillate chemistry, as well as taste and odour, to ensure blend consistency.

Ireland

Irish whiskey is a distinctive product of Northern Ireland or the Republic of Ireland and is manufactured in compliance with the laws regulating its production in Ireland. The whiskies are generally mixtures of light, column-still grain whiskies and flavourful pot-still whiskies. All components must be matured in wood for a minimum of three years, but blenders generally use whiskies that exceed three years of age.

The techniques of producing whisky distillates in Ireland resemble those of Scotland, though there are some notable differences in the manufacturing methods for the flavourful pot-still whiskies. Firstly, the mashes generally contain a mixture of malted barley and other small grains such as barley, oats, wheat or rye. The percentage of small grains in the mash bill usually exceeds that of the malted barley. Secondly, the Irish pot stills are much larger than those of Scotland and a triple distillation process is used rather than a double. The resulting distillates are higher in strength than Scottish malts, but are not necessarily less flavourful. They have their own unique characteristics because of the obvious differences in grains used and distillation conditions.

There are few distilleries in Ireland and all are controlled by one distillery group. In the 1970s many of the smaller and inefficient older distilleries were closed with the construction of a single large distillery complex. The new distillery was equipped to preserve carefully the characteristics of the whiskies that had been produced at the phased-out distilleries. Many different pot-still whisky types are produced here using the same methods as were practised at the original distilleries. Column-still grain whisky is also produced at the new facility.

The Irish blender is therefore faced with some unique challenges. There are a minimum number of new whisky types available to him and many of these are produced at the same location. In order to ensure that individual blended whisky brands have their own particular characteristics, blends will contain wide variations in the ratios of grain and pot whiskies contained therein. The blends can therefore range from extremely full-bodied to extremely light. Blend variations are also achieved through the use of older or younger whiskies, mash bill modifications for flavouring whiskies and the use of different barrel types. Whiskies may be matured in casks previously used for sherry, rum or bourbon in order to impart slightly different flavours to a whisky. New oak casks are also used to some degree.

As was evident with both Scottish and Japanese blended whiskies, it is not difficult to appreciate why Irish whiskies are unique to the industry.

United States

Whisky produced in the United States is classified as 'American-type' whisky. There are several types in the category and each is strictly defined by law according to raw material fermented, distillation proof, maturation proof, ageing period and cask type. In the case of blended whisky, the blend composition is also regulated to some degree.

The basic definition of American whiskey describes a product distilled from a fermented grain mash at less than 190° US proof, so that the taste, aroma and characteristics generally attributed to whisky are retained. Within this broad definition are listed two precisely defined categories that serve as flavouring whiskies in American blends.

The first category includes bourbon whiskey, rye whiskey, wheat whiskey, malt whiskey and rye malt whiskey. All of these must be distilled at less than 160° US proof from a mash containing not less than 51% corn, rye, wheat, malted barley or malted rye respectively. The new whiskies must be stored in new charred oak containers at not more than 125° US proof. Only the bourbon whiskies and rye whiskies in this category are used to any great extent for blending.

The second category of flavouring whisky commonly used for blending is corn whiskey. This product must also be distilled at less than 160° US proof, but the mash bill shall be comprised of not less than 80% corn. Corn whiskey must also be matured at not more than 125° US proof in new or used oak cooperage, but the new cooperage must not be charred. Both categories of flavouring whisky may be further designated 'straight whisky' if they are aged for more than two years. These 'straight whiskies' are to the American blender as malt whiskies are to the Scottish blender. They are the character-building influence in American blended whisky.

As lighter grain whiskies are an integral part of Scottish blends, so are light grain whiskies and/or spirits a necessary part of American blends. The American blender has a choice of three such products for mixing with his complement of flavouring whiskies or 'straights'. Light whiskies, grain spirits or grain neutral spirits may be used individually or in combination with one another to give a blend a degree of lightness and balance that would not be achievable with a mix of full-bodied straights alone.

Grain neutral spirits are distilled from a fermented grain mash that is generally corn-derived, but not necessarily. They are distilled at greater than 190° US proof and as such are not classified as whisky. When aged for an unspecified period of time in previously used oak cooperage, they may be designated grain spirits rather than grain neutral spirits. Light whisky is a product distilled from a fermented grain mash at between 160° US proof and 190° US proof and aged for a period of at least two years in previously used cooperage. Since it is distilled at below 190° US proof, light whisky can be classified as American whiskey. The quantities and proportions of light whiskies, grain neutral spirits or grain spirits to be used with the straights will depend on the blender's objectives for taste and aroma or perhaps the degree of sophistication he wishes to give his blend.

The regulations covering 'blended whisky' in the United States permit the use of certain 'blending materials' or 'blenders' up to a maximum of 2.5% by volume. Sherries and blending wines are commonly used in this application. Their inclusion in a blend does provide some unique and interesting characteristics.

The above insight into the complexity of some of the major blending components and their terminology enable a better understanding of the following brief account of whisky blending as it occurs in the United States.

American 'blended whisky' is a mixture of at least 20% by volume 100° US proof straight whiskies and other whiskies, grain neutral spirits or grain spirits. Although the component materials utilised in American blends are vastly different from those used in Scotland, the practice of mixing flavouring whiskies and lighter base whiskies or spirits is similar in both countries. American blends will often contain up to 35% straights and 65% light whiskies or spirits. The straight whiskies selected are generally well in excess of the minimum age requirement. As in Scotland, a wide variety of distinctly different whiskies may be used to create a desired end-result. The whiskies available to the blender are not, however, the products of as many distilleries. Rather, many different types of straight whiskies may be produced at a given distillery through mash bill variations and differences attributed to maturation periods. So despite having access to only a limited number of distilleries, the blender does have a very good selection of whisky flavours from which to choose. Given this, and some skill and experience in knowing how the various components react with one another, the American blender is in a good position to maintain uniform and well-balanced quality blended whiskies for the consumer.

There are some specified types of blended whisky in the United States that are not as popular or as predominant in the market as the general 'blended whisky' category discussed above. For example, there are blends of 'straights' that are designated 'blended straight whiskies', or if they are a blend of one particular type of straight such as bourbon, the designation is 'blended straight bourbon whiskey'. A blend of straights and lighter base whiskies or spirits that contains more than 51% of one type of straight, can be called 'blended corn whiskey' or 'blended rye whiskey' depending on the identity of the straight that is dominating the blend.

Canada

Canadian whisky is a distinctive Canadian product. It is manufactured only in Canada in compliance with Canadian laws regulating its production for consumption in Canada. The distillates contained in Canadian whisky are grain-derived (excepting a small percentage of special flavourings) and are aged for a period of not less than three years in small wood (see below). Canadian whiskies are light-bodied blends. They are considered premium blends in that all the component whiskies are matured, and usually for a period of at least six years. The age declaration must be satisfied by all whiskies in the blend. The blenders, or 'flavourings' as they are known in Canada, may include wines and sherries, rum, brandy, bourbon or malt whisky. Canadian whisky for consumption in Canada may legally use as much as 9.09% flavouring on a litres of absolute alcohol basis, although this level is not usually achieved in practice.

In the regulations covering the manufacture of Canadian whisky, the federal government has virtually taken a 'no-control' position in matters relating to grain formulations, distillation procedures and barrel types used for maturation. The attitude of the Canadian government in this regard has provided the Canadian distiller with an opportunity to operate within certain reasonable guidelines to fulfil a consumer need. The success of the product internationally attests to the wisdom of this approach.

Canadian whiskies are blends of light, column-still grain whiskies and more flavourful column-still or pot-still whiskies. The practice of blending lighter base whiskies with a selection of more flavourful whiskies is in keeping with the blending methods of Scotland, Japan, Ireland and the United States. In Canada, the lighter base whiskies or grain whiskies are most often produced from a mash of corn saccharified with barley malt and/or microbial enzymes. Some distillers located in close proximity to the wheat and rye producing regions of Canada have elected to use rye or a combination of corn and small grains almost exclusively in the production of grain spirits. Grain spirits are distilled at between 180° and 190° US proof, so that only some delicate grain flavours are retained in the distillates.

The heavier-bodied flavouring whiskies are usually combined with the lighter base whisky at a 10–15% level, which is low by comparison with Scottish practice. This relatively low flavouring whisky level is in part responsible for the lighter-bodied nature of Canadian whisky.

The selection of grains used in a flavouring whisky mash bill, and the proportions thereof, is a point of difference between distillers. The Canadian distiller is not restricted as to the choice of cereal grains used to produce his whiskies. Grains of choice are often a combination of rye, barley malt or rye malt. The mash bill formulae are usually well-guarded secrets in the Canadian industry, since the grain selection can have a significant effect on the character of a single flavouring whisky resulting from the fermented mash.

Distillation procedures are not restricted by law, so Canadian distillers are at liberty to operate continuous or batch distillations as is necessary to separate out undesirable congeners while retaining desirable ones. The strength at which distillates are produced is therefore another area of difference between distillers that is used to achieve some diversity between whiskies. Distillation practices are closely guarded by Canadian distillers, as they can impart unique and appealing characteristics to a product that a competing brand might not enjoy.

The Canadian government has also taken a liberal stance on the cooperage types permitted for the maturation of Canadian whisky. The only restrictions are that Canadian whisky shall be matured for a period of not less than three years in small wood – small wood being barrels having a capacity of less than 681 litres (150 gal). In practice, however, Canadian distillers use matured oak cooperage of 182 litres (40 imp. gal) capacity. Some distillers will elect to use a small percentage of new barrels in order to achieve a desired effect for a particular component whisky, but in general maturation in previously used barrels is the norm for Canadian whisky.

Matured barrels are of various types and are used in preselected ratios to impart a desired flavour effect to a final blend. Freshly drained American bourbon or Canadian whisky barrels are employed to create softer and more delicately flavoured whisky. Some distillers re-char previously used barrels and use a percentage of these in their barrel mix to contribute a degree of woodiness to certain blends. A few years ago, a proportion of freshly drained sherry barrels was commonly used to impart a hint of sherry to Canadian whisky. Although this practice is not common to the industry today, some distillers have simulated the tradition by adding a small percentage of sherry to a final blend or by rinsing barrels with sherry prior to filling.

The Canadian blender uses whiskies in his blends that are produced within one organisation. Distillers operating more than one distillery within Canada may choose to use one distillery for producing a particular flavouring whisky and deliver that whisky to another blending location, but generally speaking, all of the whiskies used in a blend are produced at the same location. Competing companies do not exchange whiskies as is the practice in Scotland.

We have previously discussed the benefits of having a diversity of whisky types when blending for product consistency. The Canadian blender, by not having access to a wide selection of whisky types through either an exchange programme with other distillers or by purchasing unique whiskies from competitors, has had to create the whisky variations on his own premises. This has been accomplished in a number of ways. Some of the techniques employed are mash bill variations, unique cooking procedures, the use of specialised yeast strains to impart subtle differences to distillates, variations in still design and distillation procedures, the use of a wide variety of barrel types and the selection of various maturation periods for whiskies. Distillers utilising all of these techniques would certainly have a selection of whisky types that would more than accommodate their needs. Consequently, most distillers will specialise in the use of a limited number of these options for producing whisky variations, and become very proficient in their particular area of specialisation.

There are two distinctly different approaches to whisky blending practised in Canada today. The first is the conventional method of draining pre-selected matured component whiskies from barrels into a common mixing vessel. This method is referred to as 'post blending'. The second approach is to blend selected whiskies prior to maturation so that the mixture is matured in wood as a 'pre-blend'. A particular pre-blend may be matured in a choice of different barrel types, in order to provide additional variety for the blender's post blending function. Pre-blending is not practised by all Canadian distillers; however, at least one major distiller does make use of the technique on a fairly large scale.

The method of mixing whiskies to prepare a final blend is the same in Canada as it is in the other countries previously discussed. The matured component whiskies or pre-blends are selected from warehouses in the correct proportions, and drained from barrels into stainless steel troughs. From here, the whiskies are pumped to a stainless steel blending tank where flavourings are added. At this point, the blend will be evaluated organoleptically and final adjustments made if deemed necessary by the blender.

Deproofing water and caramel are added as a final step before filtration and bottling.

Filtration

The objective in filtering whisky is to remove particulate material in order to present a clear, bright product to the consumer.

Cloudy or hazy whisky is not appealing to the eye, nor is whisky containing barrel char.

Whisky haze is a phenomenon which manifests itself more in some whiskies than in others. Heavier-bodied older whiskies and whiskies matured at higher strengths, when reduced in proof, are generally more susceptible. Certain higher molecular weight lipids and esters that are soluble in barrelling-strength matured whiskies, become less soluble when deproofing water is added. These insolubles are known throughout the industry as haze. The reaction is usually a slow one and consequently, the haze will often develop in the finished product if it is not encouraged to form prior to filtration.

The haze problem is generally controlled by chilling whisky to between 0°C and 10°C, depending upon the nature of the product, and holding it at the reduced temperature for a specified period of time. Since haze formation is a time-temperature relationship, there may be some flexibility in selecting chilling temperatures provided that adequate holding times are established to permit the completion of the reaction. The chilling process accelerates haze formation by encouraging water insolubles to come out of solution. The insolubles are then physically separated from the whisky by filtration at the reduced temperature. The opportunity for haze occurring in the final product is thus greatly reduced when chill filtration is practised.

Whisky chilling, however, is a practice that deserves a great deal of care and attention at the production level. The filtration step that follows chilling may remove some congeneric value, and if removed in excess, can subtly change the character of a blend. It is therefore not advisable to arbitrarily select a chilling program for a particular blend without first researching the effect that such a programme might have on the character of the whisky. Each whisky blend should be researched independently to determine the degree of chilling that can be accommodated without jeopardizing the character of a carefully blended whisky. Some blenders have simply refused to chill their premium whiskies for fear of changing them.

The whisky haze phenomenon has been tackled, by some distillers, by the use of activated carbon. When added to a blend at the rate of perhaps 1–2 kg per 10 000 litres of product, some success has been achieved in reducing haze, but activated carbon treatment is not as effective as chilling in preventing haze development. However, it can be a convenient partial solution when a major capital expenditure for chilling equipment is not practicable. The use of activated carbon also has several disadvantages that might negate any benefits gained by only partially solving a haze problem. It is an extremely dirty material when handled in bulk and it does offer some new challenges to filter operators. Carbon will substantially reduce throughputs, and can migrate through filter media if an operator is not familiar with its filtration properties. Filtration operating conditions have to be modified to accommodate it.

The importance of proceeding cautiously, when considering either activated carbon treatment or chilling for the prevention of whisky haze, bears emphasis. If these treatment methods are abused and not approached scientifically, the organoleptic properties of a whisky can be altered. In the United States this can have very serious legal repercussions. If a whisky that is designated as a specific type is sufficiently altered organoleptically by means of a treatment process, the distillery may lose its right to designate the whisky within the intended category.

The style of filter most frequently used to clarify whisky is the 'plate and frame' filter. This is one of the most expensive depth filtration systems available today, yet it is the most popular. Convenience of operation, trouble-free performance and consistently good results make this system as popular as it is, despite higher installation and operating costs. Other filtration systems that basically utilise bulk filtration media such as cellulose and diatomaceous earth can achieve quality filtrations but with more difficulty. With such systems, bulk medium is pre-coated on to thin cellulose backing sheets that are applied singly over horizontally oriented screens and plates stacked within a stainless steel housing. Bulk media filters are effective and attractive from a cost standpoint, but they must be constantly monitored by qualified operators during the filtration process. Hydraulic shocks can create particle 'breakthrough' with these filters, more so than with the more positive preformed pads used in the 'plate and frame' system.

The preformed pads used in the plate and frame filters are cellulose or cellulose either impregnated or pre-coated with diatomaceous earth. Asbestos pads are no longer used in the industry. Pads most commonly used are 60 cm square and 3 or 4 mm thick. They have particle retention capability of approximately 5 to 7 μm. The pads are inserted vertically between alternating stainless steel plates and frames that are suspended within a stainless steel chassis. The chassis may be portable or fixed. When the pads have been tightened down within the chassis, between the plates and frames, whisky pumped to the filter inlet must pass through a pad in order to reach an outlet port in the plate backing the pad. Each plate receives filtered whisky from pads positioned on both sides of the plate. Thin cellulose backing sheets are sometimes inserted between pads and plates in order to discourage cellulose particles from migrating off the discharge side of the pads into the filtered product. In the plate and frame filter the preformed pads serve as both the filter medium and as the gasket between plates.

The operational parameters selected by blenders for the filtration of their blends will depend to a large extent on batch size, urgency (i.e. filtration rate required for scheduling) and the nature of the product itself. A given volume of one blend may be processed through a given area of filtration surface without developing too much of a pressure differential across the filter. Another blend (or whisky type) of similar volume may, by its nature, require a much larger area of filtration surface in order to avoid unmanageable pressure differentials across the filter near the end of the run. For these reasons, it is not practicable or even possible to quote a general set of filtration operating parameters that would satisfy the wide variety of whisky types we have discussed. Each whisky type and its particular situation will determine the filtration parameters selected by the blender. Some generalised operational parameters for the filtration of Canadian whisky will be offered, however, as an example of how a typical filtration might be approached.

Through experience, a blender of Canadian whiskies will acquire a knowledge of his various blends and gain an appreciation as to how they should be handled prior to and during filtration in order to effect an efficient filtration. It is not simply a matter of producing a sparkling clear finished product at any cost. The filtration must be accomplished in a cost-effective manner, and within a period of time that is compatible with the objectives of the production planners and bottlers. The ideal situation is to filter a whisky at flow rates that are near design rates for the equipment on hand while not exceeding pressure differentials across the filter that might jeopardise the quality of the filtration. When calculating total filtration area required to filter a given batch size, within a specified period of time, the individual responsible for this task should aim to finish his filtration with a minimum amount of life left in his filter. It is very tempting to set up a filter with excess filtration area as a margin of safety against having to terminate the filtration because of blinding. This is most certainly a sensible strategy to follow if it is done conservatively. However, if one becomes too liberal with the margin of safety, the luxury can become a costly one. Filter life, and consequently the volume of product passing through the filter (i.e. 'throughput'), is more effectively extended through the practice of 'body-feeding' with diatomaceous earth. Diatomaceous earth is simply added to the unfiltered whisky in the blending tank, held in suspension with agitation and filtered along with the whisky. As the particles of diatomaceous earth become deposited on the inlet side of the filter pads during the filtration, they become an integral part of the filter medium and extend the life of the filter by discouraging surface sealing of the preformed pads. When cellulose pads are used in a plate and frame filter, a coarse grade diatomaceous earth used at a body-feed rate of 100 kg per 40 000 litres of product would be more than adequate. Cellulose pads precoated with diatomaceous earth would not require the same degree of body-feeding to accomplish the same end. A considerably lower rate of body-feeding with diatomaceous earth would be necessary.

The Canadian blender has learned that there are a few key variables that can greatly affect whisky throughputs for a given surface area of filtration medium. These are the strength at which whiskies are matured, the temperature at which they are filtered, time in the barrel, the strength at which the whiskies were distilled and the barrel type used for maturation. Any other factors, by comparison to these, are of limited importance. For example, to illustrate the effect that barrelling strength can have on filtration, it has been shown that a whisky matured for six years at 66% alcohol in matured cooperage typically requires 1 m^2 of filter area to filter 5000 litres of room temperature product reduced to 40% alcohol. The same whisky matured in wood at 77% alcohol, would require 2 m^2 of filtering area to process an equivalent 5000 litres of product under the same filtration conditions. A similar reduction in whisky throughput has been observed when Canadian whiskies matured at 66% alcohol are reduced to 40% alcohol and filtered at 5°C as opposed to room temperature. Generally speaking, lower throughputs are also experienced with older whiskies, whiskies matured in new wood or recharred matured cooperage and blends that contain higher levels of heavier-bodied flavouring whiskies.

When a filter has been set up with an area of filtration surface that the operator believes will adequately accommodate his batch size, product is circulated through the filter back to the blending tank. Initially, flow rate will be established not to exceed 12 litres m^{-1} min^{-1} and may be considerably less than this – perhaps as low as 6 litres m^{-2} min^{-1}. When the clarity of the whisky reaches the desired specification, the filtered product is directed 'on stream' to the final product tank or 'bottling holding' tank. As the filtration proceeds, naturally the flow rate will drop as the filter pads begin to blind and the pressure differential across the filter increases. With a plate and frame filter, it is advisable not to exceed a pressure differential across the filter of 276 kPa (40 psi). Beyond this, there is a high risk of particulate material migrating into the filtered whisky. Pressure leaf filters employing bulk filtration media can generally approach a 414 kPa (60 psi) pressure differential before there is a high risk of particle 'breakthrough' occurring. If a filtration is nearly complete when a critical pressure differential is being approached, it is recommended to reduce flow rate to the filter for the duration of the run in order to stabilise the pressure differential. If time permits, this is certainly a more attractive option than terminating the filtration and setting up the filter again to complete the batch. At any rate, pressure specifications across a filter should not be exceeded for the sake of expediency.

Whisky clarity is measured in a number of ways. Distillers have their own individual clarity specifications and methods of analysis. Some simply look at the product as the consumer would do, and determine thus if it has been adequately filtered. The majority, however, use a more sophisticated approach to clarity measurement with instruments such as nephelometers and a variety of turbidimeters.

Bottling

The product is received from the blending and filtration process via stainless steel lines into the bottling plant's holding tanks or directly into the filler head of each line. It is common practice to use a 'polishing' type filter of about 10 μm retention, just prior to the filler. With this last consideration of product, we can give full attention to the bottling functions.

COMPONENT HANDLING

Incoming components

In North America it is the practice to receive glass bottles in their shipping cases, by palletised load on wood pallets or on slip sheets, which are merely semi-rigid, thin composite board, sized to hold a normal pallet load. The sheet edge is gripped by a special attachment on a fork lift truck, which enables load transfer. Slip sheets can be either disposable or reusable depending on material, construction and cost considerations. Bulk handling for all incoming containers is more common in Europe for both glass or plastic and is used with either low- or high-volume line capacity.

The performance of the incoming supplies, especially the glass and label components, will be affected by climatic conditions. Label moisture variations and cold or bottle surface moisture will affect mechanical performance unless these factors are given consideration. Climatising package components to bottling room conditions is advisable if quality packaging and high line productivity levels are to be attained. In the case of plastic containers, static charges build up surface dust and cause increased container contact adhesion. Incoming ware to the line is usually blanketed with deionised air to remove the static charge. Label moisture is normally controlled at 5% to prevent curl along the grain of the label. Excess curl will make magazine loading and label retrieval difficult, and after application spring-back can be a problem. With all of the packaging component preconditions met, we can now consider bottling line equipment.

In North America line speeds for glass, round and flask bottles, commonly approach 350 bottles per minute (bpm). Any departure from these basic shapes, or exceeding 1 litre capacity, will greatly increase the difficulty in maintaining that level of productivity. The exception to this rule of thumb is the popular 50 ml miniature, which is being produced at 700 bpm in either glass or polyethylene tetraphthalate (PET) plastic. At our Walkerville plant we have bottled 500 ml, 750 ml and 1.0 litre round on a common line at speeds of 300 bpm, with minimal down-time and consistent product quality.

Bottle delivery to the line

Bottles arrive, packed upright in their inverted shipping cartons, palletised in units of approximately 70 cases on either wooden pallets or slip sheets. The pallets of cases are automatically, semi-automatically or manually depalletised, and the individual cases move to the uncaser via an overhead conveyor. Bottle discharge from the case to an accumulating conveyor provides four minutes accumulation time. This specific type of conveyor uses plastic rollers, fitted loosely on a powered shaft. The rollers turn with the shaft to move the bottles forward, but stop turning when the friction coefficient caused by the bottle accumulation is overcome. In contrast to highly synchronised bottling lines with minimal accumulation time, European bottlers favour larger accumulation capacity at multiple line locations, giving more line operational flexibility and minimal down-time. This practice dictates a larger bottling line area, which often is not available in existing plants where modernisation is being considered. In new plant construction it is the cost-effective option to follow.

Once the bottles are dumped, the empty reshipper cases are conveyed overhead to the packer operations area. Meanwhile, bottles exit the accumulating conveyor and are single-filed through the cleaning stage, where extraneous surface dust is first brushed and vacuumed off.

Bottle cleaning

The bottle cleaner receives the bottle. The bottle is inverted, and any extraneous material inside is disturbed by an air jet and removed with an applied vacuum. The bottle can also be washed but since delivery is in shipper cases direct and bottle reuse is not practised in North America, bottle washing is not required.

FILLING, CAPPING AND LABELLING

Filling

Bottle filling equipment falls into several categories:

1. vacuum;
2. vacuum-gravity;
3. pressure;

4. pressure-vacuum;
5. gravity.

It should be noted that all are pressure-dependent, as a differential in pressure is required to provide product flow. The relationship between the single or multiple systems always produces an operating pressure differential of 35 kPa (5 psi), neglecting pipe frictions. The fillers as listed are all common to our industry and all give the consistent visual fill tolerance that is required.

VACUUM FILLERS

This type of filler requires a seal on the bottle finish in order to function, after which the filling stem employs a product tube and vent or overflow tube. Product is sucked into the bottle by the applied vacuum and the vent position determines filling height relative to the top of the container.

VACUUM-GRAVITY FILLERS

Filling is accomplished by gravity in a closed vacuum system. This means that the actual flow of liquid into the container is accomplished by gravity flow after a vacuum has been established in the container equal to that already established in the filler tank. The system claims to reduce aeration during filling and to prevent the filling of chipped or cracked bottles.

PRESSURE FILLERS

These fillers are widely used to fill flex-walled containers in order to ensure fill point integrity. Unlike the vacuum fillers, they permit direct venting to atmosphere hence avoiding the risk of the container wall flexing or distorting. Every manufacturer incorporates special features into their equipment, which must be considered in the light of one's specific requirements.

Types of cappers

Closures are delivered to an overhead hopper manually, mechanically or by vacuum. They are then oriented in an upright position and single-filed as they enter the delivery chute. The closures are conveyed by vacuum, an airstream or gravity to the pick-up point where the chute terminates. A transfer arm removes the closure from the chute and swings into the position where the closure will be picked up by the chuck. Closures are applied by a friction or mechanical type of chuck to the bottle finish at the desired application torque.

MECHANICAL CHUCK

Mechanically controlled torque functions by applying a pre-set torque to a set of jaws which are released when the desired application torque activates a torque-sensitive trigger.

FRICTION CHUCK

This type of chuck is either activated by a spring or air pressure loading on a clutch plate. When the desired torque is achieved, the clutch slips, releasing pressure in the gripping ring and the container is discharged.

ROLL-ON CAPPER

The popularity of this closure is evident in Europe and it is used, to a lesser extent, in North America. The metal blanks are delivered to the bottle finish as a 'sleeve', after which the torsion side pressure springs in the head apply controlled thread rolling action, as a downward pressure assures an effective seal. Application can be equally effective to glass or plastic containers.

Labellers

Labelling technology has undergone extensive development during the last decade, basically motivated by the need for increased line speeds and productivity. It has also been subject to the external influences of rapidly changing package designs and component materials. Plastic containers, multi-layered label stock, and a new generation of adhesives have all influenced labeller design. There are three basic concepts used in today's wet-glue labellers. These are picker in line, rotary in line, and rotary. In addition there is pressure-sensitive and 'in mold' labelling now used with 50 ml PET miniatures.

PICKER IN LINE LABELLER

This type of labeller is still widely used by the industry in North America, but will be subject to replacement as faster line speeds become more necessary. It depends on cam-activated reciprocating action to provide motion to the glue transfer roller, the label pickers, the label grip fingers and the label wiping action. Typical features of this type of labeller are:

(a) The option of adding or deleting labelling stations, with a maximum number of six, at 60 rpm.
(b) It provides for a longer wiping cycle.
(c) It can accommodate multiple labelling but both front, or both back, labels must use the same adhesive.
(d) Glue application by roller transfer requires sufficient glue presence in order to pick up the label from the magazine.
(e) Glue film thickness on a label will vary from top to bottom according to the number of transfer roller revolutions.
(f) Glue film thickness also varies in bands from the centre of the label across to the sides.

ROTARY IN LINE LABELLERS

Straight line bottle flow through a series of rotary labelling stations, each of which is capable of single or multiple application on one plane, is the basic operating principle of this type of labelling machine. Label pick-up from the magazine can be accomplished by vacuum, mechanical gripper or adhesive contact, depending in the manufacturer's preference. Pre-set label compression by the moulded compression pads, which travel linear with the containers through the compression section is a feature of this labeller. Other features worth noting are:

(a) Multiple labelling stations permit more flexibility in adhesive type and label composition.
(b) Less adhesive can be used.
(c) In line wiping provides prolonged compression of labels, ensuring good adhesion.
(d) The rotary principle is conducive to higher labelling speeds.

ROTARY LABELLERS

These labellers utilise rotating motion. At each label station the number of individual labels applied can be increased by adding new pallets to existing shafts and adding appropriate new magazines and gripper cylinders. The machine range of capacity (minimum to maximum) is determined before manufacture; increased capacity is achieved by increasing the number of pallet shafts at this stage. Typically, a glue roller transfers adhesive to a rotating gluing pallet, which in turn picks up a label from the magazine. The label is mechanically transferred to a gripper cylinder and on rotation the label is applied to the container as it travels past mechanically synchronised. Complete label adhesion is accomplished as the containers pass through a series of brushes in the wiping station. Label pick-up from the magazine is generally by glue contact but vacuum pads are utilised by some manufacturers. Back label application requires controlled bottle rotation prior to entering a back-labelling station. Some additional features of the rotary labeller are:

(a) eliminates all major reciprocal motion;
(b) lower maintenance;
(c) lower noise levels;
(d) higher speeds;
(e) improved adhesive control;
(f) allows for variation of label materials at a single station.

PRESSURE-SENSITIVE LABELLERS

Labels are pre-glued and affixed at predetermined intervals on a quick release web or roll. Positive non-slip feed of the roll is critical to label placement on the container and a major consideration in design. The availability of dual roll systems enables continuous production. Label removal occurs as the roll travels over a reduced diameter roller, lifting the front edge of the label, and transfer to the container is made. Label adherence can be accomplished in any of the normal methods. Multiple labelling requires additional in line stations. Pressure-sensitive labelling is clean, convenient, eliminates wet-glue labelling problems and can provide an easy means of difficult label location placement. Costs and required line speeds must be considered with this type of labelling system.

Stamping

The affixing of federal and other stamps over the finish of the bottle has been a practice in North America to indicate product age, tax commitment and to provide visual assurance of non-tampering. With the advent of the tamper-evident closure, the necessity for stamping has diminished. Although some stamping is done in Europe, the extensive use of the metal tamper-evident roll on closures has minimised its use. It is felt that the government-supervised age stamp does have some real marketing value to the consumer and may be used in addition to a tamper-evident closure. The stamps can be applied with cold or hot melt adhesive as they are removed from the magazine by glue contact or by vacuum. Stamp wiping is accomplished by cam-activated pressure pads.

PACKING AND PALLETISING

Packers

European practice, because of bulk delivery of containers, utilises packing systems where shipping cases are formed prior to the packer or formed around the containers themselves. In North America the container usually arrives at the bottling plant in its shipper case. Therefore, the method of receiving the container, to a large degree, dictates the type of packer system used. In general terms there are three types of packers: the drop packer, the place packer, and the wrap-around packer.

DROP PACKER

The shipper cases arrive from the 'dump' area by conveyor at the packer. The bottles are positioned by lane dividers into the required number of rows and are positioned above a grid which functions to guide the bottles into their individual case cells. The bottles are dropped into the case as the grid or their supporting rails shift into alignment. The receiving case is raised prior to bottle release in order to minimise distance of the drop and the resultant impact. The dressed bottles in dropping are subject to label damage from the case dividers and are, therefore, directionally assisted into the cells by the use of low-friction plastic guides. Drop packers are, in general, noisier but operationally simpler and faster than the other two types.

PLACE PACKER

Bottle placement is achieved by seizing the bottle neck with vacuum, compressed air or mechanically activated grippers. The containers are then lowered into the shipper carton, which can be directly below or parallel to the station and on a stationary plane or raised to shorten bottle vertical travel. Place packers cycle at slower speeds, but increased capacity is possible by increasing the number of stations. These packers operate quietly with gentle placement of container into the case. Compression of the grip head can cause wrinkling to some types of neckwrap material in combination with the as yet unset adhesive.

WRAP-AROUND PACKER

Shipping case formation around the containers, as opposed to loading into a preformed shipping case, is more widely practised in Europe than in North America. Although packing into container trays is commonly done, the distilled beverage industry normally utilises complete container enclosure. This is accomplished either by using partially pre-glued cases or total in place gluing at the packer. Pre-gluing utilises case blanks with pre-glued horizontal seams. After filling, the case and flaps are folded and sealed by either hot melt or cold glue. In place gluing utilises flat blanks which are stacked in the magazine and then partially erected by vacuum cups. After loading, the filled case is transported through the flap folding, gluing and sealing steps. The industry extensively packs the 50 ml miniature size in a wrap-around inner carton, which is either pre-glued or 'slot-locked'.

Case sealer

When drop packing or place packing is used, a case sealer is required. The case flaps are automatically positioned on entering the machine and then receive the adhesive. Cold glue application is by roller or spray, and hot melt adhesives are also used. After glue application the case flaps are sealed and held in position for the required time by an overhead compression unit.

PALLETISERS

Palletised shipment of product is the accepted method for the industry. The cases can be in-fed from below, clamping the tier of cases and raising it, or from above, with the pallet lowering after each tier is positioned. Layered case patterns are normally positioned automatically by programmed controllers.

Pallet stabilisation

Because of vibration or dislodgement, pallet loads are commonly stabilised by stretch wrapping or shrink wrapping. Application of the plastic wrap material occurs in three ways:

1. A continuous band of stretch film plastic is attached to the pallet load, which then rotates a predetermined number of revolutions, while the film dispenser moves vertically in a

selected pattern, giving partial or complete coverage, as desired.

2. The stretch plastic film can be applied by rotating the dispenser around the stationary pallet. Mechanics of the application are similar to that described above.
3. Shrink wrap material requires a heat cure stage which can take several operational forms. The polyethylene wrap can be dropped over the pallet load in the form of a bag or the pallet load can be passed through a stationary film. Whichever method of wrapping is employed it must be followed by a heat curing stage. Some heat chambers are lowered over the pallet load to induce shrink, others move the wrap-applied pallet through variations of heat tunnels. Most shrink material requires a temperature of 118°C for several seconds to activate the stretch stage.

It is also quite common, in addition to wrapping, to stabilise pallet loads by applying four spots of hot adhesive or by spraying a double pattern of non-slip adhesive to the top of each case. These methods, on depalletising, can cause minimal tearing of the cases' surface and the latter will collect dirt and discolour. The option, however, is to tie the load by strapping two or more tiers, which can also damage the cases. Stretch or shrink wrapping of the pallet loads is the preferred method. The palletised case goods are now ready for shipment to warehousing or retail.

Bottling engineering

There are many scenarios that could be described in this section. Bottling line modernisation or replacement installations are affected by multiple external factors. To recommend an engineering approach to each set of conditions would produce a voluminous dissertation. Therefore, we will approach this subject considering only a new bottling line installation without those external restrictions such as space and budget. With that said, it is necessary to consider the following parameters carefully prior to the start of engineering design.

ENVIRONMENT

Modern bottling installations have changed dramatically in this area over the past decade, due to changing regulations and the desire of management to provide a more ideal workplace. The most common problem, of course, is noise and this influences the engineer in several ways. Does one purchase quieter running equipment, isolate the line from the worker, isolate the noise sources with acoustical control or use protective gear for the worker? Quite often the final decision includes incorporating several of these options, thus giving the engineer more flexibility.

Physical safety must then be considered before equipment purchase. Exposed moving components must be guarded to meet regulations and this will favour specific types of equipment purchases.

In the modern workplace it has become necessary for both worker satisfaction and productivity to provide an aesthetically pleasing atmosphere. Colour, comfortable work stations and subliminal music are a few of the options an engineer must consider.

PRODUCT IDENTIFICATION

It is essential that the engineer knows the product's characteristics by considering chemical and physical criteria such as specific gravity, viscosity, temperature range, foaming and flow characteristics. Can the product be easily purged from the system or are special cleaning operations required? Product odour, foaming and viscosity will influence the decision.

BOTTLE SIZE AND SHAPE

Bottle size, besides dictating specialised equipment needs, determines to a large degree the potential line speed. Line speed then influences equipment choice, degree of line synchronisation and accumulation required. Bottle shape also affects all of the above parameters. In general, deviation from the standard round bottle reduces line speed and line efficiency because of handling, labelling and packing complications.

CONTAINER MATERIAL

The industry now bottles in glass, and PET is widely in use for the 1.75 litre and 50 ml sizes, although there is no technical reason why it cannot be used for the intermediate sizes. Specialised equipment is required for plastic containers, although some successful conversions of glass bottle equipment have been carried out. New lines employ dual-purpose equipment where it is deemed practicable. Plastic containers gained popularity because of their lighter weights. This has required specialised equipment to transport, fill and label the containers. For example, pressure or pressure gravity fillers are used to avoid container distortion.

PACKAGE COMPONENTS

The choice of closure system, be it aluminium roll-on, linered or unlinered polypropylene, linered steel or the traditional cork stopper, influences the choice of application equipment. Labels are now made from a multitude of materials and their combinations, each requiring specialised adhesives and equipment for application. Multiple labelling on a single package can entail multiple materials.

PRODUCTIVITY

Low- or high-volume lines have completely different equipment requirements. The lower the line speed, the less complex the equipment, handling and control systems have to be. Because of diminished required accumulation area and, in general, the smaller machines that are used, the total space required diminishes. The degree of required line synchronisation also decreases with line speed.

PACKAGE QUALITY

Quality is a trade-off item. High productivity and minimum line cost are often dual requests that are possible to achieve. The penalty incurred, however, is often package quality. This situation can be avoided by the design engineer if the container shape and package components naturally lend themselves to high-speed operations. If this is not possible package quality can only be achieved by more exotic equipment and lower line speeds. The engineer must decide which direction is most suitable to meet the established management parameters.

MATERIAL HANDLING

Some mention has been previously made of material handling systems. The industry considers bulk delivery of containers to be most cost-effective, with diminishing benefits as production volume decreases. On the other hand, palletised movement of finished product case goods is cost-effective for all producers.

Volume is the critical factor because it is required to justify the automated handling system. Irregular shapes can cause difficulty in the critical single-filing operation that is required immediately after accumulation and prior to the bottling line. At this stage, the engineer must evaluate all the critical factors that will contribute to an efficient, cost-effective system for his specific project.

Today's engineer has the benefits of recently developed computerised line studies and simulations. The first step involves the construction of a computer simulation model of the line. The computer is fed the key facts about each machine and conveyor in the line: its rated speed, the frequency of major and minor failures, and the time taken to restart after failure. The computer's own comprehensive data bank supplies the required information on equipment performance.

When all data have been entered, the first computer run is made, simulating what will happen if 1 000 000 containers are processed, giving a close correlation to actual line function. The benefit of the computer simulation is that corrections can be made in the design phase rather than by on-line modifications.

Suggestions for further reading

Bureau Alcohol, Tobacco Products & Firearms 1985 *Title 27, Code of Federal Regulations*. Office of the Federal Register National Archives & Records Administration, US Government Printing Office, Washington DC

Grossman H J 1974 *Grossman's Guide to Wines Spirits and Beers* 5th edn (revised). Charles Scribner's Sons, New York

Lord T 1979 *The World Guide to Spirits, Liqueurs, Aperitifs and Cocktails*. New Burlington Books, London

McDowall R J S 1971 *The Whiskies of Scotland*. John Murray Publishers Ltd, London

Murphy B 1979 *The World Book of Whisky*. Rand McNally and Co, Chicago, Illinois

Packlowski G W 1963 Alcoholic beverages, distilled. Kirk-Othmer, *Encyclopedia of Chemical Technology* 2nd edn. John Wiley & Sons, Inc, New York, vol 1

Rannie W F 1976 *Canadian Whisky: the product and the industry*. Rannie, Lincoln, Ontario

CHAPTER 11

Energy management

J G WATSON

Introduction

When the Organisation of Oil Exporting Countries (OPEC) met in 1973, it decided that oil was such an important commodity that its price could be substantially raised. In less than four years the price of oil increased by a factor of five. This caused all industrialised countries to review their energy policies (and to some extent, their whole economies), which were highly dependent on this imported oil. Even the United States, with its own oil reserves, was importing large amounts of oil, which meant that they too had to consider alternatives.

Although it was considered that OPEC could supply all the oil required for several decades it was realised that the price would continue to rise at a rate which was impossible to predict. Also, at the known rate of consumption and the projected increase in rate of consumption, it became apparent that, when the rate of discovery of new oilfields was considered, the supply would fail to meet the demand by the early part of the twenty-first century.

Distillers, as energy users, in common with all other industries had to react. Most companies up to this time had regarded energy, whether it was oil, coal, gas or electricity, as a cheap 'raw material'. Although no figures are readily available, it is estimated that energy contributed less than 10% of the total production costs at most plants. By the end of the 1970s, if the consumption levels had remained the same, energy costs would have risen to a figure between 20% and 25% of total production costs, depending on the size and style of operation.

Initially the attempts at energy management were half-hearted, possibly due to the lack of knowledge, and possibly because it was felt that the price increase was only temporary.

However, as time went on and it became apparent that company profitability was being adversely affected by these more expensive fuels, more meaningful action began to be taken by the mid to late 1970s. As well as the natural resistance to change, possibly there is also a resistance in the distillery industry which stems from a reluctance to depart from 'well-proven traditional methods' which may or may not affect quality, and which are also very energy-inefficient. This resistance was perhaps seen more in smaller, privately owned establishments.

Usage measurement

Before we can start to manage energy, we must know how much is being used. We must also know how efficiently it is being used and relate this to price, since sometimes it can make more sense economically to use a cheaper fuel inefficiently than an expensive fuel efficiently.

Without going back to absolute basics, energy is derived from purchased fuels by combustion. These fuels can be either solid, liquid or gaseous. Usually when we talk about solid fuels we mean coal, although wood and peat in various forms are being burnt and there is considerable discussion about refuse- or waste-derived fuels. Solid fuels can also be converted to the liquid or gaseous state if this is desired. Liquid fuels inevitably mean the fuel oils, from the low-viscosity light oils to high-viscosity heavy oils. An interesting recent addition to this range is a coal/oil slurry mixture which burns on conventional oil-burning equipment. Gaseous fuels supplied to industry range from natural gas supplied by pipeline to a range of liquefied petroleum gas (LPG) products, the main ones being butane and propane. These are usually supplied in pressurised containers in a liquid form.

The other main energy source used in distilling is electricity. Usually this is supplied direct from the national supply system, although facilities exist for on-site generation of electricity and this is actually encouraged in some areas. However, it is interesting to note that in the USA there has been a gradual decline in the number of plants generating their own power, from approximately 30% kW hours generated on site in 1920, through 10% in 1960, to an estimated 3% in 1985 (Vennard, 1979). This may be due to a combination of the poor conversion efficiency of fossil fuels to electricity and the labour requirements of these plants.

SOLID FUELS

This is the most difficult medium of which to measure the consumption. Ideally solid fuel should be continuously weighed, as it is being consumed, with a conveying system which incorporates load cells, but these are expensive and it can be difficult to justify expenditure on them.

Probably the most effective method is a mechanical counter which can be fitted to a rotating shaft. In the case of an underfeed stoker, this shaft can be the actual auger or screw which conveys the coal to the combustion area, and for a chain or travelling grate the front drive shaft of the conveyor can be used – in this case, the bed depth must also be considered, but plant records usually contain this information. Frequent 'calibration' checks without combustion can easily be carried out. If possible, when batches or loads of solid fuel are delivered to the area of usage, they should be weighed as a back-up check.

LIQUID FUELS

Compared with solid fuels, it is relatively simple to measure the use of liquid fuels. There is a large variety of dependable meters which can be incorporated in the pipelines which deliver the oil to the point of use. It is also easy to measure starting and closing stocks after each operational period, as liquid storage in tanks is easily converted to capacity using a number of types of level gauge or by simple manual dipping.

GASEOUS FUELS

In the case of natural piped gas supplies, metering equipment is installed by the company supplying the gas. It is important for them (and also the user) to know exactly the amount of gas consumed, as this is the basis for the subsequent charges. Because of the wide variety of customers, they have developed a well-regulated network of pipelines controlled at specific pressures, and this lends itself to accurate measurement.

When butane or propane is being measured it is important to control the supply pressure and temperature, as these have a direct bearing on the volume of gas supplied. If this can be done, a simple rotary vane type meter should give reliable consumption figures.

STEAM

The bulk of the fossil fuels burnt in distillery plants is for the production of steam. It is therefore an added advantage to know how much steam can be produced from a given amount of fuel, and also to know the pattern of consumption of steam in certain areas, processes or items of plant.

Simple differential pressure devices, such as orifice plates, have been used extensively for this purpose in the past, but they have several disadvantages: they tend to be inaccurate at less than 20% of the full range flow; they tend to wear with time, which gives false outputs; and because of their principle of operation, they cause a pressure drop in the steam, which means that it is necessary to run at a higher pressure, which means higher fuel consumption, so that the steam flow can be measured. There are a number of different types of sensor which claim to overcome some, if not all, of these problems. It really depends on how important it is to the consumer to have this information, as the better models are the most expensive.

ELECTRICITY

As in the case of natural gas supplies, the organisation supplying the electricity must know how much they are supplying, and they have had enough experience to enable them to become experts in this type of equipment. However, it is not too difficult to install consumption meters for each area of the distilling process, which could identify undue and unnecessary consumers. Some manufacturers supply simple clip-on current transformers (CCTs) which, as the name suggests, clip on to the cables supplying a certain area or even motor.

Most electricity supply companies base their charges on a scale which is related to a value known as the maximum demand. This is the number of units consumed in a measured half-hour period multiplied by 2 (to convert to kW hours). This value is measured on the same piece of equipment half-hour after half-hour, the equipment restarting or resetting every half-hour.

The peak or maximum demand value is retained by a pointer which does not reset automatically, but is reset manually by the meter reader after each reading.

It is convenient, therefore, to know if there is such a peak in the process, as this will be the basis for charging in any consumption period, whether it be weekly, monthly or quarterly, and this half-hour could be the only high consumption half-hour in that period. The main cause of excessive maximum demands is a large number of motors starting up within a short time of each other. These peak periods can easily be spotted with simple measuring apparatus, and there is even some equipment which can control peak 'height' by switching off non-essential equipment in a predetermined order.

Auditing, reporting and targeting

AUDITING

The consumption figures, when received, must be converted to meaningful standard units of energy. The British Thermal Unit (BTU) has been much used in the past for this, and also perhaps to a greater extent, especially in the gas industry, the therm. One therm is 100 000 BTU. Since metrication in the United Kingdom, it is logical to adopt the metric unit of energy, the joule, for auditing purposes. It is a simple matter to convert kg of coal or LPG, litres of oil, therms of gas, and kW hours of electricity to joules, or perhaps more conveniently to megajoules. For a distillation process it is quite simple then to relate the total amount of energy in megajoules to units of production which, retaining metrication in mind, are litres of (absolute) alcohol. This represents the energy efficiency of a distilling plant, i.e. megajoules per litre of alcohol (MJ per LA) and has been widely adopted and accepted in the malt whisky industry in Scotland.

At a conference/exhibition organised by the UK Department of Energy at Bowmore Distillery, Islay, in October 1984 a whole range of distilleries were questioned about their energy usage related to production levels, and Fig. 11.1 shows the range of energy efficiencies which were calculated from these returns. It was obvious from the analyses of these returns that the most efficient distilleries, in terms of energy consumption, were those

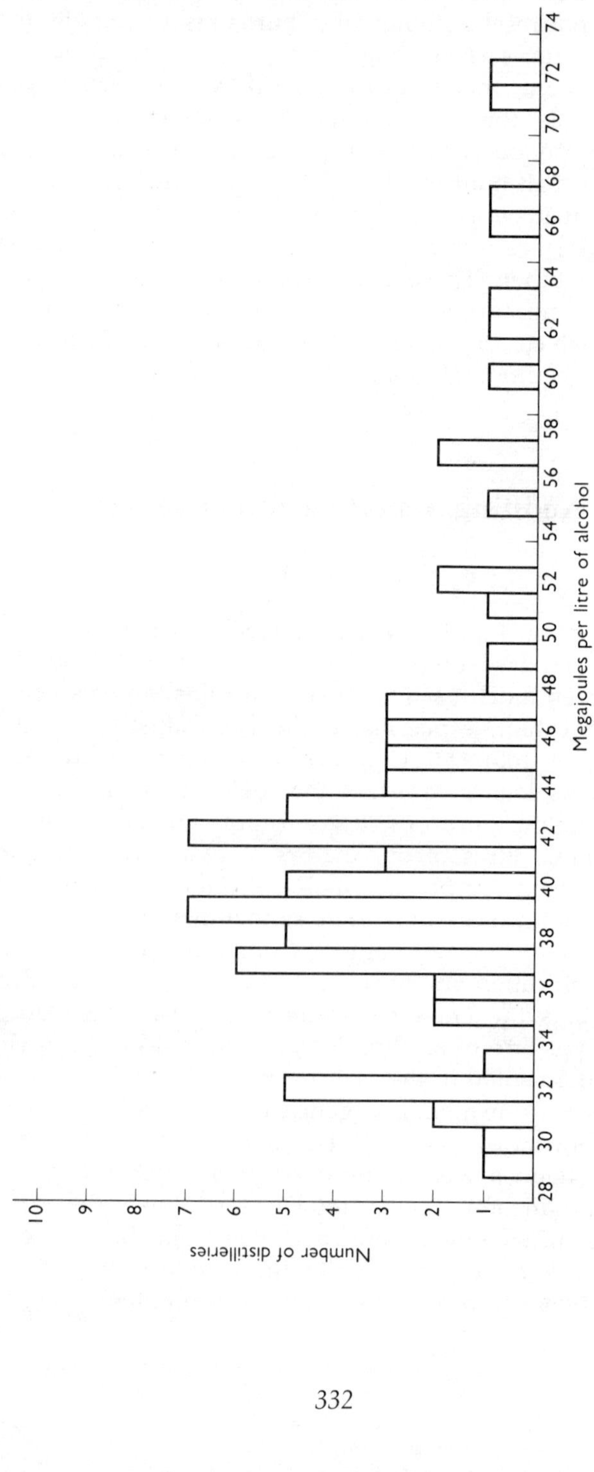

Fig. 11.1 Energy efficiencies of distilleries.

which employed steam indirectly for heating the contents of the pot stills. Those which used either coal or LPG for direct firing of the pot stills were least efficient. More detailed analyses of these returns have been made by Brown (1984, 1985), whilst Newbert (1984) at the same seminar discussed means and methods of improving energy efficiency.

On a smaller company scale, a method of auditing has been devised at Chivas Brothers, whereby weekly returns are made from all distilleries. These are then entered into a computer system which has been programmed for conversion of consumption into energy units, comparison of the total energy consumed with production, and generation of a MJ per LA figure for each location. If the total amount of malt mashed is also entered, a figure for MJ per tonne mashed can also be generated. It follows that once consumptions are known, current prices can also be entered which will give an energy price per LA produced, and also using the MJ figure give a price per unit of energy for that particular site (see Tables 11.1 and 11.2). Usually this reflects the type of heating employed on the pot stills.

REPORTING

It is important to let people know what amount of energy is being used for production purposes, and how one plant compares with another, but what really matters is how much it costs. Therefore I suggest that an integral part of energy reporting should be a section which incorporates current energy costs. I am not, in any way, underplaying how important the energy efficiency figure is to people who are actually running the plant, but managing directors respond more readily to figures related to money, whether it be pounds sterling, American dollars or French francs.

Obviously the more frequently efficiencies can be measured the better, but this depends on the time available to make measurements of consumption. This would depend entirely on the requirements of a plant, or the people running the plant. Also, as I have said, it is important that the people in top management receive information on how plants are performing, in a form that they will respond to. Bearing in mind that they have many other things to attend to, the frequency of these reports to top management should perhaps be less, but again this would depend on each individual company.

Table 11.1 Chivas Brothers (Northern Operations) Limited energy consumption for period 05/85

Distillery	Tonnes mashed	Litres Alcohol produced	Fuel oil Litres	Fuel oil Mega-joules	Coal Tonnes	Coal Mega-joules	LPG Tonnes	LPG Mega-joules	Electricity kW hours	Electricity Mega-joules	Total Mega-joules	$MJ\ t^{-1}$	$MJ\ LA^{-1}$
No. 1	320.000	131 110.4	124 309	5 133 962	0.000	0	0.000	0	56 448	203 213	5 337 175	16 679	40.7
No. 2	408.000	159 126.2	140 500	5 802 650	0.000	0	0.000	0	20 677	74 437	5 877 087	14 405	36.9
No. 3	640.000	256 189.8	231 403	9 556 944	0.000	0	0.000	0	55 690	200 484	9 757 428	15 246	38.1
No. 4	346.560	134 941.8	127 363	5 260 092	0.000	0	0.000	0	42 221	151 996	5 412 088	15 617	40.1
No. 5	829.000	331 936.9	53 553	2 211 739	159.190	4 568 753	236.727	11 836 350	72 334	260 402	18 877 244	22 771	56.9
No. 6	600.000	243 015.6	267 363	11 042 092	0.000	0	0.000	0	72 180	259 848	11 301 940	18 837	46.5
No. 7	1001.000	403 341.4	0	0	386.345	11 088 102	129.859	6 492 950	85 688	308 177	17 889 528	17 872	44.4
No. 8	580.320	229 413.1	82 368	3 401 798	253.500	7 275 450	0.000	0	67 810	244 116	10 921 364	18 820	47.6
No. 9	134.400	53 554.7	56 545	2 335 309	0.000	0	0.000	0	35 380	127 368	2 462 677	18 323	46.0

Table 11.2 Chivas Brothers (Northern Operations) Limited energy costs period 05/85.

Distil-	*Fuel oil*			*Coal*			*LPG*			*Electricity*		*Total*		*£ per*	*£ per*	*£ per*
lery	*Litres*	*£*	*£ per* GJ	*Tonnes*	*£*	*£ per* GJ	*Tonnes*	*£*	*£ per* GJ	*kW* hours	*£*	*£ per* GJ		*tonne*	*LA*	*T.GJ*
No. 1	124 309	18 360.44	3.58	0.000	0.00	0.00	0.000	0.00	0.00	56 448	1 691.600	8.32	20 052.04	62.66	0.153	3.76
No. 2	140 500	20 878.30	3.60	0.000	0.00	0.00	0.000	0.00	0.00	20 677	683.690	9.18	21 561.99	52.85	0.136	3.67
No. 3	231 403	34 178.22	3.58	0.000	0.00	0.00	0.000	0.00	0.00	55 690	1 742.870	8.69	35 921.09	56.13	0.140	3.68
No. 4	127 363	18 926.14	3.60	0.000	0.00	0.00	0.000	0.00	0.00	42 221	1 368.340	9.00	20 294.48	58.56	0.150	3.75
No. 5	53 553	7 957.98	3.60	159.190	12 042.72	2.64	236.727	54 447.21	4.60	72 334	2 269.250	8.71	76 717.16	92.54	0.231	4.06
No. 6	267 363	39 489.52	3.58	0.000	0.00	0.00	0.000	0.00	0.00	72 180	2 194.270	8.44	41 683.79	69.47	0.172	3.69
No. 7	0	0.00	0.00	386.345	21 677.82	1.96	129.859	29 867.57	4.60	85 688	2 635.400	8.54	54 180.79	54.13	0.134	3.03
No. 8	82 368	12 239.88	3.60	253.500	20 153.25	2.77	0.000	0.00	0.00	67 810	2 162.680	8.86	34 555.81	59.55	0.151	3.16
No. 9	56 545	8351.70	3.58	0.000	0.00	0.00	0.000	0.00	0.00	35 380	1 075.550	8.44	9 427.25	70.14	0.176	3.83

TARGETING

A natural progression from an energy auditing and reporting system is the necessity to improve, and this means that targets for the future must be set. It must be stressed that these targets should be realistic but nevertheless attainable. They should be set in agreement with the production personnel running the plant, thus involving these people in all aspects of the energy management of their own plant.

Production levels should be borne in mind when targets are being set. It has been observed that most plants are most efficient at 100% production, becoming less efficient as production decreases. This relationship is almost linear until production drops to about 30%, when the relationship tends to proceed at an alarming rate to infinity (Fig. 11.2). In a plant which has production fluctuations of this sort, it may be more meaningful to talk about energy efficiency in terms of LA per MJ. At least in this case, one would be starting at 0.

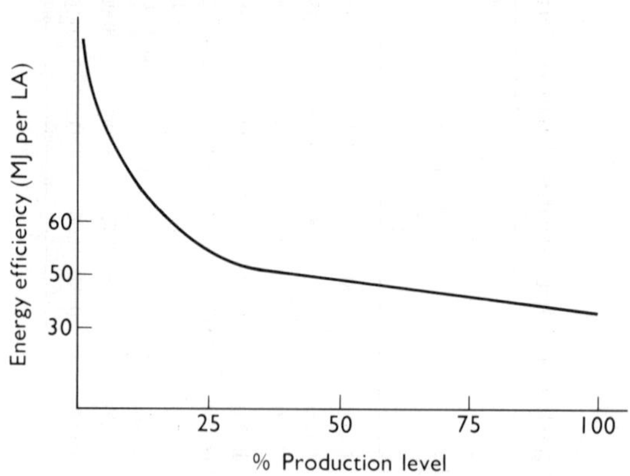

Fig. 11.2 Typical energy efficiency of a distillery as a function of production level.

Energy conservation techniques

When we have measured how much fuel is being used, discovered what the energy efficiency is, and have decided what the new target for energy efficiency is going to be, there are several separate and some interconnected paths which can be followed in order to save energy. For convenience, these paths may be split into three, which involve different disciplines and ultimately methods of conservation. These are:

1. methods of saving energy which involve no expenditure of money;
2. methods where, by spending usually less than £1000, large savings can be made (>£10 000 per annum);
3. large capital investment schemes.

OPTION 1

This is where energy conservation starts. These methods of saving energy have by far the best return on investment, since they do actually save money but hardly ever hit the headlines as most of the schemes under option 3 do. They require management to look at all aspects of plant operation to review what is being done. Questions must be asked at all levels about the current practices and methods in use in all areas of the plant. Sometimes it means that procedures will have to be rewritten, so some minimal expenditure of time and effort, and therefore money, may be involved but surely this is what management is all about.

Examples of this type of energy saving are:

(a) Does the steam really need to be turned on this tank, when hot water is not needed for another four hours?
(b) When warm water is being collected from a heat exchange process in a tank which ultimately overflows, could a valve not be closed slightly to collect warmer water and prevent the tank overflowing?
(c) After a quantity of grain has been moved by conveyors etc., does the equipment really need to run for another hour?
(d) Is it possible to increase fermenter original gravity and thereby alcohol content of charges for distillation and save on time and energy for distillation?

In addition to these examples, included in this area could be all aspects of what is referred to as 'good housekeeping'. There is no doubt that a clean plant not only looks better but also runs better. For example, where energy is being transferred, if there is less resistance it will be transferred more efficiently.

When considering procedures, it is appreciated also that

constraints are imposed by existing equipment. Sometimes this equipment is incorrectly sized, which can be wasteful of energy. Examples of this are gas burners oversized (or too many), electric motors too big and heat exchangers with too large (or too small) a surface area. This could be an area which needs to be brought to the attention of design and project engineers, but over the years the acquisition procedure has developed where it is thought to be better to be a little bit bigger just to be 'on the safe side'. Unfortunately, by the time all the people concerned have added their own little bit, items are often far too big for the actual required duty.

The summation of all these small reviews and corrections to procedures can often save a lot of energy, sometimes time and definitely money.

OPTION 2

Over the years it has become accepted that puffs of steam emanate from distilleries (and most other plants) as a sign that they are in production. These 'puffs' can be emanating from a variety of sources and whilst it is accepted that it can be very difficult to eliminate them, they represent a loss of energy.

The majority of steam (water vapour) vented from plants is from condensate collection systems or hot water storage tanks. After a heating process, which involves the condensation of steam to give up its latent heat, high temperature condensate should be returned to the boiler for re-evaporation. Unfortunately, sometimes steam traps after the heating process fail, which can result in steam passing right through the traps and into the atmosphere. A simple system of checking traps periodically, either ultrasonically, visually or by removal and inspection, can eliminate this problem.

It could be, of course, that if the steam traps are working satisfactorily, flash steam is being vented from a condensate receiver because of an excessive pressure differential. This can be rectified by reducing the pressure at which the steam is being supplied to the heating process (provided there is sufficient heat transfer area available) and this is another example of a no-cost 'option 1' action. Simple cold water spray systems can be installed in vents from condensate receivers to eliminate totally steam/energy loss, using water which would normally be used as boiler feed water make-up.

Also included in this area of conservation are repairs to leaking steam valves, minor insulation repairs and investment in equipment which can be used to optimise boiler combustion efficiencies. There are numerous instruments which can be used to set up the correctly proportioned amounts of fuel and air in order to produce the maximum amount of heat for a given quantity of fuel. The difference in running an oil fired boiler at 82% overall thermal efficiency as against 78% in a large malt whisky distillery can represent a monetary saving of £30 000 per annum. In a grain distillery, the saving would be in the region of £100 000 per annum.

As has been mentioned briefly already, maintenance of all items of combustion and heat transfer equipment must come into this area. All too often when a new plant is installed, it is properly set up but gradually over the years it is neglected until it becomes less and less efficient. It does not cost much to keep most of these items in good, efficient running order.

OPTION 3

Whereas all of the preceding comments can be applied to any style of distilling plant, the alternative energy conservation methods involving heat recovery are, to an extent, dependent on the style and method of operation of a manufacturing process.

Grain distillery processes tend to be more 'industrialised' and have been developed more or less in step with a developing industrial technology, and they can therefore be considered to be efficient industrial processes. The same can be said for by-products plants, or feeds recovery processes, which are comparatively modern developments. Nevertheless there have been several interesting and innovative developments in these areas, which will be discussed later.

Malt whisky distilleries

The area which has been developed more than most is the malt whisky production process, which had remained unchanged for centuries, and therefore offered several areas where energy conservation could be applied. In order to identify these areas it is necessary to consider a flow diagram of a malt whisky production process (Fig. 11.3).

It can be seen that the main energy requirements are for heating water required for mashing and for distillation, in both the wash and spirit pot stills. It is appreciated that there are processes which employ a triple distillation, but in most distilleries it remains two stage, so consideration will be limited to this. Basic heat balances indicate: in the mashing stage the heat requirement is approximately 5 MJ LA^{-1}; in the wash distillation stage 15 MJ LA^{-1}; and in the distillation of low wines and feints the energy requirement is calculated at approximately 10 MJ LA^{-1} produced.

Most of the sensible heat required in the mashing process is used to raise the temperature of cold make-up water to required temperatures for efficient mashtun operation. It can be observed from Fig. 11.3 that it is necessary to reduce the temperature of the worts going for fermentation. The heat recoverable at this stage is approximately 3 MJ LA^{-1}. This still leaves a net heat requirement in this stage of 2 MJ LA^{-1}. In the wash distillation, heat is required to raise the temperature of the charge to the still and subsequently boil off the alcohol/water mixture. Since water is used to condense this vapour, it follows that the latent heat which has gone into the boiling stage is transferred to the water used for condensing purposes. Similarly, the sensible heat used for heating the portion of the initial charge which has been vaporised is transferred to the condensing or cooling water, and there is still heat left in the residue in the still at the termination of the distillation. Of the initial 15 MJ LA^{-1} required for the process, a total of 13 MJ LA^{-1} is rejected in one of the three phases mentioned. In the similar spirit still distillation, of the initial 10 MJ LA^{-1} required, 8 MJ LA^{-1} are rejected and are recoverable.

It has been common practice in malt whisky distilleries for some time now to capture the heat rejected in the worts by using some form of heat exchange which transfers the heat to water which is subsequently used in the mashing process. Initially these units were composed of large bore tube heat exchangers called 'refrigerators', but these have been replaced almost universally by stainless steel plate heat exchangers.

Similarly, it has been realised for some time that there is a large amount of recoverable heat in the pot ale or spent wash discharged from the first distillation. Indeed, it has been discovered in wash stills which employ steam in pans or coils as the heating medium, that it is almost essential that the fresh wash

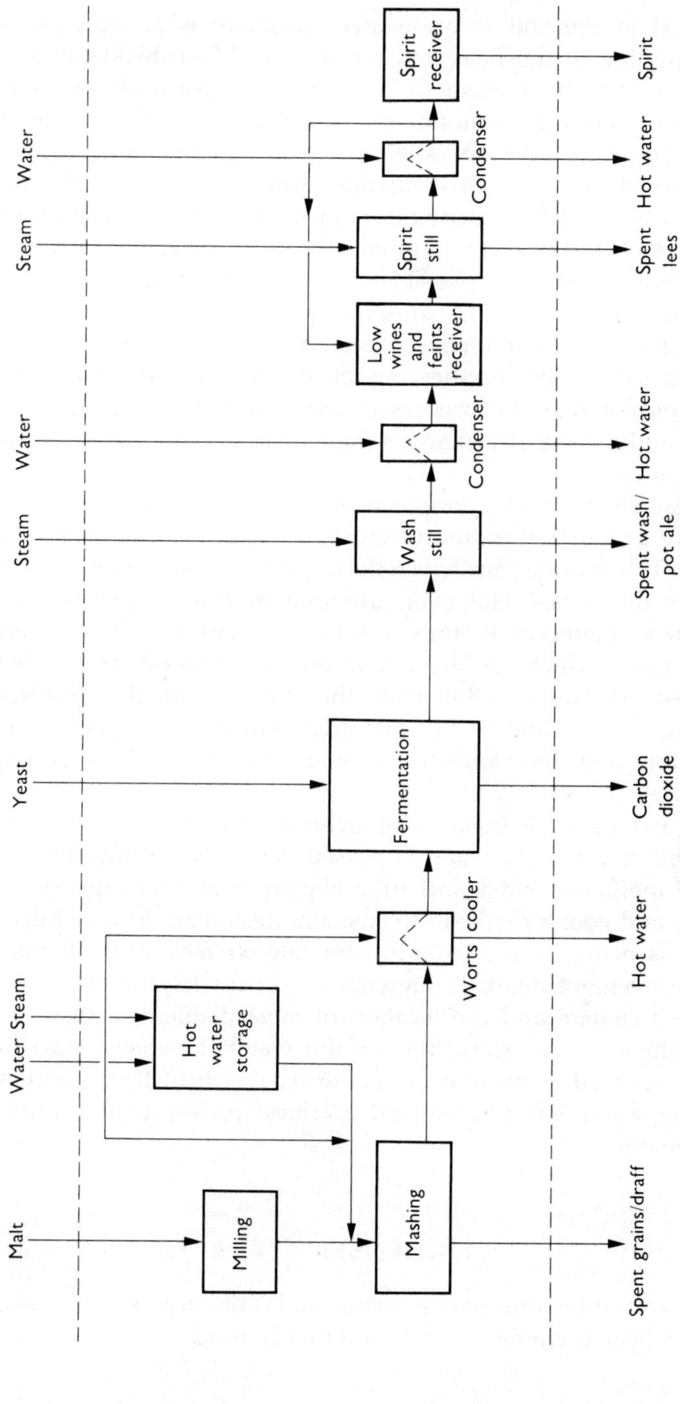

Fig. 11.3 Malt whisky production process, with no heat recovery.

charged to the still is pre-heated, or there is a large problem with fouling of the heat transfer surface. Most distilleries therefore use the spent wash to heat the fresh wash charge, which not only saves the amount of heat required for this, but also time. However, at some sites it is required that the pot ale be discharged at as high a temperature as possible if on-site evaporation is being carried out. In these instances, a portion of the condenser hot water is often used for pre-heating purposes.

It follows that it is possible to use the heat in the spirit still residue for heating the subsequent charge for the spirit still. However, this heat recovery option has not received as wide acceptance as the pre-heating of the wash still charge. It is perhaps felt that the process is too near the production of the final product and any changes may affect the quality of the spirit collected.

These then are the simple heat recovery options in the malt whisky production industry, most of which have been adopted. Figure 11.4 more accurately reflects current practice in most malt whisky distilleries. However, the total amount of heat recovered by these methods is only 8 MJ LA^{-1}. There is still another 16 MJ LA^{-1} which has been put into the process and is being rejected as waste. Obviously this heat is in the condenser cooling water, and it is this area which has received most attention and has attracted a wide variety of heat recovery schemes.

In a typical Scottish malt whisky distillery, an excessive amount of very cold water is passed vertically through the tubes of an inefficient shell and tube copper heat exchanger to condense and cool a distillate to, usually, less than 20°C, whilst the water is being heated to approximately 50–60°C and discharged to a convenient nearby watercourse. Fortunately, because of the Scottish climate and the location of most distilleries, there is no problem with the acquisition of the water. However, as I have said, the need to make an effort to recover heat from condenser cooling water has necessitated a critical review of this piece of equipment.

CONDENSER TYPES

Shell and tube one-pass vapour and one-pass water, vertical copper heat exchangers are traditionally used but even this is a

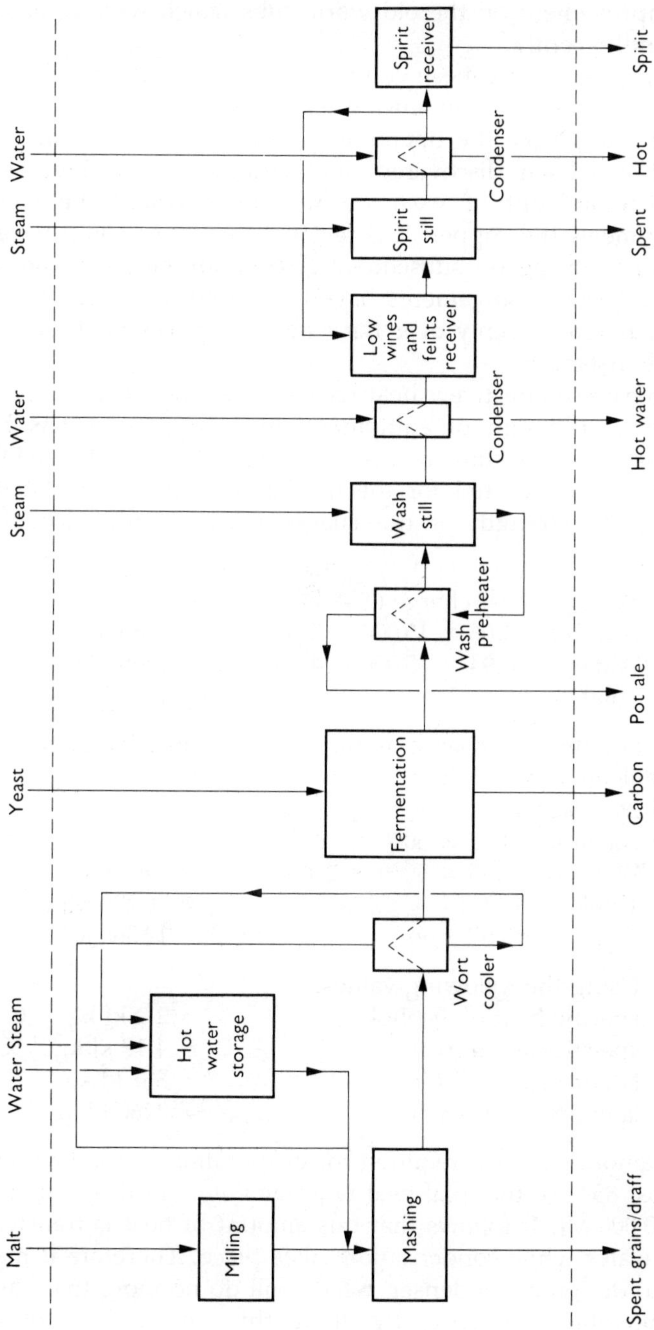

Fig. 11.4 Typical current practice in malt whisky production, showing heat recovery methods in use.

vast improvement on the old worm tubs which were commonly used until recently.

The problem with these condensers is that they are in reality doing two jobs, i.e. condensing and subcooling, and variations in heat loads vary the operating parameters of the condenser. Also it has been discovered, and accepted, that there is an inordinate amount of wear on the tubes where the vapour initially meets the copper. It is felt that this effect could have an important bearing on subsequent flavour formation in the final product. Baffle arrangements have been used to deflect the flow of vapour more evenly over the exterior surfaces of these tubes in some instances.

To maximise effectively heat recovery has meant that a certain amount of redesign of equipment and the process has been necessary. If we assume a wash still charge of, say, 15 000 litres, the total heat required for the distillation can be calculated as follows, the strength of the charge taken as 6% alcohol by weight:

1st stage: heat charge 65–95°C

Alcohol	$0.06 \times 15\,000 \times 2.4 \times 30$	= 65 MJ
Water	$0.94 \times 15\,000 \times 4.18 \times 30$	= 1768 MJ
Total		= 1833 MJ

Collect one-third of charge as low wines at an average strength of 15% alcohol by weight.

2nd stage:

Alcohol	$0.15 \times 5000 \times 837$	= 628 MJ
Water	$0.85 \times 5000 \times 2260$	= 9605 MJ
Total		= 10 233 MJ
Total for distillation		= 12 066 MJ

Using the following values:

specific heat of alcohol	= 2.4 kJ kg^{-1} $°C^{-1}$
specific heat of water	= 4.18 kJ kg^{-1} $°C^{-1}$
latent heat of alcohol	= 837 kJ kg^{-1}
latent heat of water	= 2260 kJ kg^{-1}

The amount of heat required for evaporation of the low wines is almost 85% of the total heat requirements – in this case, more than 10 000 MJ. It follows that this amount of heat is transferred to the water while condensation takes place. Therefore it makes sense to design a 'condenser' which will do no more than, as far as is possible, condense. By doing this, the amount of heat recovered can be maximised, whilst high-grade, high-temperature water can be produced and collected for a variety of purposes.

If we consider this case further, a simple aftercooler can be used to cool the collected low wines to, say 20°C. The heat or energy involved here is:

Alcohol	$0.15 \times 5000 \times 2.4 \times (95 - 20)$	= 135 MJ
Water	$0.85 \times 5000 \times 4.18 \times (95 - 20)$	= 1332 MJ
Total		= 1467 MJ

Observation of these calculated values shows that the largest amount of heat required is for the evaporation of water, which is then condensed, only to be re-evaporated in a second distillation stage. It is this recoverable heat from condensers, which can be reused in a variety of ways, that has led to the installation of various projects over the last few years.

Some work has been done using plate heat exchangers for condensing purposes. Because of the plate configuration and design, these units have much higher heat transfer coefficients. The development of plates pressed from copper sheets has attracted companies who recognise the importance of the presence of copper during the condensation stage. Also, because of the much higher conductivity of copper compared with stainless steel, which is the usual plate fabrication material, the heat transfer properties of these units are much enhanced. Because of the flexibility of plate heat exchangers, heat transfer surfaces can be easily cleaned and they can be applied to varying heat loads by addition or subtraction of plates. They appear, therefore, to be ideally suited to maximise heat recovery, and certainly can be very effective in applications where a cooling water shortage may exist. The biggest problem encountered with plate heat exchangers has been the previously mentioned wear in the vicinity of the vapour inlet. Despite regular monitoring and rearrangement of plate packs to minimise the effect, the level of deterioration has made them economically unjustifiable at some locations.

EXPORT HEAT

There have been several schemes over the years which have relied on hot condenser water for a variety of agricultural and fish farming schemes. However, most of these projects have foundered, mainly because of the lack of guaranteed water when distillation was not taking place, and partly because of a lack of specialised knowledge of these methods of production. There is one notable exception to this – the successful tomato growing scheme at a distillery in Aberdeenshire.

More obvious use of this water can be made in the distillery-related areas of by-products recovery. High-temperature water can be flashed under vacuum to evaporate the spent wash or pot ale residue from a wash still distillation to produce a syrup, which can be sold directly for animal feedstuff or can be mixed with spent grains or draff, which can then be dried to make dark grains. This scheme not only saves on the energy which would be required by a conventional multi-effect steam-heated evaporator, but also eliminates the transport costs of hauling this residue, either to disposal or to a more conventional, larger-sized feeds recovery or dark grains plant. Operating experience with these vacuum evaporation plants has indicated that the heat available from the distillation process totally meets the heat requirements of the evaporating equipment. Heating of the office blocks, workshops etc. can also be easily carried out with reject condenser water, but once again, as in the case of fish farming and nurseries, a back-up service must also be supplied.

MASHING REQUIREMENTS

As mentioned earlier, there is a shortfall in the amount of heat recoverable and required in the mashing process. Hot water from distillation condensers can be used either directly or indirectly, depending on the source of the cooling water to make up this difference.

STILL-HOUSE REQUIREMENTS

Since malt whisky distilleries work on a batch process, sometimes heat demand does not always match heat availability. A simple example of this is where a wash still charge is pre-heated, prior to charging, using condenser hot water. Hot condenser water can be used to supplement a system which already uses spent wash to effect the initial temperature rise in the fresh wash to the extent that the 'heating-up' period of the charge can be totally eliminated. It can also be used to supplement any pre-heating system for spirit still charges. However, for this purpose storage facilities must be provided for the hot water. Using a variety of storage tanks and heat exchangers, it is possible to recover a lot of reject heat, but this is only possible at high production levels where supply and demand can be reasonably well matched. Figure 11.5 illustrates the ultimate possibilities involving heat recovery in a malt whisky distillery. To minimise external pipework and storage facilities, however, it is best if waste heat can be 'recycled' to replace partly or even totally the heat requirements for distillation.

STILL HEATING METHODS

Thermocompressor

A large amount of the heat required for distillation can be reclaimed from condenser cooling water using a thermocompressor (Fig. 11.6). The importance of condenser redesign is once more stressed in this method of heat recovery. The hot water leaves the condensing equipment at as high a temperature as possible. The vacuum effect of the high-pressure steam fed to the thermocompressor changes a proportion of the hot water to steam and is recompressed to heat the still contents. It should be observed that the steam which is ultimately used for heating the still contents is at low pressure (approximately 48 kPa, 7 psig), and it will be necessary to increase the still heating elements' surface area to achieve the required distillation rates. The hot water leaving the condensers, which is not converted to steam, can of course still be used for a variety of heating duties, as previously described. The main difficulty with thermocompressor systems is possibly their inability to cope with a range of heating requirements, and it could be necessary to use more than one thermocompressor to achieve different heat input and distillation rates, e.g. running spirit and driving off feints on a spirit still distillation.

Heat pump

Heat pumps use energy to upgrade waste heat to a level where it can be readily used. Therefore it is important to make sure that the heat source is not too far from the temperature and heat requirements of the heat sink. It is also important that the relatively low-grade heat generated can be used efficiently,

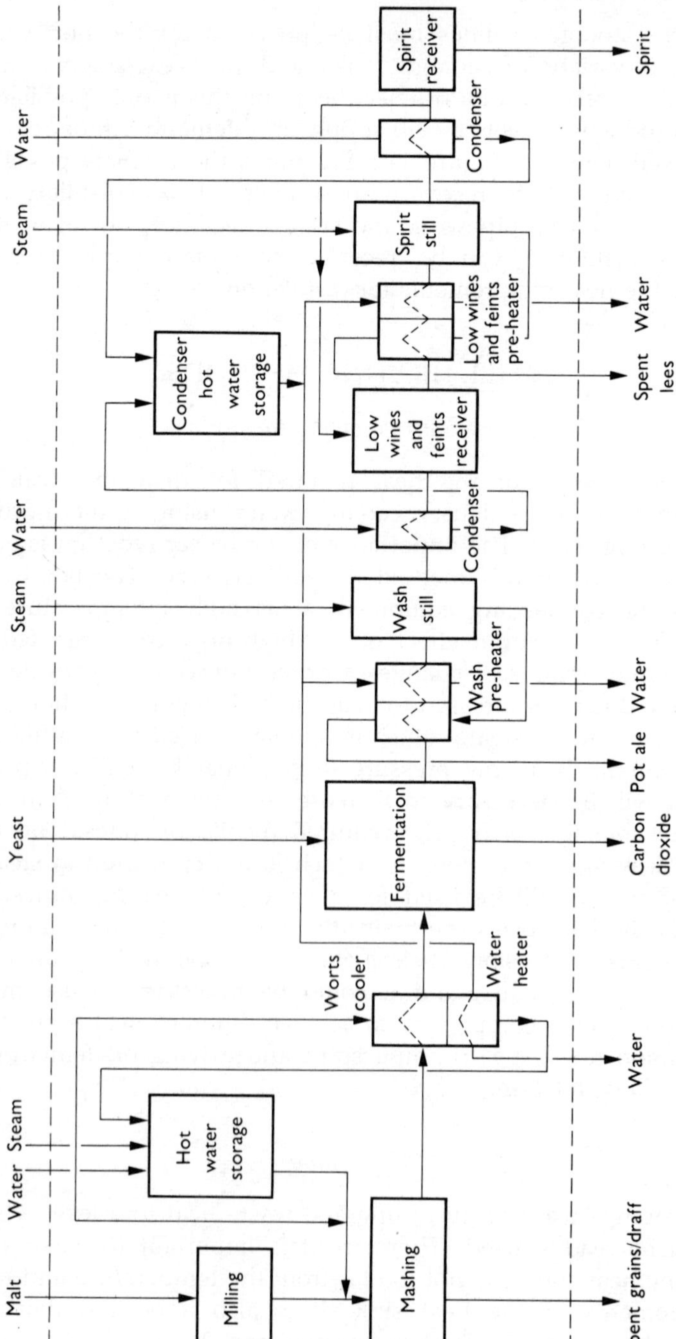

Fig. 11.5 Malt whisky production, showing maximum possible heat recovery.

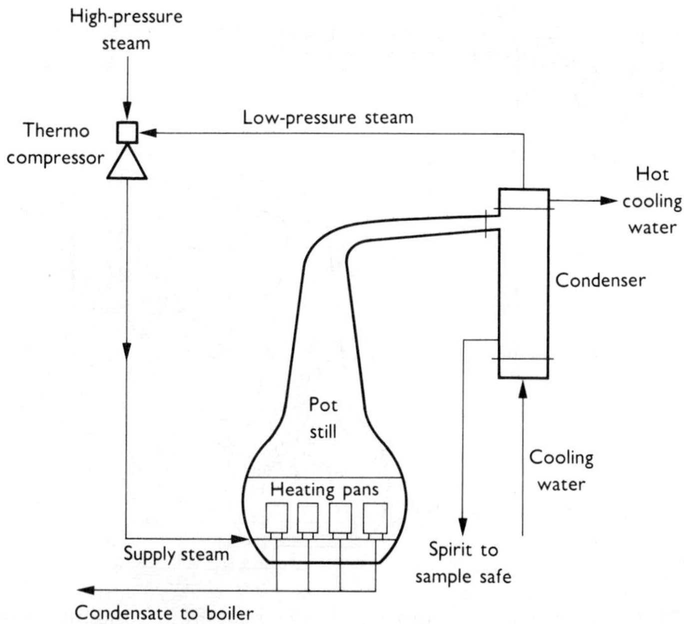

Fig. 11.6 Use of a thermocompressor to reclaim heat from condenser cooling water to heat a still.

because of the restrictions imposed by current heat pump technology. Figure 11.7 shows a proposed heat pump installation on a pot still. In this instance an electric motor has been shown driving the compressor, although it is possible to use a wide variety of prime movers, e.g. gas or diesel engines or turbines.

It would appear that this could be a completely closed cycle with all the hot water and steam from the condenser being recompressed to totally heat the still contents. As indicated, there are connections to the boiler because of start-up requirements, which demonstrate that the quality of cooling water required is boiler feed water quality. This is, of course, the case with the thermocompressor system as well.

Even more radical in design again is the old distillery 'condenser', which could probably be better described in this scheme as a 'hot water boiler'.

Obviously depending on the scale of the application, the size of the electric motor required to drive a compressor for an installation on one pot still could be between 100 kW and

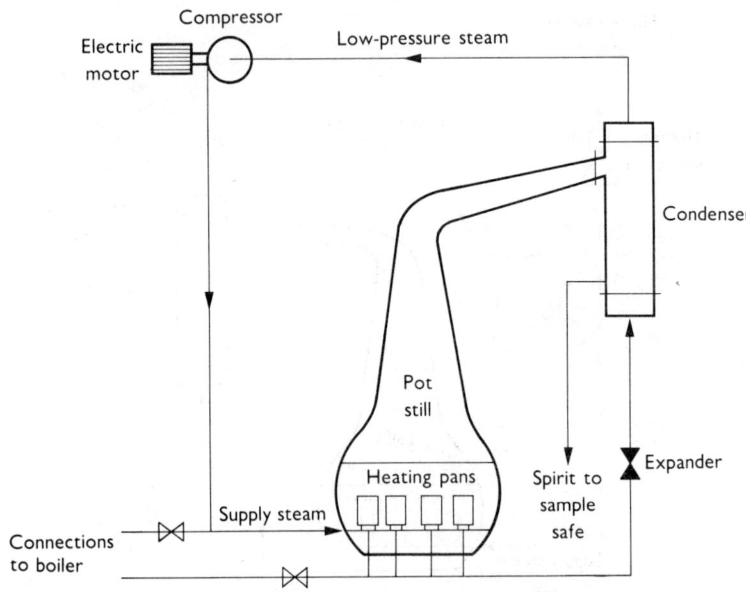

Fig. 11.7 Use of a heat pump to reclaim heat from condenser cooling water to heat a still.

150 kW. This size of electric motor is large by malt whisky distillery standards and may be intimidating. With a heat source (condenser water) available at 80°C and a heat sink needed (heating elements in a pot still) at 120°C, the compressor would have a COP (coefficient of performance/efficiency) of 4.5. This means that the heat delivered would be 4.5 times the power input. Simple calculations indicate that, whilst one tonne of steam per hour is required for a distillation, the total energy requirements of a heat pump system would be 610 kW, of which 135 kW would be electrical input energy and the remainder would be recovered heat.

Taking an average electricity price of 3.6p per kW hr, one hour's operation would cost £4.86. With a fuel oil price of approximately 6.5p per litre, the cost of a tonne of steam is very similar. Should the price of oil increase dramatically (and the price of electricity remain relatively stable), the benefits of a heat pump system would become more apparent.

Direct fired pot stills

Consideration so far has been limited to distilleries which use steam for indirectly heating pot stills. This steam is usually generated on oil, gas or coal fired boilers. However, there are still some distilleries which burn coal or LPG beneath pots to heat the still contents. This method has long been acknowledged as inefficient but is considered important in the formation of desired flavours in the final product. It is therefore unlikely that these heating methods will cease to be used in the foreseeable future.

The combustion efficiency of a pot still furnace using LPG or coal can be between 55% and 65%. Compared with an average steam boiler efficiency of 80%, this demonstrates an immediate disadvantage. The main reason for the poor efficiencies of these firing methods is the lack of an adequate surface area on the pot still bottom for the recovery of heat, which results in flue gases leaving the furnace, or wheel flue, at 500–600°C.

The overall thermal efficiencies of pot still furnaces are further reduced, in the case of old-fashioned brick settings, by air ingress and heat loss through cracks, difficulties in sealing combustion equipment and the pot stills themselves into the furnaces. Improvements can easily be achieved through the use of modern ceramic flue-insulating materials and by installing the furnaces in sealed and lagged metal 'containers'.

The high-temperature gases leaving the furnaces are ideal themselves for subsequent steam generation in a waste heat boiler. Several schemes of this nature have been implemented and have proved to be very useful, particularly in distilleries with direct fired wash stills and steam-heated spirit stills. Some of these projects are further described by Deakin (1985).

Grain whisky distilleries

A flow diagram for a grain whisky distillery is shown in Fig. 11.8. The processes are the same in as much as the processes involved are milling, mashing, fermentation and distillation. The main differences are: the different cereals used in grain distilleries (malted barley is used exclusively in malt whisky distilleries); the necessity for a 'cooking' stage in a grain distillery; and the continuous distillation in the grain spirit distillation, as opposed to the batch distillation from pot stills. The scale of production in grain distilleries is also much larger than in malt whisky distilleries, which, of course, means larger heat

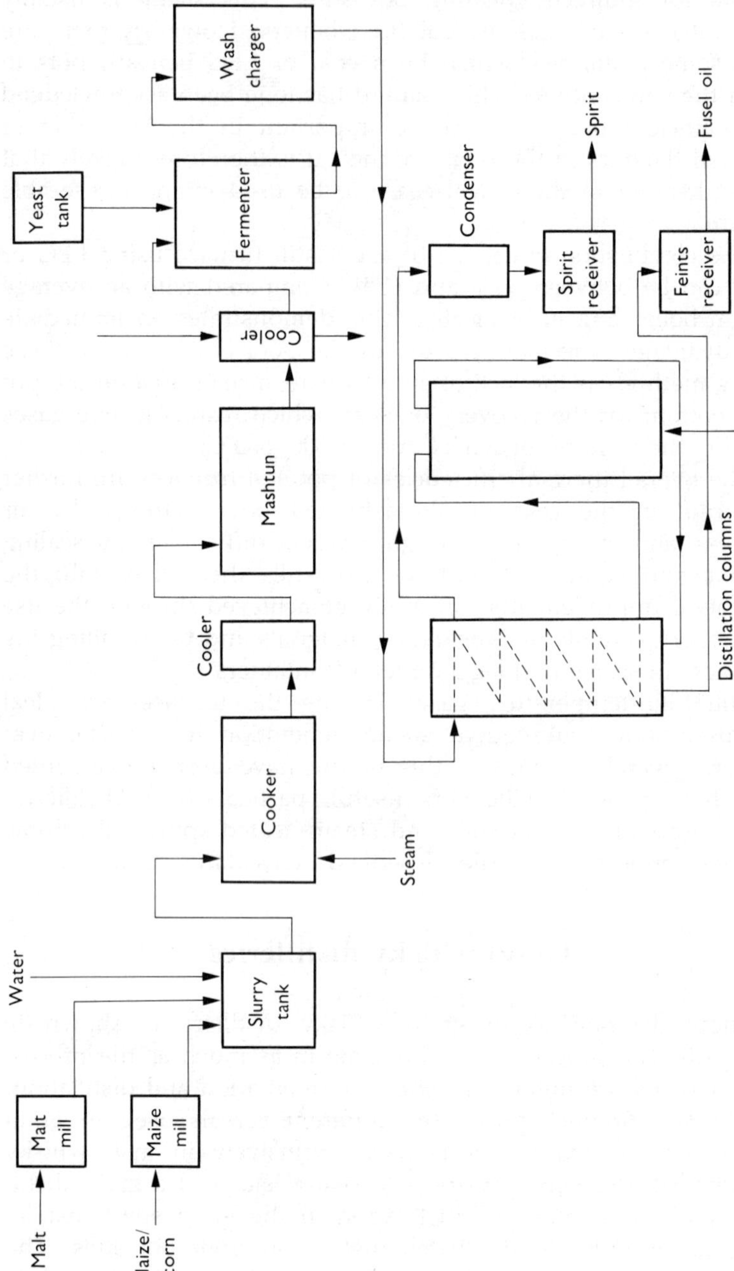

Fig. 11.8 Grain whisky production process, with no heat recovery.

loads and cooling loads. Much of the discussion about recovery of heat from worts in malt whisky distilleries also applies to the cooling of the worts prior to fermentation in a grain distillery. Similarly, hot water is recovered from overhead condensing apparatus and can be reused in a variety of different heating applications.

PROCESS INTEGRATION

Because of the scale of operation of grain distilleries, the technique of 'process integration' can be more easily applied. This has been developed as a means of identification of all the process requirements and needs, and incorporating or integrating these more effectively to improve plant management. An important area in any such process integration scheme must be energy use, and where possible reuse and/or conservation. One particular survey of a grain distillery analysed all the heating and cooling loads and the application of heat recovery techniques, and identified several areas where large savings could be made. It is important to realise that these techniques of analysis are most applicable to complex processes or plants and have not been applied to the smaller malt whisky production processes. There are opportunities for energy conservation in grain whisky production which do not occur in malt whisky production because of the differences between the processes, but these opportunities will be common to grain spirit processes in different countries.

MASHING AND FERMENTATION

In order to condition the grain for mashing, it is necessary to cook it. In most processes this is done batch-wise, but some use a continuous process. However, no matter whether batch or continuous, high-pressure steam is used and heat is recovered using a flash vessel at the end of the process. One such scheme has been publicised under the UK Government 'Energy Conservation Demonstration Projects'. In this particular scheme a slurry mixture of ground maize and malted barley is passed through a long tube and steam is injected at high pressure. The degree of cooking is controlled by regulating the amount of time for which the slurry is kept under pressure, and the process is terminated by venting into a vessel to cool the slurry. It is estimated that 80% of heat required for cooking is recoverable.

In this particular project the heat is recovered to heat the incoming slurry.

Some distilleries use the heat available from this process to heat the analyser distillation column. Because the cooker flash steam is only available intermittently and the distillation process requires a steady continuous supply, it has been necessary to incorporate a thermocompressor with a variable nozzle, which is automatically monitored and controlled to suit the varying steam flows.

In the mashing process, another significant operating method in grain distilleries is whole grain mashing. In this process the spent grains (draff) are not separated in the mashtun, as in a malt distillery, but are passed on through the fermenters and the distillation stage and only recovered at the end of the process. However, it is still necessary to cool the worts prior to fermentation, but this would be difficult to do using a conventional plate heat exchanger. Modified spiral heat exchangers are therefore used for this purpose because of their ability to pass a much higher level of solids and fibrous material. Again, this recovered water can be used to heat any cold make-up water required for the mashing process. An additional source of make-up 'liquid' for mashing is the 'backset' recovered at the end of the distillation process. This is the liquid portion of the stillage, which is recovered prior to treatment in the feeds recovery or dry-house process. This has a high heat content after being discharged from the column stills and the amount used varies between distilleries.

In the fermentation stage, it is possible to cool the fermenters to maximise alcohol production. In conjunction with this, some work is being done to develop a yeast strain which will be more ethanol-tolerant. It is hoped that the resultant 'wash' for distillation could contain 14–15% alcohol by volume. This obviously would reduce the amount of water requiring distillation in the still-house. In some countries it is possible to use enzymes in the cooking, mashing and fermentation stages, which will maximise efficiencies and increase yields. This obviously incorporates areas where energy could be saved.

DISTILLATION

Clearly, heat is required for distillation. This heat can be supplied via a reboiler fitted to a column or can come from steam injected directly into a column. As previously described, heat can be recovered from the cooking process using a thermocompressor. Another, better matched, heat source is from an overhead distillation column condenser. In one particular installation a falling film heat exchanger is used where water is evaporated at almost constant pressure and temperature, and the resultant steam is raised to a higher pressure using a thermocompressor prior to injection into the distillation column (Fig. 11.9). As in malt distilleries, spent wash/pot ale is a rich source of heat. In grain distilleries, also, it can be used for pre-heating incoming wash charges, but on a continuous basis.

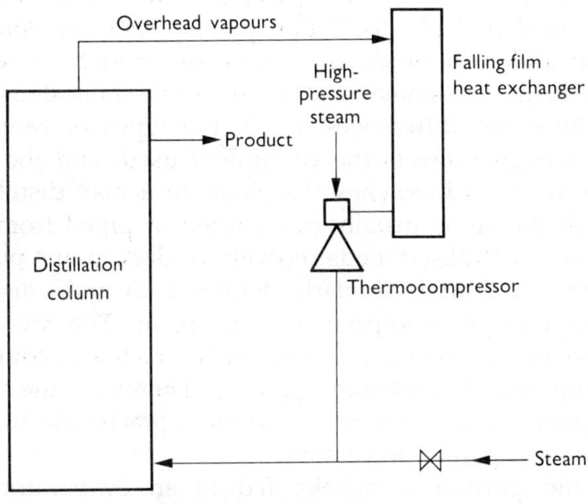

Fig. 11.9 Use of a thermocompressor to reclaim heat from condenser cooling water to heat a distillation column.

On a more radical theme, alternative distillation or separation techniques have been considered. One of these is extractive distillation, where congeners are stripped from wash using a large volume of dilution water. This method can reduce the amount of distillation effects by 50% and could lead to large energy savings. There have been doubts expressed, however, about the quality of the distillate which could be collected using such a separation technique. Some investigations have also been carried out into the use of a heat pump in the distillation process. However, instead of the mechanical compressor-driven pump, research has concentrated on the use of a chemical heat pump.

More conventionally, some distillers in the United States and particularly Canada have used hot water from cooling and condensing equipment to heat offices/workshops and more significantly barrel warehouses, which, with their sprinkler protection, could be a source of problems in the depth of winter.

Feeds recovery plants/dry houses

Although not directly a part of the whisky production process, these plants have long been associated with, and in some cases are an integral part of, distillation plants. They are common to both grain and malt production processes, and because of this the technologies are similar and often can be replicated between plants. The major differences are in the types of cereal used, which to a degree affects the equipment used, and the form in which the feed stock reaches the plant. In a malt distillery, the spent wash/pot ale is usually transported or piped from several plants to one centralised feeds recovery or dark grains plant. The spent grains or draff is similarly delivered. A grain distillery is usually big enough to support its own plant. The raw material here is usually the stillage or spent wash which also contains the spent grains from the mashing process. Therefore, the first step in these plants is to separate, as far as is practicable, the liquid and solids for separate processing.

The liquid portion is usually fed to an evaporator, which increases the solids concentration from about 4–6% total solids to anything between 35–50%, the final concentration being largely dependent on the suspended solids concentration. The most energy-efficient form of evaporator currently in use for this purpose is that employing mechanical vapour recompression (MVR) (see Fig. 11.10). These have been used successfully for several years now and, as shown, work on the principle of reusing the vaporised portion of the initial feed to the evaporator, recompressed and upgraded by a compressor, to heat and vaporise the fresh incoming spent wash or pot ale.

The measure of economy usually applied to multi-effect evaporators is the amount of evaporation compared with the amount of energy used. A 'good' figure for a multi-effect evaporator would be 7, whereas a MVR can, and regularly does, achieve more than 20. An added improvement to a MVR system used currently in a grain distillery is the method of driving the compressor. In this plant the drive is from a back-pressure turbine, which uses high-pressure steam, which, when passed out of the turbine at low pressure can be conveniently used in the distillery. As mentioned previously, evaporators can also be run using hot water generated from condensers in malt whisky distilleries.

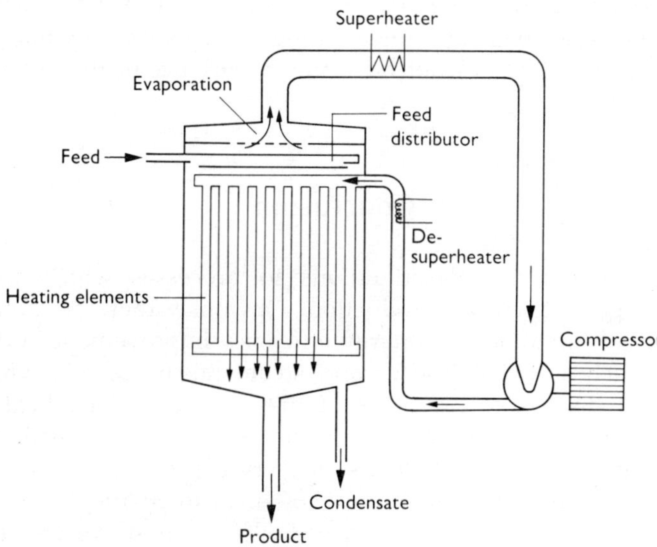

Fig. 11.10 Evaporator employing mechanical vapour recompression.

In the treatment of the spent grains or draff, it is important that the moisture content is reduced as far as possible prior to the grains reaching the drying stage, as this is usually energy inefficient. For this reason, a press is usually incorporated to raise the solids content of the grains fed to the drier to about 40%. In a grain distillery a centrifuging or screening stage precedes the press, whilst in a malt distillery the spent grains or draff are discharged from the mashing process at about 25% total solids.

In some driers which employ direct firing techniques, savings have been made by using boiler flue gas, either directly for drying or to supplement combustion air to the burners. In steam-heated driers it is possible to use flash vessels or steam or condensate from certain parts of the drying to generate low-pressure steam for subsequent heating or drying.

Using such techniques it is possible to achieve an energy efficiency in a dark drains plant of about 5 MJ kg^{-1} of dark grains produced. Although this cannot be compared to distillery production, it is interesting to note how these recovery processes have been developed efficiently in step with contemporary technologies, and have resulted in a lower net energy consumption per unit of production. No doubt the quality of the product has been considered in implementing changes but the quality of animal feedstuffs does not have the importance of the quality of distilled spirits.

Conclusion

I have limited my considerations to processes which are an integral part of whisky production. Obviously there are areas of energy conservation and energy efficiency improvement in distilleries, such as boiler houses and steam raising plants, which I have not discussed. Examples of these are combined heat and power (CHP) schemes, which can be applied to large distilleries, usually grain with on-site feeds recovery plants.

There is also an abundance of instrumentation and fuel additives which can be applied to boilers to improve efficiency. Some of these schemes are complex, whilst others are easily installed. Most distilleries are aware of these devices and schemes and, where possible, they have been implemented.

However, energy efficiency improvement has been achieved in many distilleries by implementing some of the techniques I have described. Bearing in mind the necessity of some locations to preserve a traditional appearance and the essential requirement that quality be maintained, remarkable progress has been made in a relatively short time.

References

Brown K J 1984 Energy use in the malt whisky industry. In *Energy Use in Malt Production*. Energy Efficiency Office seminar, Bowmore, Islay

Brown K J 1985 Energy efficiency in the drinks industries. *Brewers' Guardian* (February): 15–17

Deakin A W 1985 Reducing energy usage in whisky distilleries. *Brewing and Distilling International* **15**(6): 37–8

Newbert G J 1984 Why use energy? In *Energy Use in Malt Production*. Energy Efficiency Office seminar, Bowmore, Islay

Vennard E 1979 *Management of the Electric Energy Business*. McGraw-Hill, New York

CHAPTER 12

By-products

J ALSAKER

Introduction

Production of malt and grain whiskies also leads to the production of some by-products. Fortunately in the whisky industry they are nearly all recoverable and saleable by-products on which profits can be and are made. In fact, during some periods of the history of whisky distilling, producers have made more money on by-products than on the main products. The term 'by-product' is a wide-reaching term, as it covers nearly everything else which comes out of a whisky distillery apart from the main product.

As such, one can classify the by-products into two categories: those that are saleable, arising from the spent grains, pot ale, spent wash and carbon dioxide; and the unsaleable, such as spent lees and foul condensate. This chapter is concerned mainly with the first category, although some space will be devoted to the treatment of unsaleable by-products.

Characteristics of by-products

TYPES OF BY-PRODUCT

The two predominant by-products from a whisky distillery, grain or malt, are the liquid residue from the distillation, called pot ale or spent wash, and the solids from the cereals used. The solid residue, which is mainly fibres and kernels from the malt or maize/malt mixture, is called 'spent grains'. In malt distillery it will be left in the bottom of the mashtun or lauter tun and discharged from there, whereas on the grains side it may be taken out before or after distillation. If taken out before distillation, the appearance of the solids by-product will be quite similar to the malt side, apart from the fact that there may still be an additional 'dreg' from the still. In the so-called 'whole mash' grain distillation method the solids are kept with the liquid through distillation and appear as a slurry from the still. The spent wash and the grains will then be separated before the by-products are processed. In some instances the fibres and kernels may be separated before cooking, fermentation and distillation. These are normally sold in a dry state. In that case the by-products process will mainly be of unseparated proteins from the above dry separation process, and yeast residues. The residues from the second distillation in malt and grain distilleries are shown in Figs 12.1, 12.2 and 12.3.

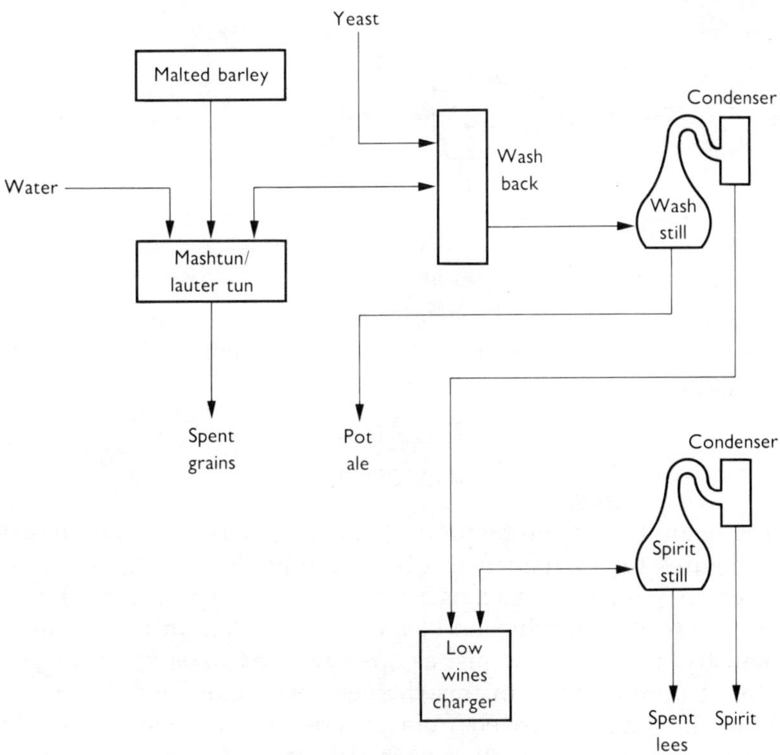

Fig. 12.1 Malt distillery process.

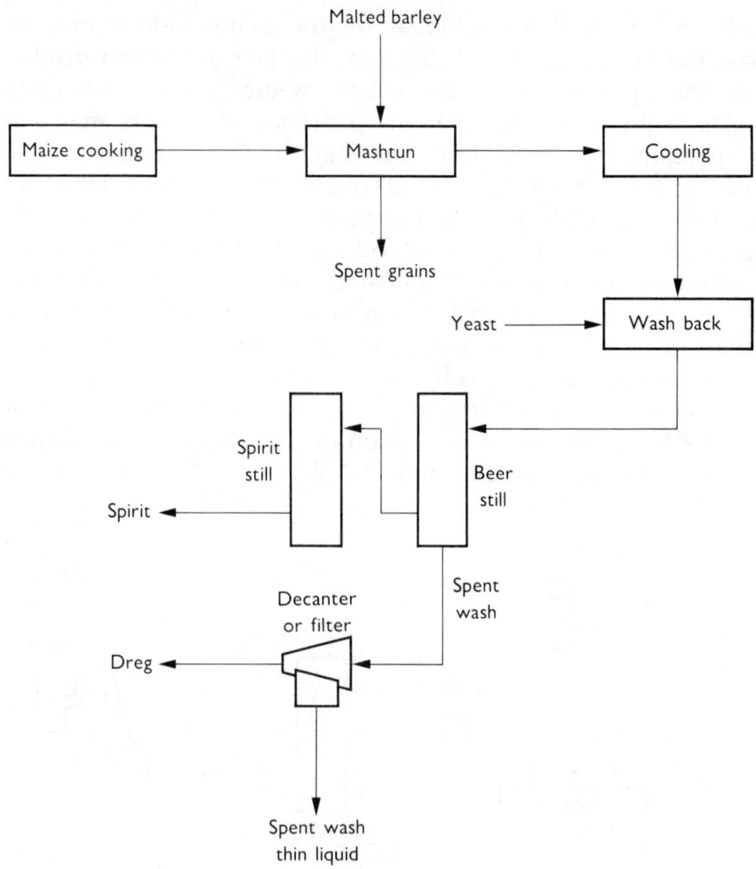

Fig. 12.2 Traditional grain distillery process.

QUANTITIES

The quantities of the by-products are normally calculated based on figures from experience. Until recently all such figures were given in pounds or gallons per gallon of proof whisky. However, with the introduction of the metric system in the distilling industry, they are now just as often referred to as kg or m^3 per litre of pure alcohol. In this chapter both figures will be given. Although figures derived from experience are most commonly used, they can be arrived at from a theoretical study if certain of the parameters are known.

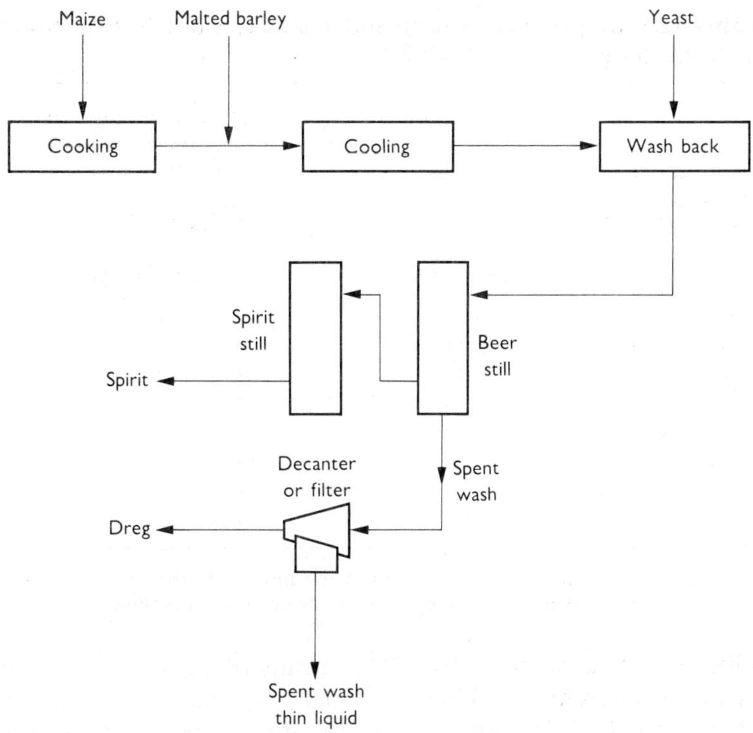

Fig. 12.3 'Whole mash' grain distillery process.

In a malt distillery the fibres, kernels and a fraction of unconverted starch will remain in the mashtun as the spent grains. If the starch content of the malt and the conversion efficiencies are known, the quantity of spent grains can be calculated. Similarly, if the alcoholic strength of the mash in the wash back is known, together with the yeast conversion efficiency, the carbon dioxide production and the strength of the spirits from the wash still, an analysis regarding the amount of pot ale and its solids content can also be carried out. However, the figures for spent grains and pot ale or spent wash are regularly available to the distillery management, since these products are either sold or processed, and so a theoretical study is seldom carried out. Typical figures in the two cases are shown in Table 12.1.

A normal size malt distillery may produce around 1 million gallons of proof whisky per year, with one mashtun, two spirit

Table 12.1 By-product output and characteristics from a Scotch malt distillery.

	lb per gal proof	*kg per litre pure alcohol*
Spent grains	17.5–20	3.5–4.0
Moisture (%)*	75–82	75–82
pH†	3.5–5.5	3.5–5.5
	gal per gal proof	*litres per litre pure alcohol*
Pot ale	4.5–6	8.5–11.5
Temperature (°C)	95–100	95–100
pH	3.3–4.5	3.3–4.5
Solids (%)	3.5–5.0	3.5–5.0
Portion dissolved (%)	60–80	60–80
Portion suspended (%)	20–40	20–40

*Moisture depends on the draining method and time in the mashtun.
†The pH depends mainly on storage of the grains before processing.

stills and two wash stills. This means that, to process the by-products from one single distillery, a plant would receive approximately 1.5 t (tonnes) spent grain per hour and approximately 3 $m^3 h^{-1}$ pot ale per hour based on a 120-hour week, 50 weeks per year.

In a grain whisky distillery the picture is somewhat different. Generally such distilleries are at least 10 times the size of an average malt distillery and the spent grains : liquid balance is different. The basic calculation is the same as above, but since maize, the predominant raw material, contains more convertible starch in relation to non-convertible, there will generally be more liquid and less solids. This has a certain impact on the by-products processes, as will be explained later.

The quantities shown in Table 12.2 are for a 'whole mash' process, where the non-soluble particles are taken to the still and are separated by means of screens, screw presses, decanters, or combinations of these later on in the process. A normally sized grain distillery with a production figure of 30 million litres pure alcohol per year (12.5 million proof gallons per year) will thus have to process approximately 4–5 t spent grains per hour and approximately 75 m^3 spent wash per hour. This is based on the same running time as for a malt distillery.

Table 12.2 By-product output and characteristics from a grain distillery.

	lb per gal proof	*kg per litre pure alcohol*
Spent grains*	5.0–6.5	1.0–1.25
Moisture (%)		
(after separation)	70–75	70–75
pH	3.5–5.5	3.5–5.5
Temperature (°C)	95–100	95–100
	gal per gal proof	*litres per litre pure alcohol*
Spent wash	9–12	16–21
Solids (%)		
(after separation)	1.5–3.0	1.5–3.0
Temperature (°C)	95–100	95–100
Portion dissolved (%)	40–70	40–70
Portion suspended (%)	30–60	30–60

*The final amount of spent grains will also vary somewhat with the grind, the distillation method and the separation of the fines. Because there is considerably more water present than insoluble solids, only small variations in the above methods will cause changes to occur in the quantities of spent grains and spent wash.

FINAL PRODUCTS

The most common final product is the mixture of spent grains and pot ale or spent wash. This dried product is called 'dark grains', due to its colour. However, the plants in Scotland also produce different by-products. A summary of these is given below.

Malt distillery

DDS – Distillers dried syrup. In these plants the spent grains are normally sold wet and the pot ale concentrated and dried.

DDG – Distillers dried grains (light grains). The grains alone are dried whereas the pot ale is disposed of by other means.

Spent grains. The grains are sold wet and the pot ale disposed of.

Pot ale syrup. This is a concentrate of the pot ale sold in a liquid state.

Grain distillery

The grain distillery may have the same by-products as the malt distillery, and in addition:

Dried dreg. This may either be recovered solids in a distillery with a non-'whole mash' distillation process, or the dried residuals in a grain distillery with a dry separation process of fibres and kernels. Dried dreg particles are finer than dark grains and generally have a higher nutritional value.

NUTRITIONAL VALUES

Information is available on the usefulness of distillery by-products as animal feedstuffs (NSCA, 1984). Typical figures for the most common whisky by-products are given in Table 12.3. Spent grains are fibre-rich proteins with a protein value of 20–25% (crude protein). They have a metabolisable energy value (ME) around 10 MJ kg^{-1} and have a lower energy content than cereals, but are less degradable. Pot ale has a higher protein value, approximately 35% crude, and also a very high ME, around 14 MJ kg^{-1}.

Spent grains are mainly used in diets for dairy cows, but have also been successfully fed to sheep, as are distillers dried grains and pot ale syrup. Dark grains are mainly sold to compounders for mixing into their compounds, but can also be used for beef and dairy cows. Distillers dried syrup is not easy to handle as a dried product, due to dustiness and hygroscopic tendencies. It is mainly sold in bags to compounders for inclusion in poultry and pig feeds.

History

The history of whisky by-products has been dealt with by Campbell (1970) and McFadzean (1979). The by-products side of the Scotch whisky industry has quite naturally lived in the shadow compared with the production of the main products. Nevertheless, it has always been a major consideration for the distiller. With the production of whisky come the inevitable by-products, which are an asset in so far as they have relatively

Table 12.3 Dry matter content, chemical composition and feeding value of distillery by-products.

	Grain distiller's grains (draff)	*Malt distillers' grains (draff)*	*Evaporated spent wash*	*Pot ale syrup*	*Grain distillers' dark grains*	*Malt distillers' dark grains*	*Distillers' dried solubles*
Dry matter (g kg^{-1})	275	258	400	400	900	900	950
Crude protein (g kg^{-1} DM)	236	198	300	350	300	265	285
Ether extract (g kg^{-1} DM)	90	82	73	22	100	69	95
Crude fibre (g kg^{-1} DM)	170	173	—	—	94	113	53
Ash (g kg^{-1} DM)	28	33	62	105	49	55	126
Ca (g kg^{-1} DM)	0.9	1.7	0.3	1.6	0.6	1.4	40.0
p (g kg^{-1} DM)	2.1	3.7	11.0	22.0	5.5	11.5	10.3
Mg (g kg^{-1} DM)	0.6	1.4	5.7	6.6	3.0	3.5	7.2
Na (g kg^{-1} DM)	0.5	0.9	0.4	1.1	1.4	1.4	0.4
K (g kg^{-1} DM)	2.1	3.7	11.5	23.0	5.5	8.0	11.9
S (g kg^{-1} DM)	—	—	—	6.8	—	—	—
Cu (mg kg^{-1} DM)	12	10	120	133	45	52	168
Mn (mg kg^{-1} DM)	—	43	17	17	22	34	53
Zn (mg kg^{-1} DM)	—	—	56	18	77	9	179
Co (mg kg^{-1} DM)	0.04	0.02	—	0.11	0.04	0.04	—
Se (mg kg^{-1} DM)	0.03	0.02	—	—	—	—	—
Metabolisable energy value (MJ kg^{-1} DM)	11.3	10.8	14.5	14.2	13.5	12.2	12.4
Degradability of crude protein	0.55	0.80	0.95	0.95	0.65	0.65	0.90

high nutritional value for animals and are therefore saleable. They are undesired because they also have high pollution values with a very high biochemical oxygen demand (BOD) if untreated. Pot ale will typically have 25 000–35 000 ppm BOD_5 out of the still, and spent lees 1000–2000 ppm BOD_5. With most malt distilleries located along popular salmon rivers and grain distilleries in urban areas, they cannot therefore normally be pumped down the drain.

The spent grains are easy to dispose of, appearing in easy to handle quantities and being relatively simple to transport; pot ale and spent wash present the problems. In the early days pot ale was often sprayed on the fields, although a treatment plant already existed shortly after the turn of the twentieth century. As the production of whisky increased after the second world war, so did the effluent and by-products problems.

The first modern plants were built in the 1960s, and they were partly spray drying plants producing DDG, partly small dark grains plants, often one at each distillery. In the late 1960s and early 1970s whisky production increased considerably. At the same time legislation on effluents became far more severe and the price of energy increased. This forced new developments on the by-products side. Therefore, to be competitive, by-products plants needed a certain capacity and energy-efficient machinery. In order to achieve this, various distilleries, often belonging to different owners, would set up one common large plant, a so-called combination plant. Smaller uneconomical plants were shut down. Transport costs played an increasingly important role in the siting of distilleries and by-products plants.

In the mid to late 1970s the consumption of Scotch whisky stagnated and even decreased. At the same time the world protein market in general, and dark grains especially, showed signs of price instabilities, which made the planning of long-term investments difficult. More importantly, the price of wet grains on local markets increased considerably. This made it attractive only for the large by-product producers, such as the large combination plants or the grain whisky distillers, to continue to make dark grains. Other by-products plants were shut down, as were several distilleries.

Efficient ways of processing pot ale were studied, and some interesting new developments, such as pot ale syrup produced entirely by the heat energy in the still condenser, have been pioneered. Other promising developments are being tested.

Description of by-products processes

MALT DISTILLERIES

Dark grains

The most common process is the 'dark grains' process (Fig. 12.4). The grains are discharged from the mashtun or lauter tun into a storage hopper, and the pot ale likewise from the wash stills into pot ale storage tanks. The grains are fed from the storage hoppers into dewatering presses, which can either be single or twin screw, where the free water is removed. This free water, which also contains some suspended and dissolved solids, is then pumped together with the pot ale to an evaporator, which will concentrate the solids to a thick syrup. Finally the syrup and the pressed grains are mixed and dried to dark grains. From the drier, the dark grains are then normally directed to a pellets plant, which compresses the grains into fine

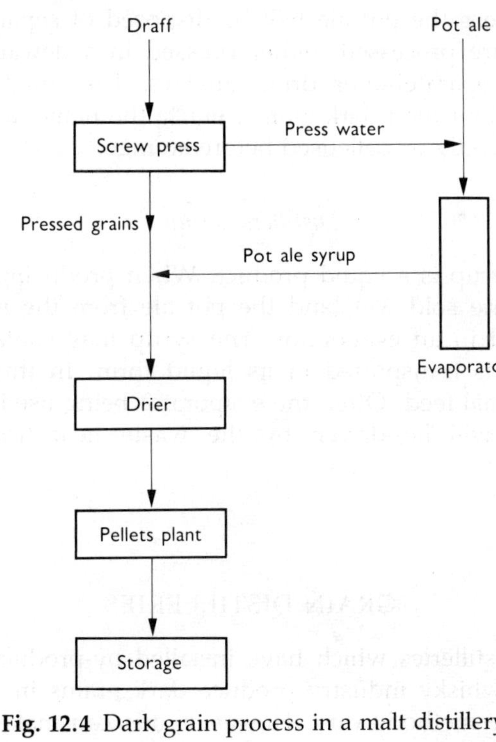

Fig. 12.4 Dark grain process in a malt distillery.

pellets. This product is easier to store and transport than the bulk meal.

In so-called combination plants, there will be an outside intake hopper facility, which will accept grains from the various distilleries feeding the plant, and likewise rather large pot ale storage facilities.

Distillers' dried syrup

When producing dried syrup, the spent grains will be disposed of separately. However, the pot ale will be pumped into storage tanks and from there directed to an evaporator and concentrated. The syrup from this evaporator will be fed to a spray drier or rotary drum drier, which will dry the syrup to a fine powder. In the case of drum driers, the final product may appear as flakes and have to be milled before bagging.

Distillers' light grains

In this process the pot ale will be disposed of separately. Only the grains are processed, either pressed in a dewatering press and dried separately, or dried directly. The dried grains are lighter in colour than dark grains, hence the name, and they can be stored directly or pelletised before storage.

Distillers' syrup

Distillers' syrup is a liquid product. When producing the syrup, the grains are sold wet, and the pot ale from the wash still is concentrated in an evaporator. The syrup may contain 35–45% solids, and is transported in its liquid form. In this form it is used as animal feed. Often the evaporator being used to produce this syrup will be driven by the waste heat from the still condensers.

GRAIN DISTILLERIES

All grain distilleries which have installed by-products plants in the Scotch whisky industry produce dark grains in some form, although the methods vary. However, one can make two broad distinctions, depending on the method of whisky production: the 'traditional' process, or the 'whole mash' process.

In the traditional process the grains are separated before distillation, as in a malt distillery. In this state the grains are pressed and the liquid is pumped to the storage tanks for the spent wash. The stills operate with wash containing very little suspended material, and the spent wash emerging from the stills will therefore also contain a very small amount of suspended solids. Some of these solids may be taken out prior to drying and evaporation, and this is called dreg. Otherwise the process is similar to that in a malt distillery. The spent wash is evaporated in large evaporators, often compressor-driven, and the mixture of the syrup from the evaporator and the grains is dried to a relatively dark product, which is the dark grains. Often dreg is sold separately, since the protein value is higher than that of dark grains.

In the whole mash process there is no separation of suspended solids before the still. Solids are transported along with the fermented wash to the stills, and are taken out from the stills as a liquid containing 5–8% suspended solids. These solids are either separated from the rest of the liquid by means of screens or, more often today, by large horizontal decanters. The remaining liquid will, depending on the method of separation, contain a certain amount of suspended solids and will be fed to an evaporator. The possible degree of evaporation is also directly dependent on the method of separating the suspended solids. If only a coarse separation is carried out, one can expect a syrup containing between 25% and 30% total solids. On the other hand, if more elaborate methods of separation are carried out, for example by means of decanters, a solids content of 40% to 50% in the syrup can be expected.

Decanters are efficient for separating solids from the liquid phase. However, they occasionally leave too much water in the solids cake, so that the drying load may be very high. Therefore, in some installations mechanical twin screw presses are combined with decanters to give an optimum grain dewatering and liquid characteristics in the liquid to be evaporated. In both cases the solid material emerging from the decanters or presses will be mixed with the syrup from the evaporator and dried. Normally the product from these installations in grain distilleries, which are relatively large, will be pelletised and stored, rather than bagged.

Unit operations

The most common unit operations which are found in the by-products plant are described below. When referring to production of saleable products, unit operations like pumping, mechanical conveying, air conveying etc., are not included. The unit operations described are: mechanical pressing; separation; evaporation; drying; and pelletising.

MECHANICAL PRESSING

Mechanical pressing is normally performed in single or twin screw presses (Figs. 12.5 and 12.6). In mechanical dewatering presses the wet grains are fed into the press through the inlet chute. The major and most important part of the press itself is its dewatering screw, which is a carefully designed heavy duty screw conveyor where the volume between the flights is reduced according to certain parameters. As this volume is reduced pressure is created, and due to this pressure water is squeezed out through the filtering screens. The solids are continuously made more compact, and are finally discharged through the outlet end.

In order to avoid rotational movement of the material with the dewatering screw, two such screws are often used, so that the

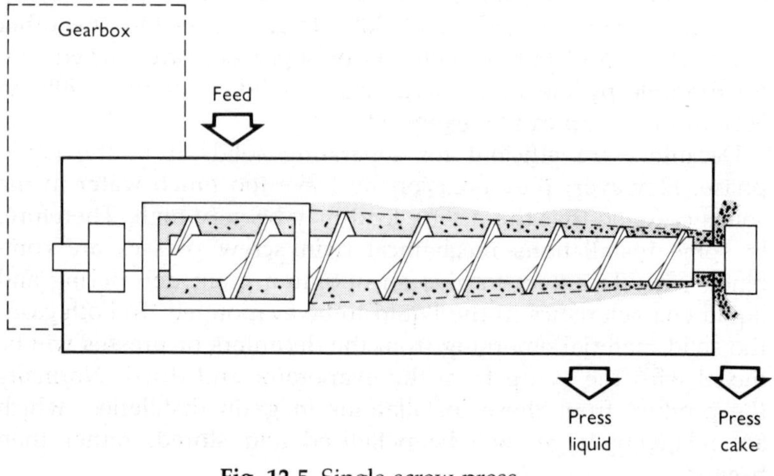

Fig. 12.5 Single screw press.

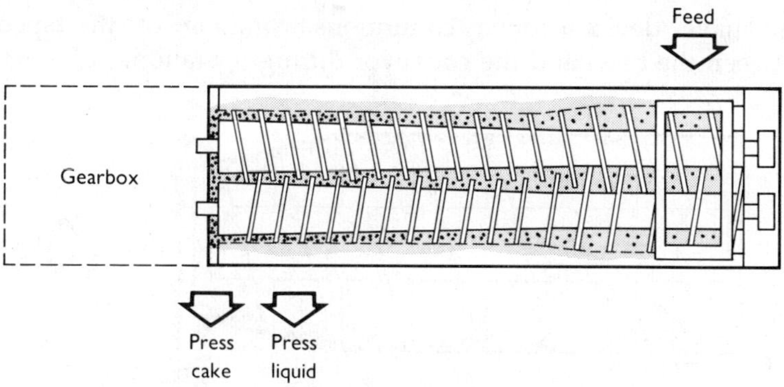

Fig. 12.6 Twin screw press.

material is transported along the press in a parallel pattern. In this way the friction between the material and the press screens is reduced, the pressure is more controllable, and the final result is normally a cake which contains less water and a liquid with less solids. Since the press liquid is normally pumped to the pot ale reservoir without further solids separation, the amount of solids in the press liquid is important.

DECANTERS

The main part of the decanter (Fig. 12.7) is the fast-rotating conveyor within the bowl, which has a speed different from the conveyor itself. Due to centrifugal force, the solids, being the heaviest particles, are forced against the bowl. The relative speed between the bowl and the conveyor also causes the solids to be slowly conveyed towards the outlet of the decanter. The liquid is discharged at the opposite end. The liquid level is maintained by using level devices on the liquid discharge end of the bowl. This level can normally be varied, although not usually during operation. If the level devices are set to make a deep liquid pool, there will be little time and distance for the solids to dry on the conical part of the bowl, the beach. With such a pool setting the liquid will have a long time to separate even fine solids. Deep pool settings will therefore give a relatively wet cake, but a good liquid separation. Similarly a shallow pool will give a long drying area on the beach, but a relatively short time for separation. This situation will give a relatively dry cake, but more suspended solids in the liquid. Some horizontal decanters also have devices for a continuous variation of the speed between the bowl and the conveyor during operation.

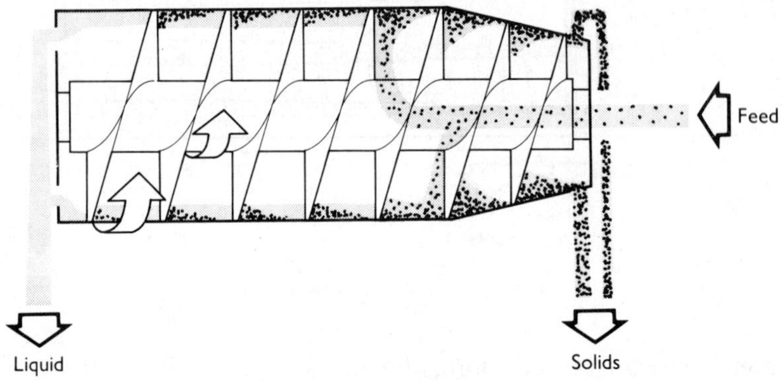

Fig. 12.7 Decanter.

It is normally not difficult to achieve a rough separation of the suspended solids and the liquid. However, as the solids become finer, they are more difficult to separate, and they will also carry more liquid with them as they are being separated out. When selecting a decanter to perform a certain duty, there will be a choice dependent on machine size, on achievable dryness of the cake and on the suspended solids in the liquid. This choice, and thus the performance of the decanting operation, is extremely important and has a major effect on the performance of the evaporator and the drier later in the process.

EVAPORATION

In the production of by-products, liquids containing 3–5% solids are concentrated to syrups by evaporation. Evaporation in this context means boiling off water vapour from a water/solids mixture. Unlike the distillation process, the liquid shows only one vapour pressure for each temperature and practically no difference between liquid temperature and saturation temperature of boiled-off vapours, hence no or very little so-called boiling point elevation.

On the by-products side of the whisky industry several different types of evaporators are found. However, they can generally be classified into two types: multiple effect evaporators

(Fig. 12.8), and mechanical vapour recompression (MVR) evaporators (Fig. 12.9). Waste heat evaporators are also used (Fig. 12.10), and though they could be classified in the first category above, they will be discussed separately.

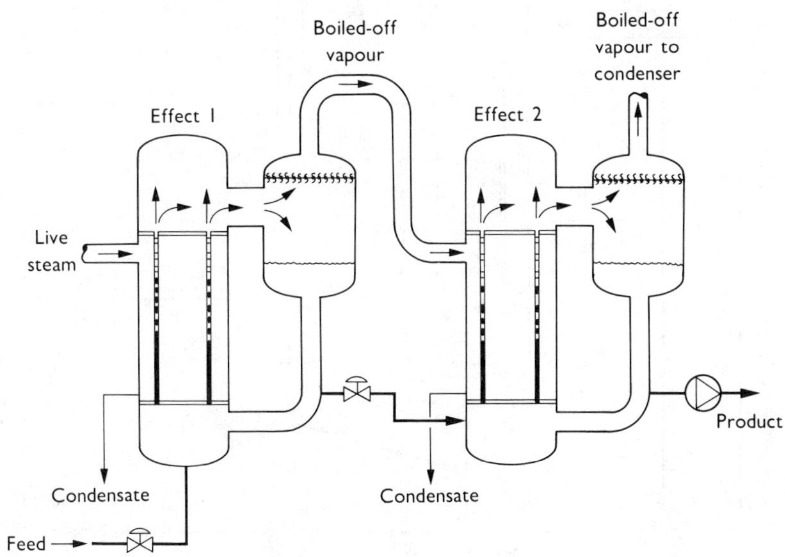

Fig. 12.8 Two-effect evaporator (principle).

Multiple effect evaporators

The multiple effect evaporator consists of several separate bodies connected together so that the vapours from the inside of the tubes from one effect serve as the heating medium on the outside of the tubes of the next effect. In this way, and by combining many effects, superior heat economy can be achieved, in practice according to the following:

No. of effects	*kg steam used per 4 kg evaporated*
1	1.2
2	0.6
3	0.4
4	0.3
5	0.25
6	0.2

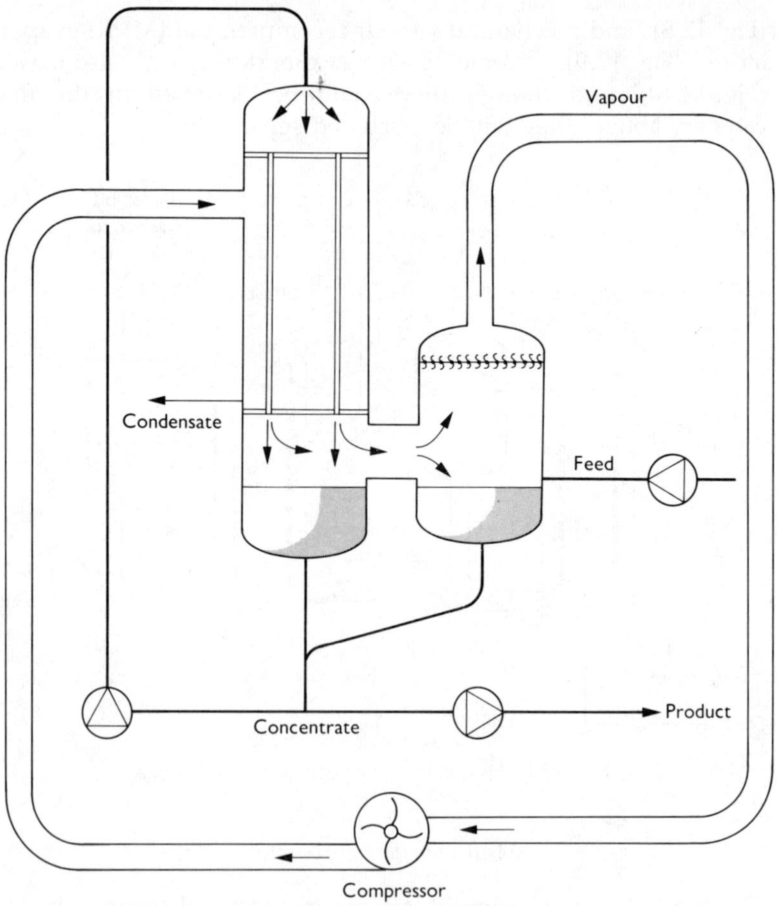

Fig. 12.9 Mechanical vapour recompression (MVR) evaporator, one effect.

Multiple effect evaporators consisting of more than six effects are normally not practicable, as the cost of equipment installation and automation will make it cheaper to look at the other alternatives.

The method of boiling can be either falling film (Fig. 12.11), rising film (Fig. 12.12), or forced flash evaporation (Fig. 12.13). In the rising film evaporator there is a liquid level inside the tubes, although the division between non-boiling, vapour formation and 'real' boiling is often difficult to observe. There is no boiling at the bottom of the tubes, due to the static height of the liquid.

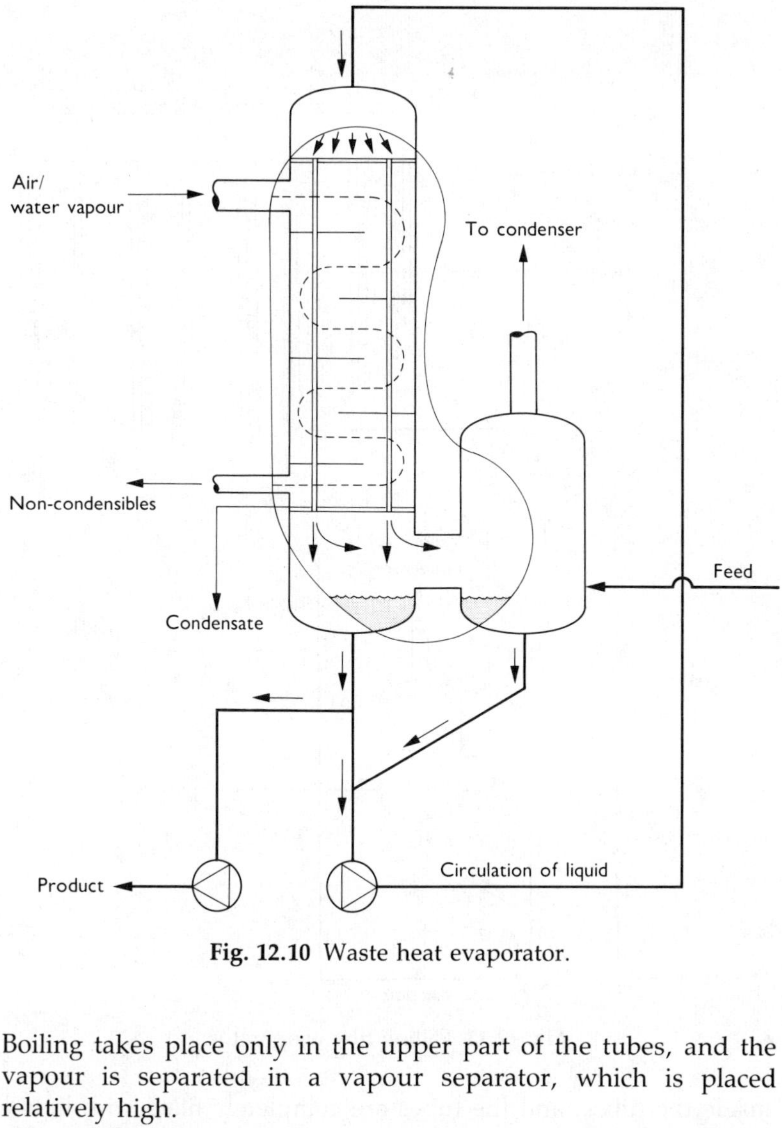

Fig. 12.10 Waste heat evaporator.

Boiling takes place only in the upper part of the tubes, and the vapour is separated in a vapour separator, which is placed relatively high.

In a falling film evaporator there is no liquid level inside the tubes. The liquid is directed from the bottom of the apparatus to the top of the tube bundle, so that on the inside of each tube the liquid forms a film, which due to the drag of the evaporation is forced down the tubes. In a falling film evaporator the liquid/vapour separation takes place at the bottom of the unit.

In the forced flash evaporation unit there is no boiling at all

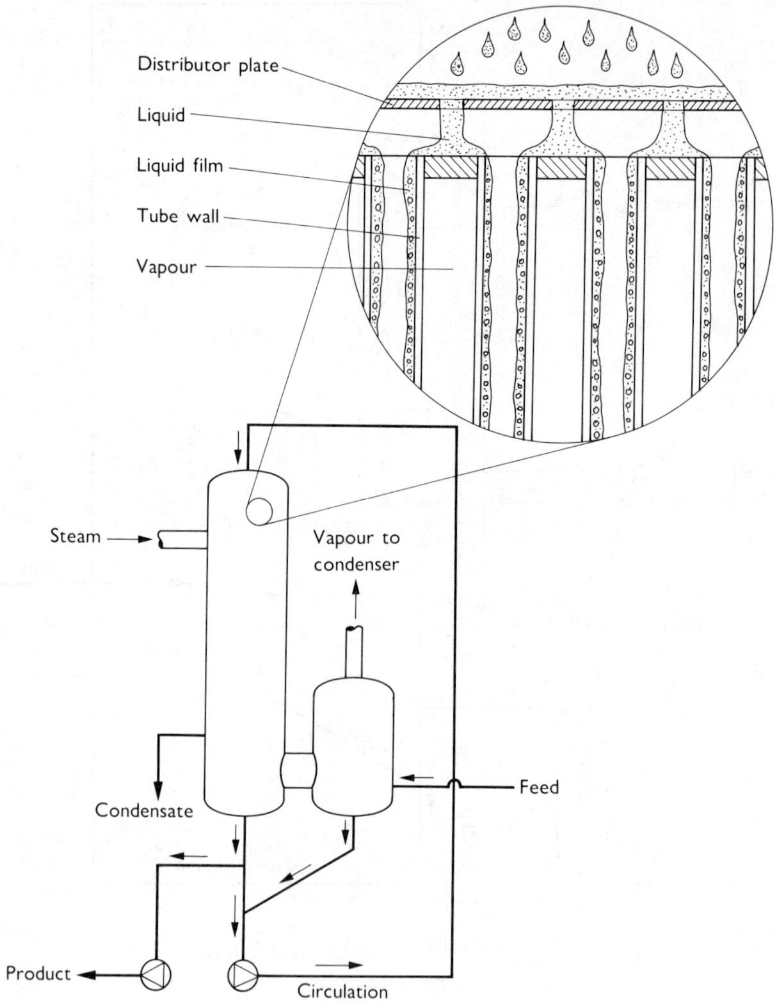

Fig. 12.11 Falling film evaporator.

inside the tubes, and the tubes are completely filled with liquid. A circulating pump creates a pressure high enough to keep the tubes full at all times. Evaporation takes place only after the pressure is released, normally just before the vapour separator.

Rising film evaporators were most common until the mid-1970s. Falling film evaporators are more used today, and the forced flash evaporator is normally used where relatively high solids concentrations are required.

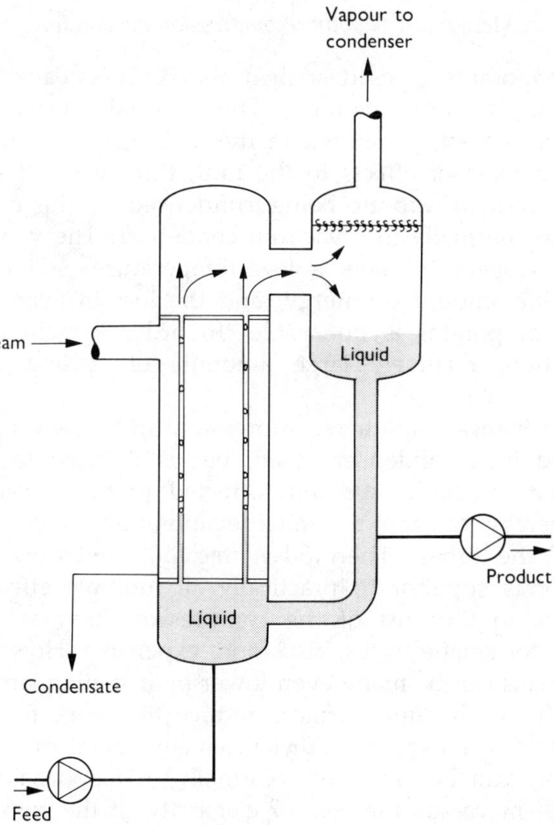

Fig. 12.12 Rising film evaporator.

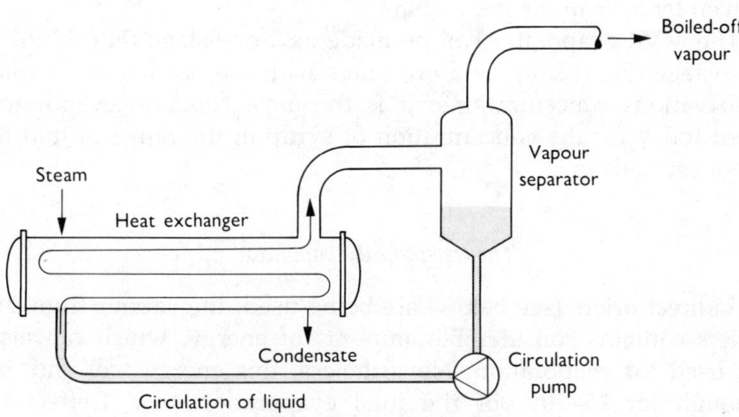

Fig. 12.13 Forced flash evaporator.

Mechanical vapour recompression evaporators

In the evaporator types described above, live steam is used as the primary heating medium. The thermal efficiency of the evaporator is then, as shown in the table, greatly improved by adding numbers of effects to the unit, but they will still suffer from the loss of vapour being condensed at the end of the evaporator, normally in a vacuum condenser. The vapour, even if it may appear to have a low temperature, still contains a considerable amount of energy, and the loss in even a five- or six-effect evaporator is noticeable. In order to achieve such a condensation, relatively large amounts of cooling water are necessary.

In a mechanical vapour recompression unit, this vapour is not condensed in a condenser at all, but is directed to a vapour compressor, recompressed and directed to the outside of the tubes, thereby causing a similar evaporation to occur on the inside of the tubes. Thermodynamically, therefore, the MVR evaporator is superior to practically all multiple effect evaporators. Due to the cost of the compressor they are normally, especially for smaller sizes, also more expensive. However, their running costs can be made even lower by installing large heating surfaces. Large heating surfaces reduce the work for the compressor, and in most cases efficiencies equivalent to 20–25-effect evaporators can be achieved. Normally, when considering the cost of steam versus the cost of electricity, if the compressor is electrically driven, an MVR evaporator will be designed to be equivalent to a 15-effect evaporator. The compressor can also be driven by a steam or gas turbine.

The MVR evaporator can be made as a one-stage (Fig. 12.9), a two-stage (Fig. 12.14), or a one-stage with the heating body split into various concentrations. It is the most common evaporator used today for the concentration of syrup in the range of mid to large capacities.

The waste heat evaporator

If indirect driers (see below) are being used, the vapour from the driers contains considerable amounts of energy, which can also be used for evaporation. Nevertheless, this energy will only be enough for 15–20% of the total evaporation as a 1-effect or 25–30% as a 2-effect unit. Such evaporators are therefore most

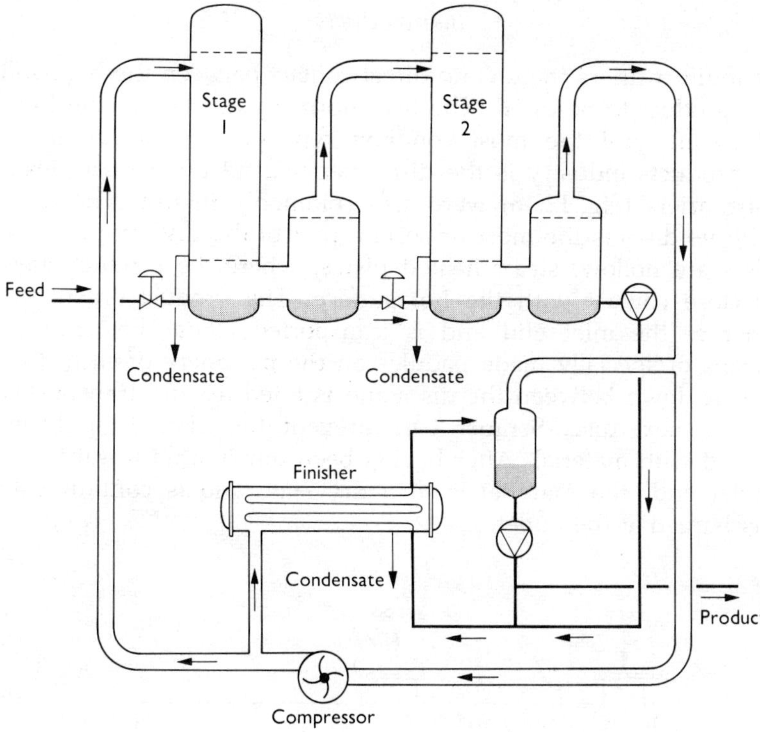

Fig. 12.14 Two-stage MVR evaporator with finisher.

commonly used as an add-on evaporator or a pre-evaporator in case the main evaporator has a shortage of capacity. In the waste heat evaporator (Fig. 12.10) the drier gases are directed outside the tubes in a special pattern, in so doing transferring their energy to the boiling liquid on the inside. These evaporators are normally made as falling film evaporators, due to the higher efficiency of heat transfer of such a system.

DRYING

Generally two types of driers are used for by-products processing in the Scotch whisky industry, indirect driers and direct driers.

Indirect driers

In indirect driers there is no direct contact between the heat and the product to be dried. The heat normally appears in the form of steam, and the most common type of drier used in the by-products industry is the disc drier (Fig. 12.15). Earlier steam tube driers (Fig. 12.16) were also commonly in use. The rotor with its discs is the most important part of the disc drier. These discs are hollow, steam-heated plates, where the product stays in close contact with the hot surface. The product enters the drier at the inlet end and is transported along the drier by means of specially made paddles on the periphery of each disc. It falls down between the discs and is lifted up and transported to the next disc. Scraper bars prevent the discs from being coated with material. After having been transported towards the outlet end, the material is in a dry state and is continuously discharged at the outlet.

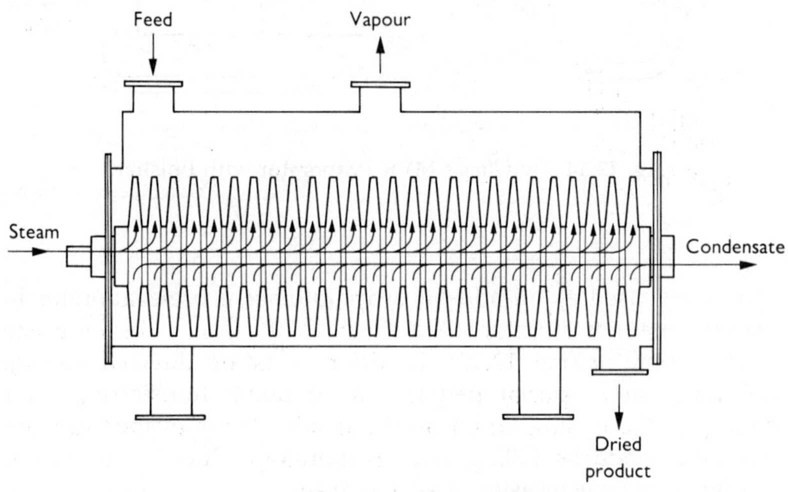

Fig. 12.15 The Rotadisc drier.

The steam tube drier (Fig. 12.16) has several specially arranged steam-heated tubes on its main rotor. The material to be dried must be pre-mixed with dry material, as this type of drier normally cannot accept the rather sticky mixture of syrup and grains. After being mixed with the dried material, the partly wet

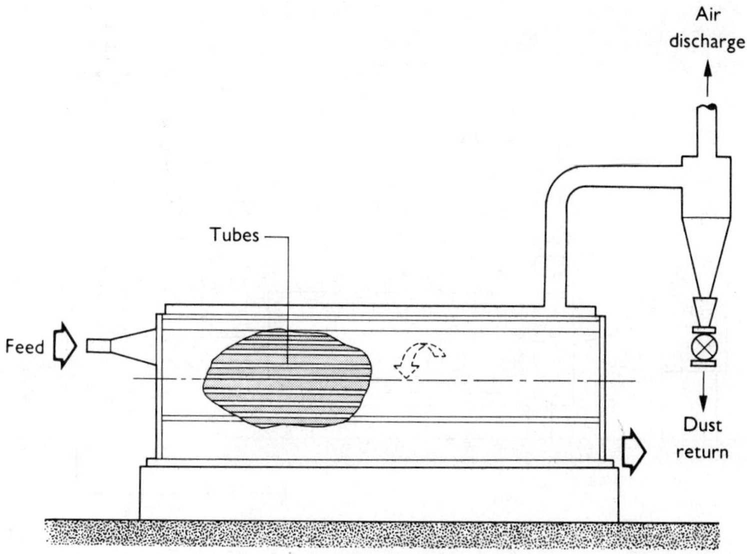

Fig. 12.16 Steam tube drier.

mixture is introduced to the drier inlet, where it comes into contact with the hot tubes. The material is transported along the drier by means of paddles on the tubes or on the structure carrying the tubes. As the material is lifted up, it falls on and between the tubes and loses water vapour when heat is exchanged. This drier also has an outlet door mechanism, and the dried product is continually taken out at the outlet end of the drier. Some of the material is back-mixed with the wet raw material, some is transported away for storage. The quantity of air used for the drying is small, compared with direct fired driers.

Direct driers

In the Scotch whisky industry today different types of direct driers are found: rotary direct fired driers (Fig. 12.17) and stationary direct fired driers (Fig. 12.18). In both cases the drier will consist of the following main items: furnace, drying zone, cyclone, and mixer.

The wet product to be dried is first mixed with already dried product in a mixing device, so that the moisture level at the inlet of the drier is kept below the level which allows sticky lumps to

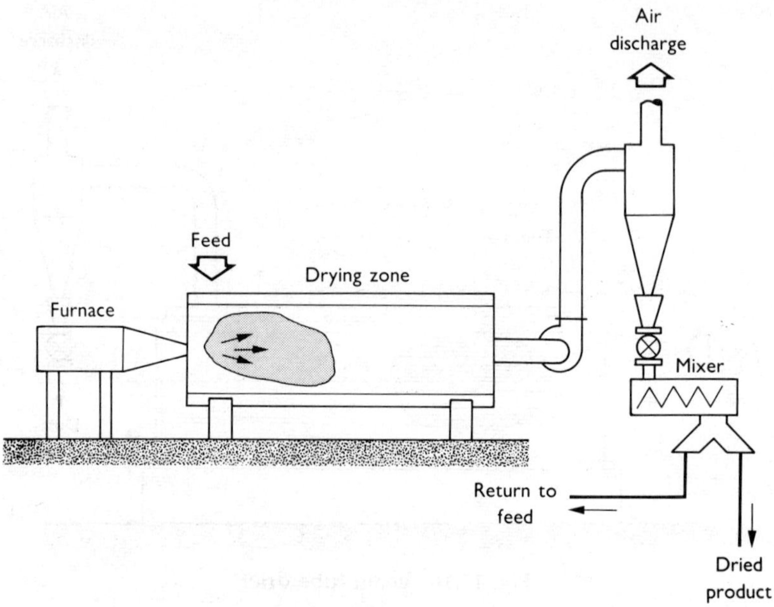

Fig. 12.17 Direct fired drier.

form. It is then introduced to the stream of hot air and combustion gases from the burner. In a rotary direct fired drier the rotor is continuously turning inside the drum. The material is continuously lifted up and dropped down to the stream of hot air and combustion gases. In the stationary types the material follows the gases up along one or more vertical tubes to a classifier at the top, which returns the solids which are not yet dried. At the end of the drier there will be large separating devices such as cyclones, which separate solids from the air. At the bottom of the cyclone the unit is equipped with a diverter valve which will direct some of the dried material back to the mixer, and some of the material away for storage.

When drying in a direct fired drier, the finer particles of the mixture to be dried can catch fire when exposed to high temperatures. Therefore the primary temperature must be kept below a certain level, which is achieved by large quantities of secondary air being introduced to the drier after the burner. This, however, affects the drying economy; the higher the temperature one can expose the product to, the better will be the economy of drying. Normally today temperatures around 400°C

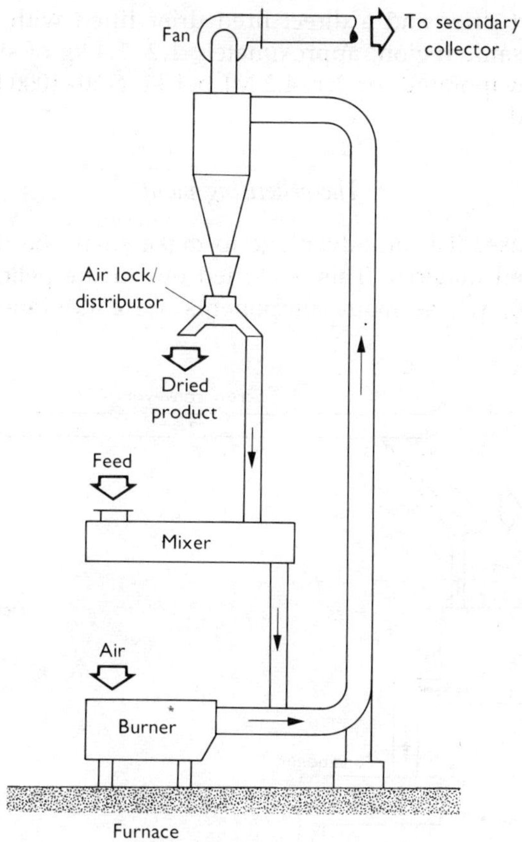

Fig. 12.18 Stationary direct fired drier.

are used. The driers will be fully automated to prevent such fires.

Direct fired driers, due to their capacity, are often found in relatively large installations, where otherwise several indirect driers would be required. They do require a considerable size of building to be erected, especially in terms of height. Indirect driers are installed both in large and small installations. They are often preferred where they can be placed inside existing buildings.

Direct fired driers can also be equipped with recycling devices for back-mixing of moist vapours from the cyclones. This will improve the economics of drying and make it possible to increase the initial temperature somewhat. The heat economy of an indirect drier and a direct fired drier fitted with recycle will be in the same region, approximately 1.2–1.4 kg of steam per kg of water evaporated, or 3.1–4.2 MJ per kg (750–1000 kcal per kg) evaporated.

The pelletising plant

In most cases it is an advantage to make small, hard pellets out of the dried material. This is carried out in the pelletising plant (Fig. 12.19), whose main components are conditioner, pelletiser and cooler.

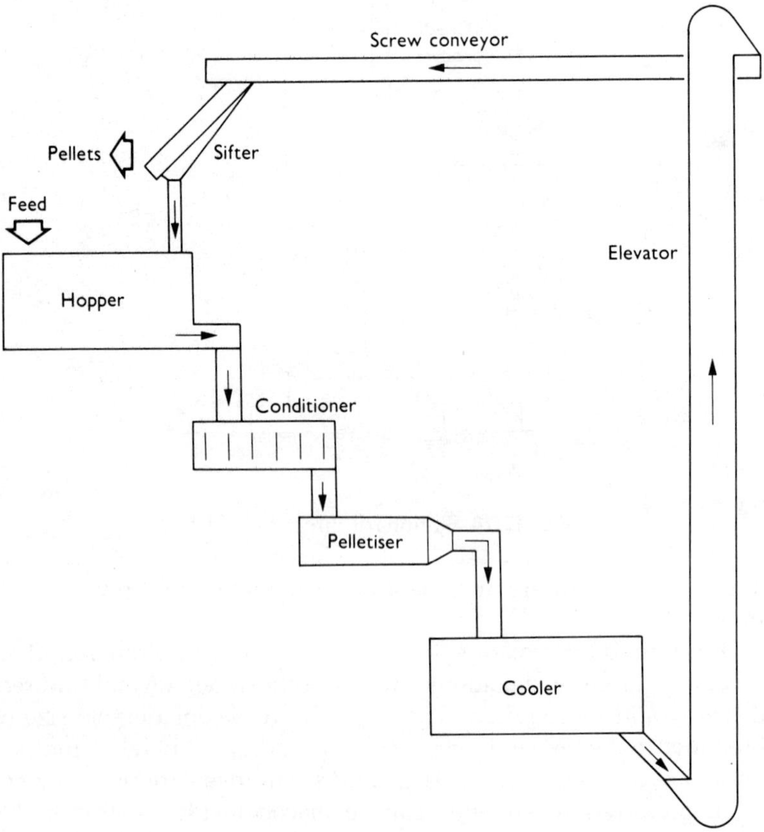

Fig. 12.19 Pelletising plant.

The material, after being stored in a feed hopper, is fed at a controlled rate to the conditioner, which is a fast-rotating mixer conveyor with a sparge tube. Normally some steam or water is added to the mixture through the sparge tube, as this improves product characteristics for pelletising. The slightly moist meal is then fed into the die of the pelletiser, where rollers press the meal through the holes of the die and thereby make it into compact pellets. Knives outside the die cut the material into suitable sizes and discharge them to the cooling conveyor, which cools the material down to approximately ambient temperature. The pelletising process will greatly improve the storage and transport characteristics of the material.

Economy in the production of by-products

It has been mentioned previously that the first small by-products plants in malt distilleries quickly became uneconomical, mainly due to their size and the methods of production. With the price of dark grains protein meal fluctuating and generally not increasing at the same rate as operating costs, and also the higher costs of energy which were experienced in the mid-1970s, the question of running a profitable by-products operation very much became a question of efficient methods of operation and size. This is clearly demonstrated in Fig. 12.20, which represents the production of dark grains from a small combination plant handling approximately 5 t of spent grains and 12.5 m^3 of pot ale per hour, and a plant three times this size. The following basic parameters are used:

Price of spent grains	£17.50 per tonne
Price of dark grains	£125.00 per tonne
Price of steam	£10.00 per tonne
Price of electricity	£0.03 per kW hour

It is further estimated that the two plants will require the same number of people per shift (2), that the maintenance and chemical costs are 2% of production for the small plant and 1% for the big plant. Compared with the price of dark grains it is estimated that the capital and depreciation costs of the two operations are based on the following:

Small plant	10% of cost of dark grains per tonne per hour

Large plant	9% of cost of dark grains per tonne per hour

It is further estimated that the small plant will make use of a traditional 5-effect evaporator driven by steam, otherwise traditional pressing and drying, whereas the other plant will make use of an MVR evaporator together with the pressing and drying.

Looking at the schemes it is clearly demonstrated that the viability of the smaller plant is highly questionable, whereas the

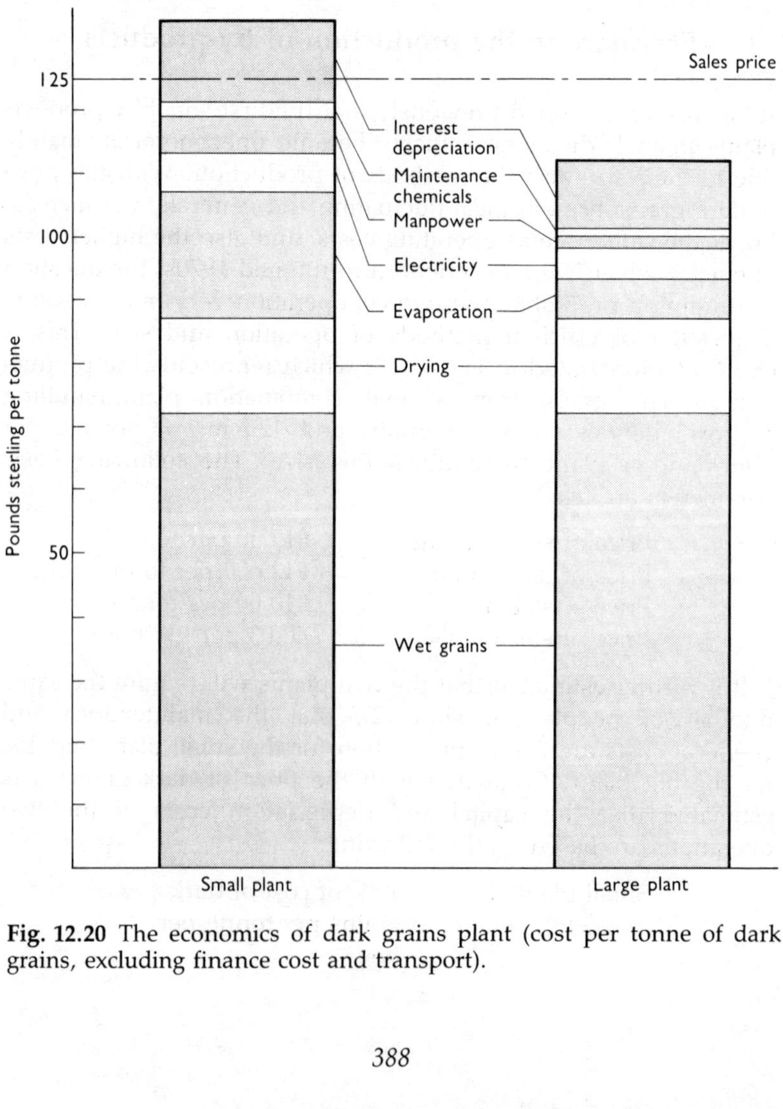

Fig. 12.20 The economics of dark grains plant (cost per tonne of dark grains, excluding finance cost and transport).

bigger one may make some profit. In this picture transport costs from the distillery have not been considered, although in some applications and for some locations they will play a relatively important part. The other major effect which can be seen, is that the cost of converting the spent grains to their 'dry state' is the most important item of expense in producing the dark grains.

The production, buying and selling of the different commodities on the by-products side is a finely balanced picture. For some time several distilleries built dark grains plants, only to find there was a shortage of spent grains on the local market. This and other factors caused some plants to shut down. The pot ale was disposed of by other means, and spent grains became available in larger quantities. This depressed the price of spent grains, so that the cost of dark grains, when considering building a new plant, therefore became less. Also the price of dark grains is a finely balanced variable dependent on other comparable sources of protein and the major movements of the main protein ingredients on the fodder market, most of all soya meal. In times when there has been a shortage of certain of the protein ingredients, the price of dark grains has increased considerably, whereas in other periods, when cheaper substitutes for dark grains have been available the price has consequently dropped.

Production of carbon dioxide

None of the malt distilleries recovers the carbon dioxide (CO_2) which is released during the fermentation of the wash, but some of the grain distilleries produce gaseous, solid or liquid CO_2. This is a fairly important by-product production process for some of the grain distillers.

In this production process (McFadzean, 1979), the CO_2 formed during fermentation is collected from the top of the wash backs and cleaned of impurities such as aldehydes, acids, higher alcohols, glycerol, furfural, glycols and hydrogen sulphide. After cleaning, the CO_2 is compressed and can be sold as compressed CO_2. Normally in the UK, however, the CO_2 is distributed and sold in a liquid or solid form. This requires further compression and cooling, and in the case of solid CO_2, expansion and preparation or sale as dry ice blocks (Fig. 12.21).

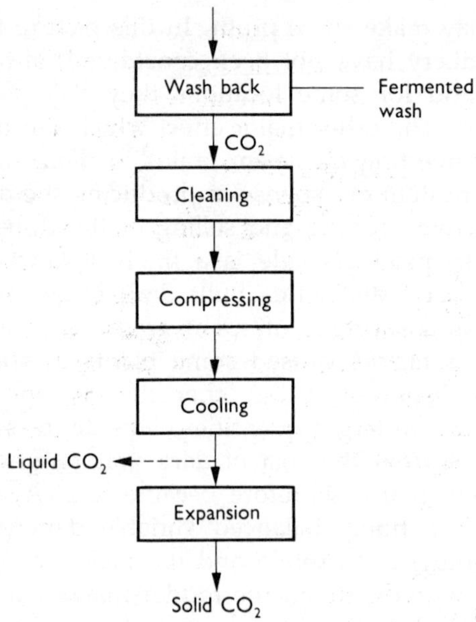

Fig. 12.21 The principles of CO_2 production.

Non-recoverable effluents from distillery

Whereas the pot ale and spent grains produce recoverable and saleable material, there are certain areas of whisky production where this is not so (Campbell, 1977). In particular this is the case with the spent lees from the spirit stills, and the various types of foul condensate which are produced during the by-product process. Amongst these may be condensate from the evaporator, and also from the drier if drier vapours are condensed.

Particularly from the mid-1970s, rather stringent measures regarding effluent output from a distillery or a dark grains plant were imposed on the Scotch whisky distiller. The statute itself (Prevention of Pollution (Scotland) Act) was passed as early as 1951, and was subsequently enforced and extended. Liquid effluents had to comply with specifications on BOD_5, and also on suspended solids. In most places figures around 10–20 ppm BOD_5 and correspondingly low suspended solids were demanded. As a typical spent lees figure for BOD_5 is 1000–

2000 ppm, and foul condensate may be anything from 1000 to 3000 ppm, it became clear that these effluents could not be discharged into rivers without further treatment. Therefore, nearly every by-products plant in the Scotch whisky industry had to install an effluent plant, in order to bring the BOD down to acceptable levels. The only exceptions were the discharge effluent from those distilleries which were situated close to the sea and had acceptably deep water discharge lines, or distilleries in populated areas which could discharge their effluent into other treatment plants.

Most of these effluent plants are built as biological treatment plants. A typical plant is shown in the block diagram of Fig. 12.22. In a biological treatment plant, or 'biofilter', as it is

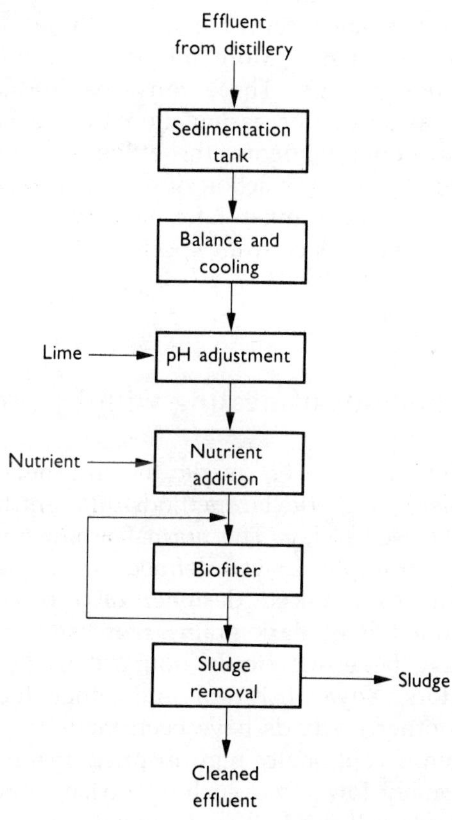

Fig. 12.22 Biological treatment plant.

sometimes called, the organic matter in the effluent, which also causes the undesirable BOD loadings, is passed through one or more filters where a bacterial growth is maintained. To prepare the effluent for the filtration itself, it normally goes through the following process:

1. sedimentation or mechanical filtration, where suspended matter is removed;
2. balancing and cooling, in order to present the filter itself with an even flow of effluent;
3. pH adjustment, to between 6.5 and 8.5;
4. addition of nutrient.

Biological filtration systems require a variety of elements to operate satisfactorily – amongst others, calcium, magnesium, sodium, potassium and iron. However, these need only be present as trace elements. More important is the addition of nitrogen and phosphorus. These must be introduced to the system readily available for assimilation by the bacteria. After adding the nutritional elements, the effluent is pumped to the biofilter reactors. During the action of the bacteria in these units, sludge is formed which must later be removed. Removal of sludge normally takes place from specially designed sedimentation tanks.

Other methods of dealing with by-products

Several attempts have been made to find new methods of handling distillery by-products, methods different from the ones most commonly used today. The normal production method for dark grains, for example, is well defined and presents very few problems to the experienced designer of a production plant. However, production of dark grains represents a process that may occasionally be economically doubtful if the prices of its main competitors, soya, fishmeal and other foodstuffs, fall. Attempts with other methods have been made in periods where energy has been exceptionally high in price or the price of dark grains exceptionally low. Two such approaches are briefly described here: single cell production of protein, and production of methane gas from pot ale.

SINGLE CELL PRODUCTION OF PROTEIN

The production of single cell protein (SCP) from pot ale was tried in the mid-1970s. Before that, production of SCP had been successfully demonstrated in pilot plants using, for example, raw crude oil or paraffin as the basic substrate. Most of these attempts were later abandoned because of various obstacles. The SCP process involves the development of certain micro-organisms like yeast, fungi or bacteria, using pot ale as a substrate. Through the cell growth which mainly takes its energy from the organic substances in the solution, the effluent strength is also reduced. The solids being formed by the cell growth would be recoverable by, for example, high-efficiency decanters, and would be removed from the substrate as high-protein solids. Test facilities on a semi-industrial scale were set up in a malt distillery in Scotland, but abandoned after approximately a year of testing.

PRODUCTION OF METHANE GAS FROM POT ALE

This is an anaerobic treatment process (Donnolly, 1980) which basically produces methane and CO_2, of which the former can easily be used as an energy source for the distillery. The first step of this anaerobic contact process is a fermentation stage. The fermentation occurs naturally in the absence of dissolved oxygen and results in the biological stabilisation of organic matter through the production of methane and CO_2. Anaerobic stabilisation of organic matter can be considered as a two-stage process, producing first acids, then the gas. In the first stage micro-organisms basically convert proteins, carbohydrates and lipids into fatty acids. In the second stage the end-products from the metabolism of the micro-organisms of the first stage are converted to methane and CO_2. A typical plant is shown in Fig. 12.34. The raw effluent is directed to the digester, where the digestion of the organic matter occurs. In order to satisfy a relatively long sludge age or solids retention time, a bioenergy separator is installed. In this unit a thermal shock is applied to the biomass, reducing the gassing within the biomass, and settling the naturally flocculent substance. The clarified liquid is then discharged and the biomass recycled to the reactor to be mixed with crude effluent. Most of the methane gas being formed in the digester is drawn off from the top, but some is recycled to the reactor. In distilleries BOD removals of above 90% have been achieved.

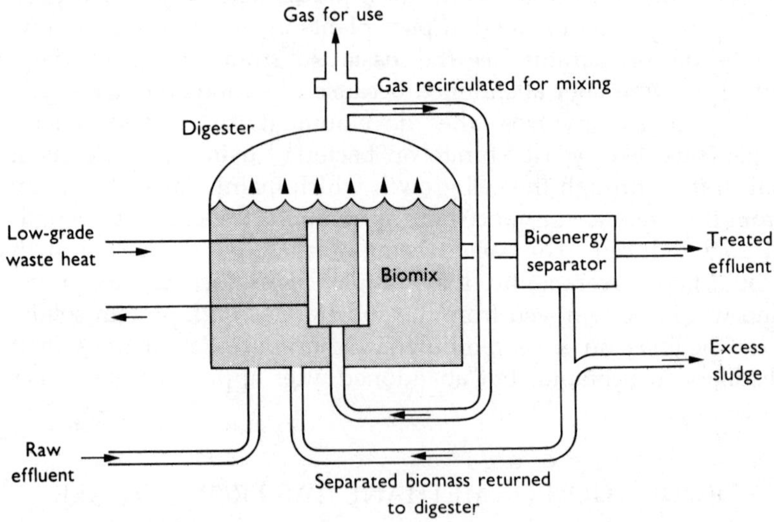

Fig. 12.23 The Bioenergy system.

References

CAMPBELL W 1970 Malt whisky distillery waste. *Effluent and Water Treatment Journal* (September): 509–13

CAMPBELL W 1977 Waste water treatment in brewing and distilling. *Process Biochemistry* **12**(5): 6–8, 32

DONNOLLY T 1980 Anaerobic treatment of waste – with 'gas bonus'. *Brewing and Distilling International* **10**(9): 78–9

MCFADZEAN W J 1979 Turning whisky 'waste' into profitable products. *Brewing and Distilling International* **9**(3): 33, 36–37

NSCA 1984 *Distillery By-products as Feeds for Livestock*. North of Scotland College of Agriculture

Index

The Professional

The Professional

Defining the New Standard of Excellence at Work

Subroto Bagchi

PORTFOLIO
PENGUIN

PORTFOLIO
Published by the Penguin Group
Penguin Books India Pvt. Ltd, 7th Floor, Infinity Tower C, DLF Cyber City, Gurgaon 122 002, Haryana, India
Penguin Group (USA) Inc., 375 Hudson Street, New York, New York 10014, USA
Penguin Group (Canada), 90 Eglinton Avenue East, Suite 700, Toronto, Ontario, M4P 2Y3, Canada
Penguin Books Ltd, 80 Strand, London WC2R 0RL, England
Penguin Ireland, 25 St Stephen's Green, Dublin 2, Ireland (a division of Penguin Books Ltd)
Penguin Group (Australia), 707 Collins Street, Melbourne, Victoria 3008, Australia
Penguin Group (NZ), 67 Apollo Drive, Rosedale, Auckland 0632, New Zealand
Penguin Group (South Africa) (Pty) Ltd, Block D, Rosebank Office Park, 181 Jan Smuts Avenue, Parktown North, Johannesburg 2193, South Africa

Penguin Books Ltd, Registered Offices: 80 Strand, London WC2R 0RL, England

First published in Portfolio by Penguin Books India 2009
This revised edition published by Penguin Books India 2011

12

ISBN 9780670085958

Designed by Joy O'Meara
Printed at Replika Press Pvt. Ltd, India

A PENGUIN RANDOM HOUSE COMPANY

To my daughters, Neha and Niti

Contents

The Professional

Introduction

Some fifteen miles from central Philadelphia, at the northern edge of the city, stands the Cornwells Heights train station. It has a large parking lot, and on most weekdays, hundreds of commuters park their cars there before taking the twenty-minute train ride to Center City, the name the residents of the City of Brotherly Love have given to the downtown area.

Since the parking lot is some distance away from the platforms, the Southeastern Pennsylvania Transportation Authority (SEPTA), which operates Philadelphia's regional trains, runs a shuttle bus between the station and the parking lot. The drivers who operate these buses have an excruciatingly boring job. Hour after hour, they wait for the trains to arrive and then drive the passengers from the station to the parking lot, or the other way around, and then again wait for the next train. As they sit slumped in their seats, most drivers look as though they would do anything to escape the tedium. Curtis Perrin is an exception.

Perrin's demeanor and behavior reflect anything but boredom. Each time a passenger enters his bus, he calls out in his gritty, hearty voice: "Hey there, buddy! How are you today?" If an elderly woman enters, Curtis says, "Hello, young lady!" Even more remarkable than his wide smile and booming voice is his memory: Curtis remembers the cars that his passengers drive. While other drivers drive along a preassigned route—stopping at Parking Lots A, B, C, and so on, Curtis usually drives each passenger to his or her own car. As the exhausted passengers alight from the bus, he calls out again, "Have a great evening, buddy!" or "Happy Friday, sir. Have a great weekend." Curtis's way has endeared him to countless passengers; they sometimes bring him gifts when returning from their vacation. Some people bring him brownies and cakes on other occasions.

Curtis Perrin is an outstanding professional first and a bus driver second. His professionalism makes him view people as individual customers he serves and not as an anonymous mass of faceless commuters. On their part, people respond and react to Curtis's warmth; they don't see him as an invisible function but as someone with affective regard for his work. People like Perrin make a real difference to their work and to the people they serve. Curtis Perrin is a professional.

The word "profession" is defined by Merriam-Webster as "a calling requiring specialized knowledge and often long and intensive academic preparation." The same dictionary defines a "professional" as "participating for gain or livelihood in an activity or field of endeavor often engaged in by amateurs."

Although this is the standard definition, I have found throughout my career that the term "professional" often seems to imply different things to different people. As someone in the workforce, you probably have your own opinions about what makes a person a "professional." Perhaps you adhere strictly to Webster's definition and consider anyone who makes money doing what they do to be a professional—whether he or she is a plumber, a hedge fund manager, or a writer. Or maybe you have a stricter definition, one that requires people to be incredibly good at what they do in order to earn the label—a plumber who starts his own business, a hedge fund manager who earns millions by bringing in clients and managing their investments, a writer who has been published in every major newspaper or magazine in the country.

We also harbor notions about how professionals conduct themselves—they dress a certain way, avoid gossip, or refuse to discuss private matters with their colleagues unless it is relevant to their ability to do their job. Even in today's modern and relatively relaxed work environment, I often hear the term "professional" applied to people who adhere to these standards.

But after almost thirty-five years of life as a professional, I have found that, in the new era of the global economy, where every person's actions have the potential to have a global impact—whether good or bad—we must redefine what it means to be a true professional.

The word "professional" has been around for a long time, and its origins are even older. Its root dates back to A.D. 1175–1225 to the Latin word *professio*, which means an oath taken upon entering a religious order.

The reason we need to appreciate the origin of the term is that we live amid scores of qualified engineers, doctors, nurses, architects, lawyers, journalists, sportspersons and accountants who believe that being a professional is merely a means of earning a livelihood, just another way to get ahead in life, to seek and create further material comfort and eventually enjoy retirement. For such people, a professional is someone who gets an educational qualification to land a job, then a coveted assignment or posting, followed by a bunch of accelerated increments and promotions.

But in my view, a professional, especially now, is something more. Today, it's not enough for someone to just be able to do a job in order to qualify for this title; he or she not only must be able to accept responsibility for their own work and actions but also must understand how that work and those actions will translate to the rest of the world. Although every country, industry and company has its own code of conduct, there is no institution large enough or powerful enough to make sure each of us is behaving in the best way possible. Today, more than ever before, we must embrace the traditional meaning of the word "professional" and take a personal oath, promising we will do our best; we will bring our knowledge, skills and attitude to make a difference to others and to regulate ourselves.

And this does not just apply to "thought leaders" or men and women in suits. You could be a software engineer, a cosmetologist, an intellectual property lawyer or, for that matter, a publicist. For some of you—the doctors, lawyers and journalists, for instance—your profession has existed for a long time and therefore you probably already operate under a set of guidelines that has been long established. But for others, your profession may not have existed a hundred years ago. Every decade is throwing up brand-new professions the world over. On one hand, professionals in their respective fields are expected to be experts in their own disciplines, to understand their nuances and, at the same time, be able to work in tandem with their counterparts in other professions. But most important, they must be able to work across borders.

There has been an explosion in the world of professions, and yet there is still no system in place that can train people how to be true professionals. No matter how much education or training you have, there is no instruction manual that tells you how to approach every problem, communicate with different people and respond to change. As a result, most of us simply bumble our way into the critical, multifaceted understanding that separates a professionally qualified, technically competent individual from a handful of true professionals.

But there is another challenge to being a professional today. This is the fact that we work globally. Even those of us who don't travel for work often have to communicate with those halfway around the world and, of course, our reliance on international trade means we can easily be affected by an event that occurs or begins abroad.

Consider Ishani Mohapatra, who studied chemistry and specialized in nonedible flavoring agents. She completed college in Mumbai, India, but today works as a fragrance development manager at a company called International Flavors & Fragrances Inc. in Utrecht, the Netherlands. She works as part of a globally dispersed team that helps create new detergent fragrances for one of her company's clients, Procter & Gamble.

Or there's my chartered financial analyst, Leslie Davis, who advises me on international tax issues. Davis lives in California but has clients all over the world who turn to him on dual taxation issues in the course of their business with the United States.

In Bangalore, India, my dentist, Vivek Nandavar, is seeing an increase in the number of expatriate clients from many different countries as they make the Silicon Alley their place of work; of late he has also seen patients come from places like the Middle

East who are turning to India as a destination for affordable medical care, often combining it with tourism.

I have been working for the past thirty years as a professional. For fifteen of these years, I have been an entrepreneur. During this time, I have dealt with customers from all around the world. Even though I'm based in Bangalore, I wake up every morning to calls from Shanghai, London, California and other places far from home. I travel almost constantly and have lived overseas. I have raised venture capital money from investors across the world, and in 1999, I cofounded MindTree, which employs people from many different national origins and operates out of offices in the United States, Europe, Asia and Australia. Our nine thousand people create software applications for companies as diverse as John Wiley & Sons publishers in the United States and ArcelorMittal steel company in Europe. We develop and create technology that is used by companies from Denmark to Korea. Through my work in the software industry and through the work of my colleagues, I routinely transact with professionals from every discipline: technology folks, end users in finance and marketing and human resources, and sometimes medical doctors, lawyers, public relations experts and real estate agents.

In the same vein, as an individual who has traveled extensively and lived abroad several times, I have been a customer of a wide variety of professional services and have relied on countless experts from medical doctors to chiropractors to airline agents and auto repairers and cab drivers to help me solve a problem or accomplish a goal that I could not do alone. Each one is a professional whom we expect to satisfy our needs in a timely, cost-effective, inclusive, courteous and, of course, competent manner.

This book is as much about people like them as it is about those who readily come to mind when we think of the term "professional." It is as much for those who earn six-figure incomes as it is for those who earn minimum wage. In short, it is for everyone who recognizes the need to behave professionally.

What defines the true professional? My goal in this book is to answer this question.

I believe, at a fundamental level, there are three qualities that make someone a true professional. These are the ability to work unsupervised, the ability to certify the completion of a job or task and, finally, the ability to behave with integrity at all times.

But there are facets beyond meeting the base-level requirements. The true professional is also self-aware and is not afraid to admit when he or she is wrong, ask for help, or refuse to do something that does not align with his or her values or goals. He understands his limits but is never afraid to speak up when he feels silence can compromise an ethical boundary. He takes care of his body, soul, and mind and prioritizes everything he does in order to create maximum value.

The true professional has a long view of time. She understands that a profession is not limited to a particular job or organization but spans the entirety of her forty-plus-year career. She makes decisions and builds relationships knowing that the things she does today can affect her tomorrow, next week, next year or even ten years from now.

The true professional is one who evolves. He embraces the new challenges that each phase of his career brings and adapts to changes in his life, company, industry or culture. He excels not only at the technical details of his job but also at the managerial and leadership aspects that come with advancement.

The true professional is comfortable dealing with things she was not trained for. As we all know, we live in times when there is a high probability that an unlikely thing—from a natural disaster to a terrorist attack—could happen. The true professional sees these events not as catastrophes but as opportunities to make a difference.

Above all, the true professional understands the imperatives of our global economy. He is culturally aware, connected and engaged with people around the globe.

A final thought before we move on: I am not a management guru. I am a practitioner. My main profession is not writing books and lecturing at business schools but running my software business. Hence, I lay no claim to profound theories born of inscrutable research. I am an observer and a doer. For this reason my book shuns complexity, and for that matter sophistication. This is a simple book; it presents practical situations, ideas and concepts and exemplifies them with everyday lessons. It is not meant to be an academic pièce de résistance. It is a workbook for anyone who wants to excel in their careers, not only today, but decades from now. I would deem it worth your time if somewhere it makes you think, find your own "how to" and someday pass it on to others. Because I deeply believe it is simply not enough to be good at your work. A true professional must have a sense of legacy and leave the core of knowledge behind so that taller trees may one day sprout from the same ground into which we all must eventually fall.

Subroto Bagchi

Part One

Integrity

Chapter 1

Burial of the Dead

Do you work in an entry-level position or even a middle-level job in a hotel, a hospital, a software company or a government organization? Or are you a self-employed professional like a doctor, a lawyer or a journalist? In all probability you are educated, know English and are working in (or have worked with) the corporate sector. Perhaps an MBA, or a student at an engineering college? You probably consider yourself a professional, or on the road to becoming one. Regardless of what you do or what stage of your career you're in, it's probably safe to say that your station in life is well above someone whose job is to, say, bury unclaimed corpses from city hospitals.

And yet this fact alone does not make you more of a professional. I know a man who has dedicated his life to dealing with dead bodies—unclaimed dead bodies. His job is dirty; it is morose; it requires no degree; yet he exhibits the true meaning of a professional more than most white-collar workers out there. His name is Mahadeva. He came to Bangalore as a child when one day his mother simply walked out on her entire village and her own family in a huff. Mother and son lived on the streets, and she worked to support him until the day she became sick. She brought herself and her son to the government-run Victoria Hospital. There she was admitted in a state of delirium, and her little son, Mahadeva, made the streets outside the hospital his home. He found many playmates among the urchins there and soon that world engulfed him. For the first time, little Mahadeva had people to play with—people who became his family—and he lost himself completely in this new world. Then one day someone told him his mother had died. Died? What was that? Where had he been when that happened?

The hospital had waited for someone to claim the dead body for the mandatory three days after death and, as often happens in countries like India where unclaimed dead bodies cannot be indefinitely kept at the morgue, had disposed of the body.

Now Mahadeva was an orphan.

A few people in the hospital ward raised money to help him go back to his village, but he refused to take it. That is when he met his benefactor, an old man who was a permanent fixture at the hospital. The man ran errands for poor, illiterate villagers who did not know the ways of the big city. He was the man who had secured a hospital bed for Mahadeva's mother. The old man took pity and asked Mahadeva to move in with him. Like Mahadeva, he did not have anyone in the world. Under his tutelage, Mahadeva grew up running errands in the hospital.

To Mahadeva, the hospital became the universe. One day, the police asked him to bury an unclaimed dead body and paid him two hundred rupees for the job. This was the day Mahadeva entered his profession, and he would go on to become the go-to guy for burying the city's unclaimed corpses. Every time the police picked up a dead body that had no claimants, Mahadeva was summoned. He had to do a turnkey job: pull the stiff body from the morgue, hire a horse-drawn carriage, put the body in it, take it to a burial ground and dig the grave all by himself—for only two hundred rupees. After completing the job, he would wait around in the hospital until he was summoned to do it all again.

Mahadeva did his work with such dedication, focus, care and concern that soon he was very much in demand. His workload increased and he bought his own horse-drawn carriage.

One day, the horse died.

People who had watched Mahadeva all these years came together and bought him an auto-rickshaw. Today, a picture of the horse has become Mahadeva's logo; it is painted on his rickshaw and appears on his business card. Mahadeva has buried more than 42,000 corpses in his lifetime, and his dedication has earned him phenomenal public recognition. Local gas stations do not charge him when his hearse is filled up, and the chief minister of Karnataka, the state in which Bangalore is located, thanked him for his selfless service to the abandoned citizens of Bangalore. Mahadeva is proud of his work, and today his son has joined him in the business.

Mahadeva: the high performer, and a true professional.

What are the two qualities Mahadeva has that make him a professional rather than someone who is simply professionally qualified?

One is his ability to work unsupervised, and the other is his ability to certify the completion of his work.

Whenever Mahadeva gets a call to go to the morgue, day or night, rain or shine, he arrives. His task is gruesome and difficult because there is usually no way to tell how or when someone died.

In his business, Mahadeva does not choose his clients. He accepts them in whatever size, shape or state they come. The day he buried the man who had taken him home after his mother died, he cried and bought the man a garland as a mark of his respect. That day, it occurred to him that he should be garlanding all the bodies he buried, not just that of his benefactor. Mahadeva believes that everyone—dead or alive—deserves respect no matter who they are or where they came from.

The police do not supervise Mahadeva, and he is not an employee of the hospital. He does not have a boss who writes his evaluation or gives him constructive feedback for continuous improvement. He reports to no one except himself. Whenever Mahadeva picks up a corpse, it goes straight to the burial ground—no place else. He completes the task with the immediacy it demands. And each time, he makes sure that all the steps are complete, from covering the dead body properly to digging six feet under, even though he knows no one is watching or would ever question him.

In most work environments, people who produce anything of economic value usually need supervision. A person who needs supervision is no professional. He is an amateur, maybe even an apprentice.

Ability to certify completion is an interesting and powerful idea. The concept is easy to understand when we look at the work of a painter, musician or sculptor. Who but only they can certify that their work is complete in all respects and is ready for the world? Who else but the painter knows that his painting did not need any more work, or that he could have done an even better job? How does the world know the musician or the sculptor truly did her best while creating something new for the world? Only they know.

That same ability to discern what constitutes completion and self-certify that the work is complete in all respects applies in equal measure to a software engineer who must write defect-free code, a young lawyer who is required to study a case and write a brief or a surgeon who must cut open a patient and make sure that he removes every instrument and every piece of gauze before stitching him up again.

Chapter 2

The Day Justice Was Murdered

On May 31, 2007, *The Times of India*, India's largest circulated newspaper, published a report that the Delhi police had sought the replacement of a public prosecutor by the name of I. U. Khan. Khan had been involved in a high-profile case in which a young man named Sanjeev Nanda had been tried for killing several people while driving drunk. The call for Khan's replacement followed a television exposé that showed how Khan had colluded with the defense attorney to bribe a crucial eyewitness in the case.

This story began in 1999 when Nanda, son of one of Delhi's high-flying families, drove a brand-new BMW through a police checkpoint and allegedly "mowed down" six people including three policemen. According to the prosecution, Nanda was traveling with his friends Manik Kapoor and Siddharth Gupta. After the fatal incident, Nanda stopped the car and got out to assess the damage. Then he got back into the car and drove off. Nanda then took the car to a friend's house and got it washed clean in an attempt to remove any evidence of the accident. A policeman, following the trail of dripping brake oil from the accident spot, came to the house and arrested the three young men. Later, in a highly publicized case, the defendants denied any involvement and maintained that the alleged driver was not at the scene and was not driving the car. Their lawyer, R. K. Anand, one of the top defense lawyers in the country, even presented theories that the accident was caused by a truck and not the BMW. Khan, the prosecutor, seemed to rest his case on the eyewitness account of a man named Sunil Kulkarni, who was at the scene of the crime. It was this man who was being sought to be bribed so that he would give false testimony. The transaction involving Khan, Anand and Kulkarni was caught on video and aired on prime-time television, causing huge public outrage.

Following this, Delhi police sought the permission of the High Court to replace Khan, and later the court found Khan and Anand guilty of obstructing justice. The High Court asked that they be barred from appearance in court for four months and that their title of senior advocate be removed. The matter went to the Supreme Court, which upheld the conviction of Anand but freed Khan from a contempt charge after reprimanding him for inappropriate conduct. Anyone with a basic sense of morality knows that bribing a witness is unethical, but what is even more shocking about this case is that the defense had help . . . from the *prosecution*!

My purpose in this chapter is not to recount a grisly story of wanton manslaughter by a drunken upper-class youth. Rather, my goal is to highlight professional ethics and the importance of displaying the highest standards of ethical behavior.

Here we are talking about two of India's top lawyers. One

of them has even represented two sitting prime ministers! The other's client list is equally impressive. These are not greenhorns in black coats. They went to law school; they were highly trained and had decades of experience to back them. How could they forget that, as representatives of the law, they could not try to unduly influence the course of justice? Did they not understand that, no matter how powerful they were, they could not compromise the interests of one of their clients?

Did Sanjeev Nanda murder those six men? The Indian Penal Code defines such killing, regardless of the number of victims, as culpable homicide not amounting to murder. Would he have done what he did if he had not been under the influence of alcohol? Probably not. A graduate of Wharton Business School at the University of Pennsylvania, Nanda had never done anything up until then that seemed to indicate he had the instincts of a killer. On the other hand, the two lawyers hired in the case did willfully attempt to kill justice. And in the process they defrocked themselves of the professional label.

Unlike Nanda, these men knew what they were doing; they cannot claim to have lost control of their basic senses due to inebriation. It was not a momentary indiscretion that made the defense lawyer conspire with the prosecutor. They knew they were breaching the professional code of ethics they'd been taught in law school. They did it in cold blood; it was a premeditated decision to buy off a star witness.

And it is not just the legal profession that should be bound by such tenets of professional conduct. The boundaries that these lawyers crossed apply in different ways to every profession.

What is professional ethical conduct and what actions may be deemed unprofessional?

What is the furthest boundary of ethics and integrity that, regardless of scrutiny and threat of punishment, every individual who calls himself a professional must think of, live by and live for? How can we learn to take responsibility for our actions and behave in the most professional way possible?

A young surgeon may not yet be proficient with the scalpel, but she must clearly understand what the basic tenets of medical practice are and what may constitute malpractice.

A software engineer, working on a client's project that requires complete confidentiality, cannot discuss his work with people not preapproved by the client (and that includes his spouse, or a significant other, or a friend at a pub). He cannot brag about his work to others because it may compromise the confidentiality requirements of a customer with his organization.

A journalist cannot curry favor with a source to get a scoop or use professional standing to seek special accommodations from the system.

A chief executive of a company cannot appoint his spouse as a contractor in his own office and do business with her.

A boss who is lending a sympathetic ear to an emotionally distressed employee cannot use her situational fragility to take her to bed.

A policeman cannot prey on a citizen who has come to seek help.

Every profession has some explicit and some implicit code of conduct, and understanding that serves as the distinction between a professionally qualified person and a professional.

Beyond the capacity to work without the need for supervision and the ability to deduce what may constitute completion of work, anyone who seeks entry into the club of professionals must understand integrity and must practice it every step of the way. Without integrity, any professionally qualified individual is actually a danger to society.

Chapter 3

Integrity Is Personal

My first few lessons in professional integrity came from my father, who was a small-time public servant in India's eastern state of Orissa. He drove home the fundamental tenet that resources provided to you to discharge your official duty do not belong to you. It started with using—or rather not using—his office staff and his official jeep for personal needs. It was, and is, customary for officers to use people employed by the government as household help. But not my father. The first time he was given a jeep by the government, it was garaged at the government housing allotted to him. I have never seen my mother in it. We could sneak in to sit in the jeep only when it was stationary and my father was not anywhere in the neighborhood. He was the district employment officer, but he would not even drive to work in the jeep. He chose to walk instead because he believed that the vehicle belonged to the republic, not him, and that it was only meant to be used on his tours to the interiors of the tribal district of Koraput where he was posted.

We studied in run-down government schools and, occasionally, some of our friends used to bring paper and other stationery to school that their parents had brought home from their offices. In small-town India, this was no big deal. But any conversation about the subject would upset my father. Not only did he have a strong opinion on such things, he did not like us to bring home gossip on the subject.

After my father retired from government work, we moved in with my brother, who had by then joined the Indian Administrative Service (IAS). As a trainee, he was starting his career as a sub-divisional officer, responsible for overseeing the law and order as well as development of part of a larger district. The state provided him an official residence as was the norm for such functionaries. In case of my brother, it was a bungalow with a large compound on which nothing grew. The land was lying unused. One day, I proposed to my mother that we plow the land and grow some lentils. I suggested we get a tractor from a family friend, and my mother seemed sympathetic to the idea. So one day, lo and behold, the tractor arrived at the gate. I perched on it excitedly and oversaw the plowing operation. The next step would have been sowing the lentil seeds. But that was not to be.

When my father returned home, he was shocked to see what was going on and he admonished me, giving me a long lecture on landownership and the meaning of land use. While it was okay to tend a small kitchen garden in the house allotted to a government servant, one could not push the envelope and start cultivating on government land. That day, I also received a tutorial on the meaning of the word "encroachment" as defined by the Orissa Revenue Act of 1951, which he read to me. I was just twelve years old at the time, but I got the message for the rest of my life.

A part of the large bunglow was designated as the residential office of the subdivisional officer. In it, there was a big wooden table where mounds of files were stacked each night for my brother to go through and that were replaced the next day. On the table, next to the files, sat a black phone. It was the first time in our lives that my family had seen a telephone at home. For someone like me who was born in a house with no running water or electricity, the telephone was an object of curiosity and pride.

When it rang, it sent electrifying energy through my entire being. I wanted to pick up the phone to answer it whenever my brother was not around. It felt so grown up. My father maintained a hawkeye on the instrument and prohibited me from even touching it. Forget about talking to friends (most did not have a phone anyway); going near it was forbidden because it belonged to the republic.

Those early lessons made it easy for me to understand later on the demarcation between what belongs to Caesar and what belongs to me. I never needed a formal course on corporate governance.

Sometimes, Father's ways seemed strange and extreme. But I realized how relevant his messages were when, back in 1990, I was the resident manager for U.S. operations for Wipro. One day, my most important customer called. He was extremely embarrassed about the subject of conversation, but at the same time wanted me to know that one of our engineers was making personal long-distance calls to Chennai, India, using the telephone in the office in Austin, Texas. He had run up close to a thousand dollars in making those surreptitious phone calls. Until then, the client did not have pass code protection on telephones and every phone in the office was equipped with long-distance calling capability so the U.S. team could be in constant touch with their counterparts in Japan. My head hung in shame at my colleague's breach of integrity. It was not generational poverty that had led the engineer to make personal phone calls from the office telephone; it was lack of basic understanding of what did not rightfully belong to him.

Over the years, working in India and overseas, I have learned to correlate professional integrity and good business, taking off from where my father's grounding ended. Today, I can say with absolute certainty that integrity in business and work is not a matter of corporate piety—it simply makes good business sense. Globally, positive examples abound in every sphere. But more about that later. Let us first understand what integrity means in the professional context.

Simply said, it means:

- We follow the rules.
- Where rules do not exist, we use fair judgment.
- When in doubt, we do not go ahead and do what suits us; we seek counsel.
- Finally, faced with a difficult choice, we ask ourselves: Can my act stand public scrutiny without causing embarrassment to me and my family?

If these four tenets are applied to our daily decisions, answers to the most daunting questions will present themselves and our subsequent professional conduct will always meet the highest standards of integrity.

In civil societies, all professions are bound by rules that usually have a basis in fairness. Sometimes, the rules do not suit the professional or business interests of an individual in a given transaction. The thing to do then is not to bend or circumvent or flout the rule that does not suit you, but to contest it and, pending resolution, follow it. This is what constitutes the basic obligation of any professional. So when a tax law does not suit me, I will protest the law, but while protesting it, I must ensure that my business pays the tax (albeit under protest) until such time as that law or its interpretation has been reversed by the system. I cannot be part of a system and flout the law at the same time. If I live in a country, my citizenship requires me to embrace the legal system in its entirety. The moment we selectively subscribe to the legal system, it will fail.

No society can, however, fully state all the norms of acceptable professional behavior through enacted laws. No corporation can tell you exactly what, as professionals, you should or should not be doing. This is why we need to use fair judgment. In fact, if we exercise fair judgment most of the time, one does not require a framework of rules.

The concept of fair judgment is subjective, but essentially it is the voice of conscience. The basic question you need to ask yourself is: How would I feel if someone did to me what I am about to do to another?

So here I am, planning to take a shortcut and bend the rules against the system. All I need to do is step back for just one moment and ask, What if I were the system? What if I were at the receiving end of things? When a moment of temptation comes, as it sometimes must, all we need to do is switch roles and place ourselves in the shoes of the potentially most-affected party. In an instant, the path presents itself.

Most problematic professional situations can be avoided if,

before committing to a certain path, a professional just picks up the phone and asks someone he trusts and respects what to do. Seeking counsel is an almost certain way of preventing problematic situations from arising in the very first place and then becoming a matter of perpetual regret.

Finally, the greatest deterrent for most of us is public scrutiny. Even if we have our moments of private indiscretion, there are only certain things we are comfortable with sharing publicly. If we simulate the possibility that what we are about to do might be uncovered, the choice will become clear. How would I truly feel in front of my family if my indiscretion was scrutinized or came to light? That one thought is enough to show me the path in most situations where I may be drawn toward the convenient, but not the right thing to do.

Chapter 4

Doctor, Heal Thyself

On February 9, 2008, *The Guardian* newspaper in London ran a report about an Indian doctor who had fled to Nepal after being investigated for his alleged complicity in running an illegal kidney racket.

Amit Kumar was accused of running a private hospital just outside Delhi that allegedly lured or forced hundreds of poor people into "donating" their kidneys. Kumar then sold the organs and made a fortune.

He was arrested in a resort very close to the Indian border. At the time of the arrest he had a bank draft for 900,000 rupees and cash piles of 140,000 euros and 20,000 dollars Kumar apparently kept a low profile in the hotel where he was staying, wearing a hat and sunglasses, but aroused suspicion when it was found that he was cutting stories about the kidney scam from newspapers. According to the police, he was attempting to flee to Canada where he owned a home. On his arrest, Kumar addressed the press, insisting that he had not committed any crime, that he was not running away and was confident he would be cleared. Investigators, however, said he had confessed to carrying out the illegal kidney harvests. He and his brother had run the racket together, selling the organs to American, Greek and British nationals.

We presume that Amit Kumar had taken the Hippocratic oath as a medical graduate before entering the profession. Irrespective of every other test of integrity, all he had to do was to ask four simple questions before doing what he did:

- Is my intended act consistent with the law of the land?
- Is what I am going to do the fair thing?
- If I were to seek counsel from someone with no vested interest, would the person advise me that it was the right thing to do?
- Would it embarrass me if the news of what I intend to do broke out?

Amit Kumar, also known as Dr. Santosh Raut, ran the racket in Mumbai and Delhi and was prosecuted in Mumbai, from where he had jumped bail. In Mumbai he conned beggars and street dwellers into selling their organs. In Mumbai and Delhi he had a network of nurses who were paid large amounts of money to get the victims to his operating room.

Doctors and nurses are professionally qualified people. But when they breach the principles of integrity, they cease to be professionals. When such people are faced with a conflict and they are about to breach the principles of professional integrity, it is not always the law but fair judgment that can prevent them from crossing the line.

Breach of integrity is not relevant for doctors and nurses alone. It is a virus that infects every profession. On April 20, 2004, a top editor at *USA Today*, Karen Jurgensen, abruptly resigned her position at the newspaper for her failure to intercept a series of fabricated stories by foreign correspondent Jack Kelley. Kelley had fabricated substantial portions of eight major articles, including one that was a Pulitzer finalist (an eyewitness account of a suicide bomber that turned out to be nothing more than a figment of Kelley's imagination).

From Amit Kumar to Jack Kelley, most of the people you will read about in this book were educated, well informed and mostly affluent. They had a choice at every step and they chose to do what was convenient, not what was right.

At the end of the day, professionalism is a personal choice people make, and for some, integrity simply becomes a way of life.

In many parts of the developing world, young people do not work part time while studying in high school or college. In contrast, in many Western countries such a thing is more the norm. While working part time as gas station attendants, grocery packers or baristas in a coffee shop, young people begin to understand the boundaries of a professional work environment. Most of them begin to appreciate the concept of company time versus personal time, the idea of serving a paying customer, acceptable workplace behavior and the difference between the penny in their pockets and that in the cash register.

Further, in many places around the world, the education system fails to provide an understanding of the idea of professionalism. There is little content on the concept of professional ethics in technical or professional curricula. And there is no conversation about the subject. In the absence of all that grounding, despite being professionally educated, most young people come into the workplace as novices.

Finally, most workplaces do not articulate their position; far fewer practice ethical behavior in any demonstrable manner. This is not to imply that they do not abide by the law. What I am saying is that most fail to consciously articulate a shared understanding of ethical behavior, what it actually means in that particular work environment and how it should be practiced by everyone at work. And in only a very few companies is ethical behavior truly nonnegotiable.

Chapter 5

The Many Shades of Gray

While there is only one pope at the head of the Roman Catholic Church, in the Hindu world, religious leadership is quite decentralized and there are many who are considered to be religious heads. Prominent among them are the *shankaracharyas*. Unlike the pope, who sits in Rome, these shankaracharyas have their seats in different places of religious significance. One such sits in the southern Indian temple town of Kanchi, famous for its ancient architectural beauty and fine handloom cloth. On December 18, 2008, a report filed by the India Abroad News Service (IANS) sent shockwaves throughout the Hindu world. It reported that charges were pressed against the shankaracharya of Kanchi along with his successor; the charges involved murder and financial embezzlement. As the story unfolded in the following weeks, there were allegations and media reports about sexual wrongdoing in Kanchi's religious high place as well.

The world over, stories like these involving the heads of religious orders have made news in recent times. These often make us question morality itself. The conduct of individuals like these rubs off on the institutions they represent. Their personal conduct becomes synonymous with institutional conduct. In difficult times like ours, we all look up to individuals beyond ourselves for reassurance, to get clarification on what to do when we are at a crossroads.

When we are children and things around us fail, we look to our parents for value clarification. Later in life, we look up to a teacher or a mentor. Beyond them all, we seek regulation and protection from the idea of the government, which to most of us is the protector of civil society, the arbiter of laws and norms that dictate how all citizens must behave. But beyond the government, we turn to our religious leaders—a parish priest, a shankaracharya, an imam or a rabbi.

In the last couple of decades, we have seen this established model break down with dismaying regularity.

In the United States, we saw an elected president compromise the sanctity of his office by giving in to the allure of sex; we saw him lie under oath. We saw big business—from New York Stock Exchange chairman Richard Grasso to Martha Stewart to the folks at Enron to Satyam Computers chairman Ramalinga Raju—fail the test of transparency. The successive convulsions in the U.S. Catholic church involving serious accusations of sexual abuse left people wondering: Where is our source of strength in times of difficulty? Who clarifies values?

To me, large institutions in particular have progressively exposed their own vulnerability. When obviously unethical things happen in places we hold in high regard, people begin to discount the subsequent messages they receive from them. It works like this: If a child witnesses a parent giving or taking a bribe or listens to parental conversations that validate such things, the child will actually begin to associate these despicable acts with those she looks up to. When the same child listens to lectures in school about the importance of being truthful or not taking what does not belong to her, she will not actually "hear" the message because she has already witnessed people she trusts disobeying it. If it's all right for them to ignore this advice, why shouldn't she?

One day, the same child, now grown up and professionally qualified, joins the workplace, already a cynic about the benefits of integrity, and will discount her organization's values. This creates significant challenges for business organizations that value integrity and expect their people to be principled in their personal and professional capacities. But how do you communicate a code of conduct to someone who is a witness to the positional collapse of values at home, in the government, in big business and in organized religion?

Posters, screen savers and first-day orientations at work during which we hear corporate rah-rah by folks from the human resource department do not work anymore. The only option left for those in positions of leadership is to talk the talk and walk the walk. They need to be transparent and explain to others what transparency means and why it is good for everyone.

While leaders need to demonstrate high personal integrity in order to deliver the message to their organizations, the truth is that value clarification is most needed not when things are going right but during dark moments, when things are falling apart. People do not appreciate values when everything is going well.

Each time there is a fiscal irregularity, or a report of misuse of power, or sexual misconduct at the workplace, we invariably get caught up in two things: First, facts and emotions get mixed up. Then, we get lost in what is called the "many shades of gray." This is the moment of truth.

People falter when they begin looking at professional misconduct not in black and white but in shades of gray. Once you let yourself enter the gray zone, you get lost, because the human eye can actually see 108 shades of gray! Once you become mired in the many shades of gray, value clarification takes a backseat, for the majority of people find taking a black-and-white stand on a breach of integrity an uncomfortable proposition. Some managers skirt the determination of facts and let emotions cloud their judgment. They get cold feet dealing with a breach of ethical conduct because they do not like to be in the position of a hangman.

The other fear people have is about organizational embarrassment and business loss. Consider the following situation.

You have just discovered that your star salesperson has faked a travel voucher. He is negotiating a huge order. The discovery has taken place at a time when he is about to close this very valuable deal, which will make the company a lot of money. And he is the only person who knows the ropes in the client organization.

What would you do?

- Keep quiet?
- Warn him and let it go?
- Warn him only after the order comes in?
- Fire him on the spot?

Think.

Chapter 6

Firing the Star Salesperson

We left our story as you were deciding what to do with the star salesperson who had forged a travel voucher. The path you choose affects the organization.

I cannot tell you what you should be doing. It really depends on your personal and organizational value system, and I certainly hope the two are not opposed to each other. On this issue, rather than give you a prescriptive outcome, let us talk about the process organizations must have in place in order to deal with breaches of integrity.

First of all, an organization needs to articulate its position on integrity. As a part of that articulation, it needs to publish the process for dealing with how a breach is reported. The organization needs to anticipate potential breaches; even in the most well meaning of places, sometimes things will go wrong. The process for dealing with a breach should consider: Who can report the breach? Who will investigate it? What steps should be followed during the investigation? The consequences of a breach of integrity must be common knowledge. At MindTree, for instance, we have articulated an integrity policy. This is not left in the wilderness of the intranet portal. It is freely downloadable from our Web site. We keep the document current with instances of breaches as they sometimes take place and the price the organization paid each time. There is a whistle-blowing process, and when a breach is reported, we set up something akin to prosecution and defense teams so that facts and emotions are separated and witch-hunting does not take place. We investigate issues without fear or favor and irrespective of who is involved and what the business implications could be.

Assuming that an organization is committed to an in-depth investigation of the reported breach, the most important area where many flounder is speed. More than anything else, speed is critical. You do not want to drag your feet because the matter is embarrassing or the potential outcome is inconvenient. The greater the delay, the greater the chance of people moving from black and white to the many shades of gray. That is what happens under the judicial system in most cases of public investigation. An organization must not fall into the same trap.

If the issue of complicity is factually settled as a result of the investigation, the organization must brace itself to absorb the consequences. If you know that a salesperson has forged a bill, and such forgery is defined as a breach of integrity by the organization, at that moment it matters very little that his firing could adversely affect the outcome of a bid in progress or subsequently affect employee morale. This is the classic moment of truth. It is true that the action of management could mean loss of business, but the clarity with which management thinks and the speed with which it acts determine the consequent social memory that guides the vast majority of employees and eventually makes organizational values tangible.

The most critical aspect of dealing with issues of integrity is the price management is willing to pay to ensure integrity in the workplace. Once people figure that out, compliance does not require constant policing.

Following definitive action, the next step is communication. This is not always easy. A senior vice president is being asked to leave on charges of sexual harassment. Would you or would you not explain the circumstances to his direct reports? What acts must be publicly and widely communicated and what acts require that the individual concerned be given a safe passage? Do not focus on getting the "right answers" to these questions. The bigger question is whether or not an organization encourages conversation and has an existing framework to deal with such issues.

Finally, all is not over just because you have decided to part ways with a delinquent colleague. It is not just a matter of firing someone. The organization needs to deal with the inevitable—the residual toxicity of such incidents. The responsibility of management does not end with handling a breach with speed and fairness; it must be very sensitive to the emotional fallouts that such acts leave behind. When a breach is dealt with, top management has to help the organization in making sense of the situation and the subsequent decision. It has to shift from blame to building collective learning for the future. It must identify individuals and groups who may need help to heal. After all, the fired salesperson could also be a much-loved colleague of many years and the senior vice president may have been a great problem solver and a role model in other ways for people below him.

Apart from such people, the process invariably affects those who have investigated the issue or corroborated facts that lead to someone getting fired. These individuals tend to develop what I call "hangman's remorse." It is important to keep communication channels open and encourage such people to talk even after the event is over. This helps to reinforce the idea that their action was not just something required as part of their job but was indeed the right thing to do.

Having done all of this, an organization should know that residual toxicity could remain within the larger organization among those not close to the incident. This can occur when a silent majority forms its own opinion and does not express it but lets it fester. Positive reinforcement, visible actions of top management and constancy of purpose are needed to detoxify the silent majority. And be prepared—the actual healing takes a long time.

Part Two

Self-awareness

Chapter 7

Where Competence Ends

Two African American men. Both born in the same decade—one in 1943, the other in 1947. Both sports stars of formidable fame. Raised by ordinary parents, both were endowed with enormous talent, and won prestigious college scholarships to pursue their game. Both worked hard, extremely hard, to reach the highest level of their professions. Both achieved prestige, acclaim and fame.

But behind the glamour, there was pain. In time, both men had their own brushes with the law. One for crimes ranging from shoplifting to murder. The other for protesting apartheid and oppression against immigrants from Haiti.

One went to the hall of notoriety and the other to the hall of fame. The former is American football player O. J. Simpson and the latter was the tennis legend Arthur Ashe.

Both had unmatched professional skills, but Simpson can never be regarded as a true professional. Ashe, a man who died of an HIV infection he received via a blood transfusion after a heart attack, lives on in the hearts of millions of people around the world who have little to do with professional tennis.

In the grueling world of professional sports where only the best survive, it takes something more than the highest batting average, the most formidable backhand, or the most MVP awards to be a professional. What separates O. J. Simpson and Arthur Ashe is self-awareness—this is the streak that divides millions of professionally qualified people in multiple disciplines from the few hundreds who inspire professional respect.

Some people are naturally more self-aware than others. But all of us have the capacity to become progressively self-aware through life's experiences. Without self-awareness, we fail to balance ourselves through the ups and downs of life, have problems in accepting reality and often do not know what is good for us. As a result, we may even end up making wrong professional choices that have a spiraling downward impact.

What constitutes self-awareness? How do we build self-awareness? What signs indicate the evolution of this idea in our journey to becoming a true professional?

In this section, I discuss some critical qualities of self-aware people. These qualities, I believe, build on the foundation of integrity and ethics, and will take you along the path of developing into a true professional in your chosen sphere of work.

Chapter 8

Knowing Who You Are

It is important that each one of us pushes our limits and operates outside our known boundaries so that we explore new possibilities. Yet it is also important not to get carried away—particularly at the height of one's career. In societies where a certain degree of sycophancy is part of the accepted code, it is easy for one to reach too far. A famous actor who is able to portray varied characters with great conviction on the silver screen begins to believe that in real life he could do as good a job, becomes a member of Parliament and then falls flat. It is one thing to play the role of a leader of people on film; it is quite another to execute that in real life. Examples abound in every other field.

Always remain aware of where you have come from and where you are today. Being grounded is a key requirement for carrying success on your shoulders without being burdened by it. I come across many professionals who shun their roots, because it brings on a sense of personal inadequacy. Sometimes they prefer to adorn their conversations with false accents, carefully donning clothes that they hope cover up the anonymity of their early upbringing. Many cling to the branding acquired through a degree or diploma of a certain institution.

Do not let yourself get carried away; do not start believing in the myth about yourself and your achievements. The self-aware understand what their true strengths are; they know exactly how much of their success is because of their inherent strengths and how much is situational. Many of us are extremely fortunate to have been mentored by the right people, fortunate to be in positions of great influence sometimes through sheer luck, or have been the children of fortuitous circumstance where we were surrounded with very competent people who did not compete for credit. None of these take away from our success. Believing that any consequent success is only due to one's inherent capacity is dangerously wrong. Once we begin to think and believe this, we have reached the beginning of the end. We lose our bearings at great cost. Once lost, it often takes a terrible personal loss to restore sanity. Showered with undeserved adulation and recognition, self-aware professionals remind themselves and others around them that their personal qualifications and abilities have only brought them so far. And they take both success and failure with the same equanimity.

Chapter 9

Being Authentic

The phone rings. The caller is someone I've known only as a casual acquaintance at best. I have not heard from him in a decade.

"How've you been? I've been reading so much about you and the success of your company. I really admire what you've achieved. I have read your blog," he gushes. "By the way, my uncle was asking about you the other day. He thinks very highly of you." The caller's uncle was a person of eminence, but I had had one brief meeting with him many years ago, and I doubted he would be inquiring about me through his nephew.

There's honey dripping from the caller's voice; I'm beginning to feel uncomfortable with all the praise and familiarity. Where is the conversation leading? I wonder.

And then comes the real purpose. The caller is looking to change jobs. "I manage a $400 million account, you know," he says. I'm wary when people make such claims. "Managing an account" usually means the person stopped doing real work a while ago. Even though MindTree did not have much interest in the domain he worked in, I asked him to send in his CV, offering to put him in touch with the head of a division who could possibly use his services.

A couple of days later I received the CV and forwarded it to my colleague. Imagine my surprise when I discovered that my so-called admirer had already been in touch with my colleague and had sent his CV even before calling me. Now why did he conveniently forget to mention this to me? The story does not end here.

A few days later, I was having lunch with a friend who is the CEO of a firm in the same sector. He casually said, "Your protégé has been in touch with us. Tell me, is he really good? He told us, if we wanted to know more about his work, we only had to ask you."

"Whom are you talking about? I haven't recommended anyone to you."

No surprises here: my "protégé" was the out-of-the-blue caller. He had gone ahead and offered my name as a reference without even asking me first!

There is no point in being inauthentic and insincere. A true professional has no need to embellish, name-drop, or pretend to be something he's not. Being authentic might not always get us what we want, but that is better than the ignominy of being unmasked, because we live in a small world where everything is connected, and the hollowness of our sincerity will eventually be revealed.

The professional world really craves authenticity; do not dismiss it as old-fashioned.

Chapter 10

Being Comfortable

As we advance in our careers, we outgrow many inadequacies of our past. We gain through the experience of the assignments we handle; we acquire fresh knowledge from others that equips us to handle things differently from the past. Yet the self-aware professional is conscious that there is bound to be some gap in his knowledge, knows that he may never bridge this gap and, most important, feels comfortable with this fact.

A long time ago, I met a customer who, while saying good-bye to me at the end of a long and happy association, told me that I should be comfortable about not being technically qualified, even though I was working in the R&D section of an IT company. During my time there, I used to attend myriad meetings with our customers and engineers and must have shown some fallibility somewhere that did not escape the customer's attention. He was giving me advice on the importance of developing comfort with personal inadequacies. After that I have never felt uneasy about being a graduate in political science working in the IT industry. More important, I do not pretend to understand things when I cannot.

Today, I work for MindTree, where highly competent teams tackle complex technical problems. However capable I might be, I simply cannot fathom the complexity and depth of their work. I may sometimes be capable of deep questioning, I may have the intuitive capacity to cut through issues, I may have the breadth of experience to bring in an external viewpoint which blinkered teams that work at the cutting edge of technology may miss—yet none of that gives me the ability or competence to write a software algorithm or understand the physics of how the alternating character of material helps us store information in bits and bytes.

I may add value in some meetings, and not in others. When technical experts speak, my silence may signal my total lack of understanding. But if I open my mouth, I may disturb the harmony. A professional does not need to hog the limelight or monopolize airtime.

If you cannot add true value, then you must not add to the problem by pretending. The more you pretend, the more naked you become. Ever since I received this piece of momentous advice from a guardian angel, I have attended countless meetings where I began by admitting that I am a complete novice or have sat quietly while others have taken center stage, and I have never felt excluded or reduced in stature because of it.

Sometimes, stating your ignorance can be the simplest solution. Others then take it upon themselves to explain complex technical jargon in easy-to-understand language. Concede the ground and wait, emotionally secure. The team will come back to you when they need you, and then you can truly add value.

Chapter 11

Seeking Help

Each one of us may have the education and training that make us competent at our jobs. These are our strengths, and we must build on them.

But as I mentioned earlier, there is always bound to be a gap in our training and knowledge to do all the things our jobs may entail, and in such situations, admitting ignorance is just one step. Actively seeking to bridge that gap is the next important step in becoming self-aware. Quite often, it is achieved by seeking help from others.

Seeking help from someone—a peer, a mentor, your supervisor, even a junior—is not a sign of weakness. We are surrounded by knowledgeable people in our own organizations and certainly outside of them. But we feel self-conscious about actively seeking their help. We are plagued with self-doubt: What if a peer uses the context to showcase himself as better than me? What if my junior loses respect for me if I ask for help? Am I exposing diminished capacity or intellect by admitting I need help?

These can be real risks sometimes. But again, if a professional does not take risks, how will he develop emotional self-confidence? The moment we seek help, we actually expand inner capacity; we create the emptiness into which others pour their capabilities.

Seeking help is not always about solving an organizational problem. Sometimes we may have a personal problem—an emotional issue, an awkward health matter—which we keep to ourselves. Sooner rather than later, it will adversely affect our work.

There is no shame in asking for help—in matters of work or in matters of mental and physical well-being.

Chapter 12

Not Suffering False Comparisons

Every professionally qualified person makes comparisons with colleagues, peers, seniors, even juniors some time or other. Such comparisons arise because of a competitive spirit that is built into us, often through a ruthless social and educational system that teaches us to falsely interpret the meaning of success in material terms.

Every now and then I meet people who constantly compare themselves with those whom they have left behind decades ago, sometimes pegging their measure of success on where former colleagues or classmates might be today.

Consider this. A midcareer doctor who has embraced the life of research is constantly nagged by the so-called material success of a colleague who was a comparatively mediocre medical student but is now minting money because he chose to open a private practice fifteen years ago. Our man loses faith in his own work and achievements, regretting the years he has lost, unable to turn the clock back. Falsely attracted to the other person's station in life, he yearns for the other man's money, not realizing that the individual in question may be leading a less than complete life. Quite possibly he has no time to enjoy his money! But no tertiary consideration yields solace. The shadow of comparison progressively engulfs the researcher so that eventually he loses focus, gets pulled in by a chimera and sacrifices that most precious thing: contentment.

It is futile to make such comparisons. They cause unnecessary pain, based as they usually are on superficial knowledge, and are certainly devoid of an understanding of the other person's journey to reach that position.

Chapter 13

Having a Reasonable View of the Future

Do we all know with certainty where we are headed?

Probably not. Yet when you look around and observe the professionals you admire in various fields, you invariably sense that they seem to know where they are going; they are in command of the situation even as they face the indeterminate challenges of the future. This comes from the ability to build a view of the future, as against being helplessly dragged into it by unknown forces.

Building a view of the future, and knowing where you are headed, requires acknowledgment of the ground reality, a statement of intent in the overall direction, and sometimes a clear destination or purpose. The sense of destination gives people what celebrated psychologist and author Viktor Frankl calls "something to look forward to" in his book *Man's Search for Meaning*. The author survived the Holocaust at the notorious Auschwitz concentration camp and others in Germany during the Second World War. This is where he studied the psychology of people who survived the worst brutality and hopelessness anyone could ever experience. Yet those who survived the concentration camp did so on the strength of their sense of journey.

They believed that they would get out of the place some day.

They believed that they had an unfinished task to do.

In that belief, they felt that it was they who were in charge despite the helpless immediacy of their circumstances.

Chapter 14

Looking Beyond Money

A professional who sees his work primarily as a means of earning money runs out of meaning very soon. After you've achieved basic comforts, the quest for material success actually erodes self-worth. It becomes most pronounced in the later part of one's life. Many among us begin to suffer from a sense of emptiness that becomes difficult to decode. The greatest satisfaction, and the more enduring one for a professional, is the admiration of people with whom we do business. At a certain stage in one's career, it is peer recognition that sustains us. But beyond it all, the ones who last the longest in the race are those who have given something back to their professions. These are professionals who are driven by a sense of legacy. There is no sustenance bigger than the power to build an intellectual and emotional inheritance.

Many midcareer professionals nurse an empty desire to do something beneficial for a larger common good. "I want to make a difference to society" is often just false piety. Do small things on a sustained basis; do things for your own profession; do things that make a difference to you, your family, your friends, your neighbors and your colleagues; do not worry about changing the world.

The day you feel empty, shift the attention from yourself to others—go and spend time with a bunch of colleagues who have just joined the organization, help an intern with her work, write a series of how-to articles based on your experiences, take on pro-bono work with your industry association. See how the pitcher of emptiness begins to fill again.

Chapter 15

Being Deeply Self-observant

During leadership training sessions, I often ask groups of people to close their eyes for a few minutes in a quiet place and count the number of sounds they can identify. It is a very interesting exercise. Participants usually pick up the rustle of paper, footsteps, the sound of a distant automobile, a chair softly creaking as someone shifts position, probably a conversation in the corridor, the ping of an incoming message on someone's BlackBerry.

When the exercise ends and people recount what they have heard, most people invariably miss two things: sometimes the soft hum through the air-conditioning duct, and almost inevitably the sound of their own breathing. When you can hear your own breathing, you are beginning to observe yourself. But this is just the surface of self-observation. Just as you can observe your body, you can learn to observe your own mind.

Learning to be self-observant is learning how to watch yourself as you think, as you work and interact with others.

I learned to be self-observant in my professional life when I was a rookie salesman. As I went from sales call to sales call, I learned not only to make the call but to observe what was happening. From the time I opened the customer's door, I observed the way we greeted each other, the smallest details of the office, how we settled down, how I initiated the conversation, the questions I was asked and the answers I gave, the tiniest reaction of rejection, query or approval, how the conversation flowed, and how we finally concluded.

The capacity to watch your own actions and utterances is the starting point of being self-observant. This can take you to a state where you can observe thoughts, feelings and actions. For example, if I cannot see myself when I become angry, if I do not play back such a moment, frame by frame, I will never be able to identify whether the cause, the degree of my reaction and its effect was appropriate, adequate or over the top and unacceptable.

Being self-observant is an amazing way to help you perform an "action replay" each time. Record events—who said what, who did what and how things unfolded—and play them back later. Watch yourself at work; you will be amazed how much you will learn about yourself, and this will help you move forward. In the process, you become both player and coach.

Observing yourself on a daily basis and replaying your actions and reactions does not mean you cease to be natural, or lose spontaneity. Just as the player on the sports field is unconcerned with the camera recording the game, the professional is able to go with the flow, be natural and normal, but has the capacity to record events and faithfully reconstruct them down to the smallest detail. The camera is in the mind. Self-observant people do not suffer from illusions.

Chapter 16

Reining in Reactions

Once, I was talking with a bellboy at the Taj Coromandel Hotel in Chennai. He had worked in the same hotel all his life. During our conversation, I asked him what was the worst situation he had ever faced in his job in all the years he had worked there.

After a long silence, the man recounted an irate guest who, for no reason, had flung a bag at him as they were going up in the elevator.

"What did you do when he did that?" I asked.

"Nothing, sir. I felt very bad. Then I thought something had probably made the guest angry." That "something" of course had nothing to do with the bellboy.

However much we try, there are those inevitable moments when a strong emotion gets hold of us. We lose control of the situation. Only a great person can express a negative emotion for the right reason to the right degree at the right time. Unfortunately, most of us are not that perfect. Even if the reason seems right most of the time, the degree of reaction is excessive. What follows is an undesired consequence, regret and often some damage—even if we originally had just cause to be angry.

The bellboy had serious cause to be angry at the abusive behavior of the guest. But at that moment, the bellboy had the greater power between the two: the power to rein in his reaction.

Imagine a policeman being taunted by a crowd. The crowd waves fists, pushes him around, someone calls him names.

How does he decide, when he has the power of violent retribution in his hand and justifiable reasons to use it, what to do? Stare down the crowd, use a baton or use a bullet? How does he know how much is enough?

In extremely high-pressure situations, often the best emotion to express is control. And a true professional has a calibrated thermostat that prompts the degree of reaction and control required in any given situation.

Chapter 17

Welcoming Feedback

Most feedback, by its very nature, feels uncomfortable. This is because feedback is packaged as, or is perceived to be, personal criticism.

It is interesting to note that a sound technician treats all feedback as unwelcome noise. That is exactly how we feel when a customer or a senior colleague provides feedback. We gently thwart it in the sound studio of our minds. In the workplace we covet praise but gloss over most other feedback, even when we say we want to hear it for continuous improvement.

People are taught that every customer complaint presents scope for improvement. Yet scores of employees would rather not deal with an irate customer. How often do people read customer feedback in the raw and not in the aggregated form? How often do we read between the lines of a summary? How often do we travel to the customer's location and discuss negative feedback? Some never even know what their customers look like.

A very common instrument of feedback in many organizations is the 360-degree appraisal. In it, an employee is asked to choose a set of at least six peers and six subordinates, apart from the supervisor, to whom a questionnaire can be sent for feedback. We all have the urge to seek feedback from people who are more likely to say nice, reassuring things about us than make critical observations. That in itself is an insecure act and indicates a lack of maturity, of self-confidence to remain open to ideas and thoughts.

A professional craves real feedback. The ability to freely seek feedback, and more important to take feedback and act on it, is something that can only be learned over time. And it requires constant effort to master. But without developing this ability you cannot become a true professional.

As you practice the art of listening to feedback, you begin to pick it up when it is a whisper, like the sound of your own breathing, and you pick it up from unusual sources.

Chapter 18

Not Suffering False Attractions

You have just started working in your current organization. While applying for the job, you had indicated that you have made enough career changes and now want to settle down. At the interview, you went to great lengths to establish why this was the job and the kind of organization that you were really looking for.

You got the job. Three months into it a headhunter calls you. There is a new multinational in town and they are looking for someone with the same profile as yours. The job pays 50 percent more than what you currently make and according to the man at the other end of the line, it has great professional challenge.

That night you have trouble falling sleep. A lady named temptation visits you.

She whispers into your ear all the things wrong with your current assignment. She reminds you of noncooperation by your new colleagues. She eggs you on by reminding you that you have suggested pathbreaking ideas in the short time you have been there, but no one really listens to you in this new organization. She brings up the fact that you felt shafted when, after joining, you found out that some of your colleagues were earning more than you for doing the same job. Why then do you feel this false sense of virtue—as if you owe the organization something for hiring you?

It is not the false virtue, but the false attraction of which one must beware.

The fact of the matter is that you have not paid back your organization for taking a risk in hiring you just three months ago. The fact is that you did not anticipate the amount of effort you would need to make before being accepted by peers and subordinates. The fact is that, to the headhunter, you are just another head to hunt, to make his cut and bonus.

Flirting with false attractions makes us lose affection for what is on hand. If you do not have a serious need for the offered job or assignment, do the professional thing and resist the temptress.

Chapter 19

Never Lose Touch with Your Work

David Ogilvy, the founder of Ogilvy & Mather, one of the world's most respected advertising agencies, once beautifully described the idea of maintaining contact with your work when writing about his first mentor in the professional world. The man was the head chef at the Majestic Hotel in Paris. He had a retinue of assistants to cut, wash, cook and do all the things that were needed to make the restaurant's signature dishes. But this gentleman did something else to ensure that he earned respect as the boss. Occasionally, he cooked a dish, just to show his assistants that he still could.

As we advance in our organizations, we lose touch with the real work that goes on in them. As you start dictating notes, you lose your typing skills. As a senior surgeon, you stop operating and can no longer tie a suture as well as an intern can. As the head of a business division, you become so big that someone has to change your slides while you make presentations. You no longer do your own research; you cannot design your PowerPoint presentations. You stop making cold calls once you become VP of sales.

These are all signs of decay.

A professional does not let go of the basic ability to work because it is like losing your fingers. There are some things you must continue to do at any stage of your career. Not just the cerebral strategizing, but the actual work. For the chef, it's not about the presentation of the dish, but about how to handle the kitchen knife to make the perfect cut.

A few months before he retired, my father started what we thought was a weird practice. He picked up his own plate after every meal and washed it. He started washing his own clothes and cleaning the toilet. As a result, we had to follow suit. When asked why he was doing all this, he simply said, "Soon I will be a retired man." My mother, though blind for the greater part of her life, always washed her own clothes, made her own bed and swept the floor.

Doing what are often thought of as menial tasks has a calming effect on us. Doing the small things in life yourself is very important. These have greater significance than you know. Sometimes, the most profound ideas come not when you are in the boardroom but when you are washing dishes. So do not stop walking when you have a car. Do not stop driving yourself because you can be driven. Make your own slides. Write your own speeches. Go and make a cold call every now and then. Make an omelet or a cup of tea for yourself. And sometimes, get into the weeds at work.

Chapter 20

Being Proactive

Valmiki, the sage who wrote the Ramayana, counted the many essential qualities that make a great king. He saw them manifest in the personality of Lord Rama. One among them was the quality of being what he termed *agravashi*, or the one who initiates conversation. Imagine when two great individuals meet. Invariably, one of them makes a subtle, often brief, but observable, proactive stance—sometimes it is the first extension of the hand or the first spoken word. That seemingly trivial motion is the sign of a king. I believe that is also the sign of a true professional.

Being proactive is an amazingly wonderful attitude and a behavior that earns us memorable relationships in both life and business. Everyone loves to deal with a proactive individual. Every boss prays to have such a person on her team. What makes someone proactive?

First and foremost, like the king, a proactive person is self-confident. People who are unsure of themselves are unlikely to extend their hand first. This self-confidence is not about who you are; it is about where the conversation could lead. What if by extending my hand I end up committing to something? What if you ask me for something I cannot give or do not have? Fear of commitment makes us stay put, keeping the hand and the words to ourselves.

A proactive individual is genuinely interested in the well-being and welfare of the other person. One day, I got a call from my family doctor. A week before I had complained to him of back pain. He had prescribed a course of medication and now I was feeling fine. "Subroto," he started, "you have not been a good patient." He was admonishing me for my silence. In that moment, he was not really concerned about his fees or his medical practice. He genuinely wanted to know how I was faring and why I had not made the time to call.

Proactive individuals are not worried about creating work for themselves as an unwanted consequence of reaching out. Why fix something that isn't broken? Why nudge a customer who has not called for months about how the machine you repaired is working? What if it opens up a Pandora's box?

For all you know, the reason he has not called is that your competitor has offered to take back the bad equipment, scrap it and install his latest model—and has worked out a great financing deal your company has not even thought of!

Proactive behavior is about our small dealings in everyday life. It is about being the first. You could be the first to send a clipping of an article that could help a peer working on a related project when she is snowed under. You could be the first to reach out to the boss and request time to discuss your performance appraisal. You could be the first to analyze the price performance of a solution you installed for your customer a year ago. The first to write a point-of-view document for the firm, or the first to apologize when you butt heads with a colleague in a meeting.

In other words, you could be the first to be king! That is the power of proactive thought and action. Sometimes, being proactive can and will land you in trouble. But what kind of king is afraid of trouble?

Proactive people think on behalf of others. Sometimes it is also about thinking ahead of others. This attitude is then followed up by thoughtful preparation. If you analyze the reasons for attrition in the company before the bosses have woken up, guess who is the one spending nights researching the data? If you are about to suggest to your customer a new payment model that is good for her and good for you, who is the one spending time on the weekend doing spreadsheets?

There is another critical side to proactive behavior. In the world of business and professional dealings, people do not like to be taken by surprise. Anything that may cause grief if discovered must be brought to light by you, voluntarily, ahead of time and put in front of all the stakeholders who could be affected by it. It is uncomfortable while you do it but it builds lasting relationships.

Proactive behavior extends to our personal lives as well. Someone who is proactive at the workplace is more likely to be proactive with his health, and in his relationships with friends and family.

Chapter 21

Taking Charge

In 2003, as the chief operating officer of MindTree, I wanted to take my direct reports on an annual exercise to some place away from Bangalore, where the company is headquartered. The purpose of the exercise was to set out the objectives for the various departments I was responsible for. I wanted to tell them about my priorities for the future so that they could align their objectives with mine and with one another's. In the process, I also wanted them to share their functional goals with one another so that everyone was aware of what each department was doing.

The stated plan was to drive down in a small bus to a place in the middle of nowhere where my friend Captain Ravi, an ex-army officer, ran a highly successful outbound learning center called Pegasus Camp. There I wanted the team members to listen to one another, question, ideate and return with solid alignment and a collective vision for the year ahead.

Along the way, I also wanted to expose them to a situation involving extreme pressure and uncertainty so that they became emotionally ready for what I foresaw as an unpredictable time ahead. With Captain Ravi's help, I hatched a plan unknown to my colleagues.

The group assembled on the appointed morning for the bus ride. They were told that I would not be joining them as I had decided to go ahead by car. Only the bus driver knew what was to come.

An hour and a half into the journey, the bus turned off the highway onto a barely navigable road. The bus had to go past a tiny village before coming to the Pegasus Camp, which was located down a hill by a lake. When the bus approached the village, the managers, still somewhat sleepy and now weary from the rough-and-tumble journey, were accosted by angry villagers who had barricaded the road. The driver had no option but to stop, and as soon as he did, the mob surrounded the bus.

Apparently a car ahead of the group had mowed down a goat and sped off. The irate stick-and-sickle-wielding mob demanded that everyone get down from the bus and stay right there until the car owner returned to compensate the villagers for the loss of the goat. The group was stunned. The description of the car and driver suggested that I was the one who killed the goat. The villagers were relentless, their tempers rising by the second. They refused to engage in conversation and kept demanding that the group get down from the bus.

Away from the scene, inside the village school, I sat glued to a small window and watched the reactions of the people on the bus: fear and uncertainty. That bus had some very qualified, competent and senior managers onboard. People who reported to me were usually stars. They were trained to handle very difficult corporate situations; they were engineers and MBAs and trained negotiators.

But this situation was different. Here were villagers who did not speak English, were armed and angry and seemed like they could and would do anything if slightly provoked. I wanted to see how the leaders would react, who would take charge. The first reaction of the group was to stay put in the bus. No one stepped out.

After a while, a few of them tried reasoning with the mob from inside the bus. But it was futile because the group did not have a nominated leader and it is impossible to negotiate with a cacophonous bunch. Most of the managers just stayed in their seats, anxiety writ large on their faces. It was soon evident that the villagers were determined to get everyone out, and who knew what they would do next.

Then in the melee, I saw two men, Nagan and Sripad, slowly get up, come to the door of the bus, open it and get off. These two were not the most senior among the group; they were not the stars who would wow everyone with their presentations in the boardroom. They could best be described as somewhere between solid citizens and unsung soldiers. In that instance, however, where others were courting safety, these two men had waved aside all caution, putting group before self. As they cajoled the villagers, their captors got more menacing. Surrounded from all sides, they kept trying to pacify, to reason, standing between the villagers and the bus door, negotiating on behalf of their colleagues. At this stage, I came out of the village school with a goat in tow, and seeing me, the villagers stopped the premeditated drama. The mood in the bus changed from fear to jaw-dropping disbelief to relief and happiness. Later, we sat down to decode what had happened to appreciate the nature of unscripted events in the workplace.

Several years after the incident, many of the stars in the bus have either become dust or gone on to shine elsewhere; the two men who climbed out of the bus that day continue to serve Mind-Tree well. But to me, the goat that was never killed settled the issue as to who among the busload could be counted upon in case of a future crisis.

Faced with a potentially dangerous situation, people freeze. They are afraid to push their way into a crowd because they feel powerless. They justify their inaction by thinking that not being in the front, not taking charge of a dangerous situation is the wiser (and safer) thing to do.

We have often heard that no one can give you power, it has to be seized. While the first part of the statement may have a tinge of truth to it, the second is simply untrue. Power is not something material that you can seize, like a gun or a piece of land. Power is never seized. It is always generated from within. Whether you are powerful or powerless, you must feel these qualities before you can become them. Developing the power within, to have the confidence to take charge in the most difficult and potentially dangerous of situations, is the hallmark of a true professional.

Chapter 22

Courtesy and Humility

The chairman of Infosys, India's much talked about software services giant, N. R. Narayana Murthy entered the room with a young colleague and a guest. There were only two chairs in there. That morning, Murthy was somewhat preoccupied with his own thoughts. As Murthy and the guest settled down for a conversation, he took one chair, while the guest took the other. Only after the meeting was over did Murthy realize that his colleague had been standing the entire time, and Murthy felt terrible about the insensitivity he had shown.

In our discussions, Narayana Murthy has always stressed the importance of courtesy and generosity as a marker of professionalism. More important, you must show generosity and courtesy to people when you are in a position of power. Nearly two decades ago Murthy was nominated to receive an award instituted by a well-known charitable organization in western India. The evening before the award ceremony, he noticed a famous woman columnist and walked up to introduce himself. The lady simply looked through him. The next day, after he received the award, the lady told him that she had not known *he* was the recipient of the award when he had approached her the evening prior. Murthy simply replied, "Does it really matter?"

Many years later, a friend of Narayana Murthy's won a major award, and the same columnist happened to be at the party in his honor. This time she made it a point to walk up and talk to Murthy, who had by now become an internationally known figure, to inquire if he remembered meeting her a long time ago. Murthy politely said yes but also distinctly remembered how rude she had been when she was the better known of the two. Murthy sums up the issue by saying, "You're Amitabh Bachchan and you show grace and courtesy—that is when it matters. Not when you are just an aspiring actor." Generosity, grace and courtesy become truly valuable only when shown to others at the height of your professional career. (For those who do not know, Amitabh Bachchan is a living legend of Bollywood cinema and one of the greatest actors of all time.)

Professional adulation is like the beacon of a lighthouse—it falls on you for a few moments and then it must move on. People who think the spotlight will always remain on them, and are often rude, insensitive and self-absorbed, soon discover the loneliness of being in the shadows. Professionals must know that humility is critical to enduring success. When we are humble, we can listen to others. When we do, we remind ourselves of our weaknesses when others are discussing our strengths.

When you suspend humility, you only hear the adulation and it becomes regenerative until you turn deaf. Often it takes a personal tragedy to bring you back to normalcy, and that is an avoidable price to pay.

Murthy practices something very interesting to retain his

humility. Every night, before going to bed, he stays awake with the lights out for a few minutes to recall the mistakes he made during the day. During this time, he speaks to his conscience. Sometimes, when he has made a regrettable mistake, this process of recounting the incident is cathartic, because it makes him relive it and only then can he let go of the pain. Before he falls asleep, he prays that he should be given the strength not to repeat the same mistake.

There is a proverb in many Indian languages. There are three leaves—the one on top is falling, the one in the middle is trembling and the young one beneath is awaiting its time. In every profession, the ones on top should be aware that theirs is only a temporary accommodation.

Given the constant ebb and flow of our professional lives, it is important not only to have this humility but to have an appreciation for the potential in people below us—to recognize and nurture that special someone who right now is not quite there but may go further than you have. In Eastern culture, a certain respect for senior people comes rather naturally. A professional does not take this respect for granted.

Chapter 23

The Big Picture

I was experiencing a mild but pestering pain in my left shoulder for a few days. Then one morning at work, I saw a colleague with a cast. What happened? I inquired. Lo and behold, he had a similar but more serious shoulder problem and had undergone surgery. Hearing him, I asked if I could see his doctor. I was visiting India for a few days from the United States at that time and requested an appointment right away.

The doctor overwhelmed me in the first meeting. He had read every single article I had ever written, as well as my books. After a brief examination during which he convincingly concluded that I had some kind of degeneration, the doctor informed me that a famous cricketer had the same problem, indicating that I was in exalted company and that surgery was in order. "How lucky I am to be able to operate upon you," he gushed.

Just to be sure surgery was needed, I had an MRI done the very next day. When I called my wife in the United States and informed her, being a very balanced person, she asked that I meet our family doctor in India before rushing into things, and I did that. Since

I was already under the spell of the scalpel, the family doctor sent the MRI reports to a sports doctor, who in turn confirmed that I had the same problem the famous cricketer had. Convinced of the diagnosis, the first surgeon I had met was now eager to put me under general anesthesia and get started.

My wife asked me to come back and get another opinion before doing anything.

Back in the United States, a land famous for commercially minded doctors, I made an appointment to see an orthopedic surgeon who ran a private practice near my home in New Jersey. He saw the scans I had done in India. Then he looked me straight in the eye and asked, "Mr. Bagchi, how old are you?"

"Fifty," I replied.

He went on to say that his job was to cut people open and get paid for it. And yes, the MRI did show some degeneration of the shoulder. But if at the age of fifty my body did not show signs of degeneration, when would it? He advised against any surgical procedure. I had to accept the signs of my body's aging.

"But will I be able to ride my bike, Doctor?" I feebly asked. "What if it pains?"

"No pain, no gain," he replied, and I was ushered out of his clinic.

It has been almost four years since this episode and the shoulder pain no longer bothers me as it once did. But through its discovery and resolution, I learned an amazing lesson about professionalism. As we look at data, we must also look at the big picture. Only the big picture, the context in which we live and work, makes the facts relevant. And it is only when we understand and actively look at the big picture that we will develop into grounded professionals.

Part Three

Professional Qualities

Chapter 24

Of Time, Body and Soul

I started working when I was barely nineteen. That was a good three decades ago. In the intervening years, I have grown up professionally, married, raised two children, started my own company, written three books, acquired hobbies like teaching, photography and biking, learned yoga and meditation, made friends, attended the occasional social event, and yet I have never felt that I have less time on hand. This may sound unusual to many people who groan, "I don't have the time to do the things I once liked," or "I cannot cope with my work and have problems with my work-life balance."

One cannot be a great professional unless one masters time. This is not a book on time management; there are countless books on that subject. While you should read some, I do not believe that you can learn to master time by simply reading about ideas. It is probably more effective to watch and learn from people who actually seem to manage their time well, and who feel productive and satisfied. I have met many such people, and if there is one primary quality they all share it is self-discipline. If you lack basic self-discipline, you will soon find that poor time management is just one of your many problems.

There are deeper issues that you must address before you can achieve success with time management.

Most important is the status of your health. Good health is critical to productivity. Poor personal health affects concentration and leads to a feeling of exhaustion. The people who complain most about lack of time ironically are people who are in the prime of their life, but probably in poor health. The key is to ensure that you pay attention to what you eat, how much you sleep, the exercise your body gets and the ways you unwind. Strange as it may seem, these basic things go a long way in expanding the time available to do things.

Pay attention to how much you eat. Most of us eat way more than is necessary. We have so many options now to eat out that we are not careful. I will never forget a conversation I had with the famous cardiac surgeon Dr. Devi Shetty, who told me that of all people, Indians have the highest risk of heart attack because they overeat. Genetically, the average Indian is designed to make do with much less food. But now there is abundance and, as a result of both availability and affordability, many Indians eat excessively, leading to coronary complications. For many people, consumption is what defines success and it begins with food.

Then there is the accepted notion that the best way to do business or bond with colleagues is to go out for a drink. Nothing is further from the truth. In my work, I often have formal dinners with visitors and clients. If a client likes to drink a glass of beer or wine over dinner, that is fine. I have never lost anything by abstaining. Never use your work as an excuse to drink.

Next to eating in moderation is the need to sleep well and

sleep adequately. Most young people who come to the workforce bring erratic sleeping patterns from their student days, and over time get into the cycle of sleep deprivation. Sleep is as important as diet and exercise to produce great work.

The forties are usually the peak years of a professional's work life. It is then that you are snowed under with work pressure, apart from increased expectations from your family. Very often, your day starts early and ends late. You probably travel ten days a month. You do not realize it, but all this consumes you from within. Begin to pace yourself, slow down. It is a great time to use techniques like yoga and meditation to sharpen the ability to do more by doing less. In other words, you need to know how to burn less energy to deliver the same level of productive work. Sometimes prayer, or a quiet period for reflection early in the morning or before you sleep, is just as helpful.

Chapter 25

Doing More by Doing Less

As young managers, people tend to be all over the place. Some simply revel in multiplicity. They have no time to pause and think. They thrive on multitasking, constantly checking e-mails, sending out text messages and calling their voice mail number while rushing to and from meetings or frantically trying to finish a project on time. In the process, many develop a state of mind in which volume of activity becomes more critical than the value one adds.

Doing more, being engaged 24/7, does not mean you are achieving more. As a professional, the trick is to do more by doing less. To do this, you have to disengage from doing too many things at the same time and prioritize based on where you can make a larger impact. The practice of disengagement from distractions becomes very important, particularly for midcareer professionals.

Strange as it may sound, a way to disengage from multiplicity is to learn something new—a sport, a foreign language or a musical instrument. Pick up something that regenerates you and revives the spirit of curiosity and learning in you. Give it undisturbed time.

As you move into your forties, it is time to become your own benchmark. It is time to bury past rivalries and disgruntlements about who got which promotion ahead of you. Make the time to actively mentor someone and groom a replacement—this enables expansion of roles, encourages thinking beyond the obvious and gives you more time to focus on larger issues. As you mentor someone you find yourself energized and are thus able to do more.

As we approach midcareer we find ourselves slowing down, not being able to keep up because we begin to develop a knowledge gap. Make an active effort to bridge this gap. In a knowledge-intensive world, the only way you can remain ahead is by learning. The sources of learning themselves may change as the compression of time takes place in your work life. You may have to pay attention to what you read, participate in industry bodies and interest groups and find newer ways to accelerate learning and absorption. Look for newer sources of learning as well as newer ways to learn. Meanwhile, cut down distractions. You know very well what they are in your own case.

Chapter 26

The To-do List

I cannot remember a time when I did not make to-do lists. Each morning, I look at what I need to accomplish that day, and in the evening I check off all of the things I did. It is my small victory over time. When we started MindTree, I decided to go a step further. I told myself that I would save the notebooks that contain my never-ending to-do lists and my diary for each year. As a result, I have accumulated the notebooks and the diaries—these have pride of place in my office. I do not think they are more precious than the work I have accomplished but they egg me on; they help me to know that I was in charge and that I did my best.

Many tasks on the list do not get done for days or sometimes weeks. Inevitably, when I have accomplished them I cross them out with a sense of victory. My to-do list is not just about work. While I cluster work-related things in a group, I also have a personal to-do list and it gets the same treatment. I write both lists on the same page of the notebook. It is unbelievable how powerful this simple tool can be.

The to-do list is nothing but a set of goals—not the big goals of life, but the small day-to-day ones that we have to get through on our way to achieving our vision. Goals have a powerful ability to shape our behavior, motivate us, create energy. Once you build goals, you enter the virtuous cycle by which when you are close to achieving a goal, newer goals appear, and this process keeps you going. This is the essence of the purpose-driven life. During a conversation with the Dalai Lama on where he got his power from, the great man told me, "Your purpose determines your power." The humble to-do list is the beginning of purpose and prioritization.

The issue some people face is not the absence of a goal, but rather the presence of too many and the lack of prioritization. I find many young people wanting to learn a new language, enroll in a personal development program, join a gym, read a book, write a blog, run a marathon and socialize, all at the same time. It is important to prioritize one's life. Ask what is really important to you. Also ask yourself why that thing is so important and whether you are willing to commit yourself completely to it.

Chapter 27

Saying No

It is surprising how the inability to say no results in time wasting, poor prioritization and the feeling of always being rushed and behind on deadlines. I am constantly surprised by the number of people who attend company meetings in which they have no interest. It is a criminal waste of human potential. Do not attend meetings just because you get invited to them; say no when you know you have no value to add to the discussions. You will be surprised how your forthrightness will eventually evoke admiration.

When you learn to say no to unimportant things, you have the time to take on a task in whose outcome you have a serious stake and in which you believe. Attend meetings in which you have a solid viewpoint and can add or receive value. And undertake travel that is necessary. Filling up an airline seat to fill time is not a good idea.

The inability to say no to their superiors is a major problem for many young professionals. Often it comes with unnecessary awe for a person in a position of power. A young doctor may feel compelled to drink with a senior surgeon who likes company. Young salespeople who look up to their older managers feel obliged to go along with the senior person's eccentricities. I would advise two things: never do anything you do not want to do, and learn to say no to additional chores or requests outside of work. If the senior is a professional, he will appreciate your candor. As a professional you must set the limits, and when you do, people will respect you for it.

Chapter 28

Quit Whining

Jobs are not meant to satisfy us. Jobs do not know who we are, what we are seeking and what our special needs could be. You may think this is a mere philosophical statement. On the contrary, I believe it is the most practical and rewarding way of looking at a professional's career. When I see scores of successful people around me, I believe that their achievements are due largely to such a perspective.

Some people dislike their work, some like the salary but not the work, some dislike their boss and yet others dislike their colleagues. In some cases people dislike the idea of work itself. A professional realizes that work is a blessing and, most of the time, it is therapeutic. We are lucky that we have work to do. Every day, increasing numbers of people in this world feel left out, unlucky, because they have woken up to a morning without any work. Go to villages anywhere in the world and you will realize what unemployment really means and how scores of people are grateful if they can manage even seasonal employment.

Many professionals argue that it is not the work they do that bothers them; they are bored with the routine. "I seek challenge" is the headline on most résumés. The world goes around because of its capacity to be predictable. The bees and other insects pollinate the crop so that we get our food with "monotonous" regularity. Imagine the sun getting bored with its work or the Earth feeling bored with its axis? Routine things done well make life livable.

And what is not routine about a cardiac surgeon who operates on people every day? Or a pilot who flies airplanes? At Bangalore's famous hospital, Narayana Hrudayalaya, doctors operate on newborn babies and infants who suffer from heart ailments. These small children must remain for days in the intensive care unit where every baby is assigned a dedicated nurse who keeps a watchful eye on the baby at all times. The hospital boasts a 95 percent operation success rate—not because of the doctors who operate and leave, but because of the nurses who take care of the babies after the operation. The job of the nurse is as routine and as predictable as it gets. But it is her work that ensures 95 percent of the children operated on would never have to return to a heart hospital. That is the definition of a 95 percent success rate at Narayana Hrudayalaya. The great challenging job is often an illusion; making the routine interesting is the real challenge.

Sometimes, we have very genuine problems at work. We may be underemployed or stymied in our quest for advancement, or have toxic bosses and difficult colleagues. The way out is to face up to the problem. Speak to someone in the organization or even an outside trained counselor. Discuss the issue, find a solution, be open in examining how much of the problem is because of you or your own insecurities. Try to correct yourself. If you find the organizational fit is wrong, find another job—but do not whine.

It simply does not help. When you whine about your workplace troubles, chances are two kinds of people listen—those who do not care anyway and those who are happy that you are in trouble.

If the solution is finding another job, be sure you are not going to an organization where you will encounter similar problems. Many people end up in the wrong profession because of lack of prior knowledge about it, and then they complain for the rest of their lives. It is never too late to start all over again. There are enough resources available to help you understand your aptitude. So it is never too late to find out what you are cut out for and re-align your job, rather than remain stuck and complain about your work.

There are situations, however, when high achievers tire out or simply outgrow their jobs. When the latter happens, there may be no visible external sign. No one is aware of the situation except you; only you know when you have outgrown your shoes because they pinch, and the same is true when you outgrow a job. When you see the signs of mental pinch, consider expanding your organizational role. Ask for additional responsibilities. Walk up to your boss and ask to take on more work. Consider spending time with younger colleagues. Build into your work elements of service—that will invariably make your job, and your achievements, more enriching.

Chapter 29

A Long View of Time

I first started writing for *Dataquest*, India's leading computer industry magazine, way back in 1982. It started with an occasional piece every now and then, but I took it seriously enough for the magazine to eventually offer me three consecutive columns to write. In the process I built personal visibility and a professional reputation that not only helped my work but also got me a job at Wipro Corporation. In the later years, as I grew in my professional life, I maintained my commitment to writing and eventually wrote a column called "Arbor Mentis" for *Businessworld* magazine. This column drew the attention of my publisher, and that is how I was given my first contract to write *The High-Performance Entrepreneur.* Around that time, Prashant Hebbar, former journalist at *Dataquest*, and Indrajit Gupta, my editor at *Businessworld*, joined *The Times of India*. Their goal was to revamp the business page of the number-one newspaper in the country. As part of it they wanted a human-centric column that would make the nonbusiness reader "stray into and stay" on the business page. Hebbar had followed me for a decade and

Gupta had interacted with me closely while at *Businessworld*. They asked me to write a new column, "Times of Mind." It soon became one of the most talked-about columns in the Indian media. When Indrajit Gupta left *The Times of India* to become the first editor of *Forbes India*, he asked me to write "Zen Garden," my biweekly column in the magazine. Similarly, after writing *The High-Performance Entrepreneur*, I was commissioned to write *Go Kiss the World*, and the success of both books has now led to *The Professional*.

The reason I am sharing this with you is twofold: I believe that one must take a long view of time in building any professional relationship. It goes well beyond the work you produce at your job. My relationships with Hebbar, Gupta and my editor at Penguin are not just about writing but about how, over time, we have engaged with each other, how we have ideated, ironed out differences and co-created valuable work. If working with each other was not a pleasure, something we were jointly committed to for the long term, we would not have continued our relationships, and would have moved on to other choices. It is the power of association that builds formidable value over time, which the myriad other choices cannot provide.

The second point I want to make is the importance of taking every assignment seriously. Whether I wrote a regular column, an occasional newspaper piece or for that matter a book, I gave it all I had. To me, my writing commitments were no different from commitments at work. This attitude called for many personal sacrifices because it was always like having two jobs. Yet the rewards have been huge.

Every professional's footprint must be larger than his daytime job. Our professional identity goes well beyond the employment badge we wear at work. So while dealing with people who may be suppliers, customers, industry associates and other stakeholders, one must build a long view of time and treat every small engagement with them with all seriousness, as if life depends on it.

Chapter 30

Mavens, Connectors and Evangelists

Among the most powerful things you can have as a professional is a strong network. A good network, cultivated wisely and used well, is a great expander of time. It can achieve great things for you. To make it work for you, you need to know two things: First, you have to contribute value to a network before you can get value from it. Second, its primary job is not to give you the ability to meet your targets in the short term. In other words, a network needs to be cultivated with a long view of time. One needs to understand how to use it wisely and well.

A good network can bring you in minutes what hours of individual research may not. In many organizations today, leaders encourage people to form voluntary communities. There are stories galore in knowledge-based companies of how someone in the United States reaches out to the global community of fellow professionals late at night with a request for help and solutions arrive in his mailbox the very next morning from people far away in Australia or India. When you acquire the reputation of being a contributor of value to a community, you are invariably flooded with high-quality help when you need it.

I have my own short list of favorite professionals—doctors, journalists, economists, filmmakers, architects, young entrepreneurs—whom I reach out to routinely, seeking their advice on dozens of matters. You may think for a moment, What is so great about this? Isn't this the same thing as using your world of contacts to help you get things done?

To me these people are not "contacts," they are "connectors." The world of a professional consists of what we call mavens, connectors and evangelists. A maven is someone who produces, consumes and trades knowledge. A group of mavens is usually connected to an expert who is called a connector. A connector provides clarification, assistance and linkages to other connectors. A group of connectors may link up to an evangelist, who is the last word in a particular field of knowledge. In a world of mobile devices, the Internet and information overload, these people can substantially cut down research and consulting time, which is precious to me. Without them I would have to do hours and days of work, whereas these people can guide me effortlessly to the precise point of knowledge or expertise, or provide me with a counteropinion. You do not build a network of such professionals through the low-level principle of trading favors. In fact, such people would dislike the very idea. Interaction with them is governed by the principle of mutual respect and affection which is built over years.

Chapter 31

White Space

Every day, as I hit Bangalore's busy Inner Ring Road, I look for my favorite sight: a three-wheeler converted into a delivery van with the words EGG VAN written on it. Presumably, the owner is in the egg business and uses this vehicle to deliver eggs to retailers in the city. But when I see the Egg Van, it is not loaded with eggs. It has a shutter at the back which is open a foot above the fender. Through it, the sun filters in and I see two young girls, dressed in school uniforms, sitting cross-legged with books open on their laps. Between them, I see their schoolbags. I have never seen their faces but I can imagine their eyes. The picture uplifts me in an instant. On the same road, I see countless office buses that ferry bored young men and women with tired looks on their faces, some sleeping, others switched off from the world via a pair of white earplugs.

I call the two girls my "angels of white space." I do not know how the term "white space" originated, but in telecommunication, it denotes frequency allocated to a channel but not used. Typically, broadcasters are provided additional frequency that is not meant to be populated so that adjacent broadcast stations do not overlap. In print, a white space denotes emptiness so that we can read the characters that form a word and a group of words that form a sentence.

In our professional lives, white space is a train or a bus ride to work, it is the time waiting outside a client's office, the time spent on long flights. We have all been given huge white spaces and we simply let their power go to waste. I am not suggesting that we cram work into every available moment. Quite the contrary, these white spaces should be an oasis of recreation and rest. Mindless use of time that neither helps get work done nor rests the mind is a waste; use your white space for getting things done in a planned, proactive manner. Find your own angels of white space.

Chapter 32

Creating Reuse

Whenever you create something useful, always consider how you can reuse it.

You're preparing an internal presentation to your sales team on sales productivity. A little bit of tweaking and generalization would probably make it a great training resource for the sales school your company runs or a valuable download to offer on your Web site. Offer the piece to a trade journal. Convert the idea into a white paper.

An internal memo to the organization on how to protect customers' intellectual property could become part of a point-of-view paper that could be used as a handout. While Mind-Tree's integrity policy is an internal document, it is also a great thing to showcase each time someone asks us about corporate governance.

By looking at what you are creating—an essay, a PowerPoint presentation, a policy document—as something that can be re-used, you can expand time and get more done in your day.

Reuse is about building components at work wherever possible so that we do not do the same thing over and over again. It is also about recognizing that everyday artifacts have multiple purposes.

The downside, however, is twofold: First, wanton reuse builds a "cut-and-paste" culture in which advice and resources are created with one party in mind but sometimes get reused on another party that might not benefit in the same way. Second, unless one is careful, there can be issues with intellectual property.

Reuse of work artifacts is a great way to expand time. Vast amounts of knowledge just sit in the content management system of an organization, yet professionals spend enormous amounts of time reinventing the wheel.

Chapter 33

What Is Your Touch Time?

In a factory, amid the countless activities that keep a manufacturing person busy, all that truly matters is when the raw material physically touches the machinery. It is only at that point, as the material converts itself into the next higher stage in production, that value is added. While a factory head may feel proud of the state-of-the-art machinery in his factory, what truly matters is not its capacity but the touch time the factory is able to achieve. Similarly, every professional must know the equivalent of touch time in his field. We are all busy doing things that fill up our workday, but these do not necessarily add up to touch time.

Once, I ran a study involving more than three dozen salespeople over a number of months during which they agreed to maintain logs of their activities for each day. I decided to conduct the study because they were all complaining about how stressed they were, how they did not have enough time to make new sales calls, or submit expenses, and certainly no time to feed customer information into the order processing system or chase accounts receivables. What was their principal gripe? That they had to perennially follow up with the manufacturing unit for product shipment. They complained that they were always on the phone speaking to the factory about delivery and short shipments. So we agreed that everyone would log their activities to see how much time they spent "following up with the factory" every day and how big this issue really was.

After the data was processed, the results surprised everyone. To begin with, the follow-up with the factory was taking less than 20 percent of their total time. Even though this was high and could be significantly reduced with systemic changes, it was nowhere near the amount of time the salespeople had imagined it to be. The study turned up other interesting data as well. The actual face time the salespeople had with potential customers was less than 12 percent of their total time—they all thought it was three times as much! For salespeople, face time with a prospect is the real touch time. Salespeople often forget to focus on this touch time, believing that time spent traveling, waiting in office reception rooms or making small talk with the receptionist is core sales activity.

Time expands only when we measure our value-added activities. The amount of time every white-collar worker wastes on unproductive activities adds up to quite a bit at the end of a day, week and month. Anyone who sings the refrain that he works really hard must step back and measure his touch time.

I work with high-achievement individuals to help them build their leadership capacity by working on self-awareness. An essential aspect of the process is an analysis of how they are spending their time. I urge them to look at a hard copy of their previous ninety-day calendar at any time. Most people who complain about not having enough time have many "not-so-hidden" time wasters.

People come to the office only to immediately go for a coffee break; smokers spend on average an hour a day taking nicotine breaks. People also attend unnecessary meetings. They surf the Internet and answer personal e-mails while at work. While some may argue that none of these things may be in the way of someone's productivity and may actually help some, we must be aware of the amount of time in a day we spend doing these things, and how they could be taking away from the time we need to do more valuable tasks.

When I ask people what they do when they are not working, many draw a blank and some say they watch television. Ask them what they watch and how much time they spend in front of the television, and you get a very evasive "This and that, not much time" in reply. Ask them to quantify the time in front of the television and it can be six to eight hours a week. Watching television can be relaxing and informative, but so can reading a book, or meditating—doing things to gain a complete body-mind-soul makeover.

Senior professionals who are productive and get things done on a sustained basis do four things particularly well. They get briefed, they seek help, they use commute time effectively and they periodically take a mental shutdown.

Given the information abundance we all experience, there is a pathological need to consume information by reading reports. As one goes higher up the ladder, the effective use of aides not just to take instructions but to absorb information and synthesize on your behalf, and sometimes even offer a viewpoint, can be very useful.

Seeking help is a great way to expand time. It is amazing how much idle time is available in any organization. In MindTree, there are people who are always waiting for a project to come through, and every time I need research done, I invariably reach out to such people. The call for help energizes them as much as it expands my time. Organizations that promote the culture of a "capacity market" in which people can notify one another who is available at any point in time actually increase productivity and at the same time build thriving communities that bond.

Apart from all this, it is very important to take periodic shutdowns. Many people complain about lack of personal time but the idea of a shutdown gives them withdrawal symptoms. Professionals who take compulsory breaks at regular intervals are able to get more work done, in less time, because of the power of a relaxed mind and body.

Chapter 34

When Paths Diverge

In today's world, it is unrealistic to think that you will remain a one-job or two-job professional through a forty-year career span. I have made my fair share of job changes but I find that the places where I have made a recognizable impact are the ones where I spent five to ten years or more. Such a thing obviously requires a high degree of alignment between the individual and the organization. When we talk of that, usually people emphasize the organizational priority to get the right people in. But I believe that it is as much the individual's responsibility as it is the institution's. Good people tend to go to good organizations.

If you are any good, you should know that you have the right to outgrow your job after you have made an impact, you have the right to occasionally get bored and you have the right to covet more space. There's nothing wrong with seeking and making a job change. But a true professional makes sure he has thought through the consequences, sees the big picture, knows how to make the change and finally does not burn bridges while parting ways.

There is a great deal of similarity between a job and a marriage. You never simply marry someone, you marry into that person's family. The same is true of a job. It is not merely the job, but the organization that you are going to join. A great job in an organization that does not align with your values is a waste of your time. Another way to look at this connection is to think of the job as the house you want to move into and the organization as the neighborhood. A great house in a bad neighborhood may not be that great of a house after all.

If the neighborhood is great, you can progressively move to a better house someday. If you have really outgrown your job but are value-aligned with your organization, you may just need to find another place in the same company in order to make an even larger impact.

Sometimes it still doesn't work, and one must change jobs.

It's important to understand that most job changes do not change you. If there is an inherent incapacity that is not getting you the raise and the promotion in your current job, you will not solve the problem by simply switching organizations. Fix the gap, and if you cannot, come to terms with it. Do not fool yourself with the idea that a mere job change, just because it is possible, is going to solve your woes.

Always do a reference check. Reference checks are not the employer's prerogative alone. Ask the recruiter for the names of people who might be willing to talk to you and help you understand the new organization. Reach out to customers, suppliers and ex-employees. Look for patterns that prove or disprove what you read on the Web site. Say that you would like to spend a day at the proposed workplace before signing the acceptance letter.

While you do your due diligence, also know the trade-offs. If you so badly want double the compensation overnight, do not complain about office politics at the new place.

Once you have landed your new job, make sure you set it up for success. Try to blend in and not blend out in the initial days. Build value before seeking recognition. And do not make comparisons. As in any new relationship, so in a job, it is downright irritating for the new "spouse" to hear how heavenly your ex was.

Do not prey on your last employer. Everyone has a hard time not hiring an ex-colleague who has been their absolute favorite, or staying away from an ex-customer. To build your future at the cost of the previous organization, destabilizing it, leaving it hollow as you move on, is not professional conduct.

And most important, professionals always take the goodwill of everyone with them when they leave. You never know when you may need to come back as an employee, a customer, a supplier or simply a mutual reference.

Part Four

Managing Volume

Chapter 35

The Power of Vision

K. Srinivasan was a village postmaster in the erstwhile Madras State in 1957. A contented man in all respects, he had just one personal regret—he had never attended college. When his first child was born, he had the dream that his son would one day become an engineer. In the years ahead, there were two real constraints: the villages he got posted to often did not have a good school. He decided his wife and children must stay in one place where there was a good school for the sake of the children's education, and they would not accompany him on his transfers. This brought up the second constraint—having to pay for two homes. In order to avoid the additional expense he decided that, instead of renting a house for himself whenever he was traveling, he would simply sleep in the village post office, which was usually nothing more than a mud hut. The post office benefited as much as Srinivasan did. Even in his sleep he could hear the telegraph machine, and if the Morse code chatter revealed bereavement in some family, he would immediately get up and go convey the message, even in the middle of the night.

Srinivasan's son Janakiraman did well in the vernacular school he was studying in, and won a scholarship to the Regional Engineering College at Trichy, which is now among India's best engineering schools. From there, he went on to do his master's at the Indian Institute of Technology, Madras, before starting life as an R&D engineer. Eventually he became the chief executive of R&D at Wipro. That is how I met Janakiraman (affectionately known as Jani), who was to become a cofounder of MindTree.

Jani's story is about a father's dream, a vision. That vision made him sacrifice his own personal comforts and the joy of seeing his child grow up in front of his eyes, but he did it in order to realize his biggest dream—that of seeing his son get a great college education and grow into a successful adult.

A personal vision is an inherently powerful thing. The word "vision" has its origin in the Latin word *videre,* which means "to see." But at a deeper level it refers to an unusual competence in discernment or perception, intelligent foresight; it is the manner in which one sees or conceives of something. It is also said to be a mental image produced by imagination and a mystical experience of seeing as if with the eyes of the supernatural.

Vision is not about the past. Vision is not about the present. Vision is, almost invariably, a distant image of the nonexistent thing we call the future. It is the result of a mental process that telescopes a person into an image that simply does not exist in current reality. From then on, that image becomes the guiding factor. It helps build energetic action. As if tied to it by an invisible string, the person is slowly pulled forward by it, until the imaginary picture and life's reality become one.

Most visions bring with them what is called the "burden of

dreams." This idea is most appropriately depicted in the 1982 Werner Herzog movie *Fitzcarraldo*. The movie is about a man possessed by the idea of bringing the opera to the rain forest, which is considered sheer madness by everyone around him. He had no personal means, but undertook a hazardous journey up the Amazon to claim rubber estates so he could fund the building of a European-style opera house in Iquitos, Peru. His dream was a burden that brought him despair, banishment, rejection and unimaginable obstacles, including brushes with death. Yet he persisted until his dream was fulfilled. If you get a chance, see the movie.

Dr. Govindappa Venkataswamy was a gynecologist. He suffered from rheumatoid arthritis and watched helplessly as his fingers slowly became twisted and permanently warped, making him unfit to deliver babies. Instead of ruminating on this cruel twist of fate, he reeducated himself, this time in ophthalmology. He trained his twisted fingers to handle surgical tools and became so deft as an eye surgeon that he could perform a hundred operations a day! Then, at fifty-seven, his retirement orders came. But instead of getting shelved in a corner and slowly fading away, he opened an eleven-bed eye hospital in a rented house in Madurai, India, because he wanted to do something to help the rural poor who were becoming blind due to completely avoidable reasons.

From this humble beginning grew the legend of Aravind Hospitals. Today, collectively, the five Aravind hospitals see 2.4 million patients a year and perform an astounding 300,000 operations a year, of which 70 percent are done free of charge. Dr. V, as he came to be called, said in an interview with *Fast Company* magazine, "If Coca-Cola can sell billions of sodas and McDonald's can sell billions of burgers why can't Aravind sell millions of sight-restoring operations and, eventually, the belief in human perfection? With sight, people can be freed from hunger, fear and poverty. You can perfect the body, then perfect the mind and the soul, and raise people's level of thinking and acting."

Dr. V died in 2006. The power of his vision is embodied in the fact that its beneficiaries are twice, three times, even four times removed from him. None of the 300,000 people annually who have their sight restored are related to the man. That is exactly what guarantees the continuity of Aravind's success.

To me, Dr. V symbolized the ultimate professional.

I call people like my cofounder, Jani, children of vision.

Most of us begin life with the vision of a parent, a relative, often a teacher—these visions inspire us to study a particular subject, choose a profession or take up something that becomes our calling. We all grow up as children of that vision. Yet countless among us lose our power to dream and act as we subsequently settle down into a well-paying career.

Our professions are instruments to build and deliver value to others. What good is an instrument in the hands of a person who has no vision?

Vision is what vision does. American futurologist Joel Barker tells us that it is not enough to have a dream. You have to act on that vision. Vision without action is far less noteworthy than action without vision. Every day countless people imagine possibilities for our future—from a world free of hunger and poverty to acing a test to owning an island—but very few act, thereby reducing a potential vision to a mere daydream.

Vision must be acted upon; it must be externalized and articulated. And if it involves other people, it must encompass and include them, building what Barker calls a vision community. In the end, the worth of the vision is in its executed reality.

Many professionals, themselves children of great vision, lose the way somewhere and give up on their visioning capacity, which then becomes a recessive gene. Some lose it out of lethargy, some allow material success to submerge it. Yet there are others who abandon it as they face obstacles and blame their environment for it. All of us must understand one vital truth: it is the job of the environment to present obstacles. That is the way nature is built.

Nature designs things in such a way that dreams must die in the same proportion that wild salmon breed. Out of the 2,500 eggs a wild coho salmon lays, 2,125 die. Which is to say only 375 eggs hatch. Of these, only 30 survive. From this lot, 25 are either eaten by predators or die due to environmental reasons. Eventually, only 2 survive to become fully grown adults, to return to their source with the burden of their dreams and spawn. The same ratio applies to anyone who builds a vision that is larger than life. Just because you are a hatched salmon does not mean you are guaranteed safe passage to the ocean and back.

Professionals who have a profound vision are able to uplift not only their own lives but also the lives of those around them. As a professional, bring the power of vision to your work, and act upon it. Do not be fazed by the size of your adversary; the size of your adversary determines the size of your success.

Chapter 36

Affective Regard

The American Heritage Science Dictionary defines values as "beliefs of a person or social group in which they have an emotional investment." The same source defines values in sociology as "the ideals, customs, institutions, etc. of a society toward which the people of the group have an affective regard. These values may be positive, as cleanliness, freedom, or education, or negative, as cruelty, crime, or blasphemy."

I think both definitions merit attention. The first one tells us that values are personal. They are an emotional investment.

The second definition talks about affective regard. Values are not imposed to enslave us, and must be embedded in groups to which we belong. The profession we are trained in and the organization we work for are the two foremost groups in which we need to invest our values and build affective regard. A profession is not to be treated as an ATM or a gambling den. We need to build a value-centric view of our profession. A value-centric view can be built only by people who are capable of emotions.

According to Peter Drucker, the father of modern management, values provide us with constancy of purpose. There can be no profession without a set of professional values. When those values are transgressed, the professional mantle is disfigured, maimed and left in a state of sacrilege, as we saw in the examples of the doctor who sold kidneys and the lawyers who attempted to buy a witness in the earlier part of this book.

When we subscribe to a profession, it is to the exclusion of other options and it signifies a social membership. As a result, people who belong to a professional group must agree to abide by a set of values that makes the group both cohesive and worthy of societal respect. That is why every profession articulates its own set of professional values that are basic and form its core—these values do not vary across organizations related to that profession. If these values are violated, it amounts to desecration of the covenants of that profession. Thus for a medical professional, nonrefusal to treat a person in need is an inviolable covenant. For a lawyer, client confidentiality is nonnegotiable. For an army officer, the safety and honor of soldiers at war is a core covenant of that profession.

Values sometimes get internalized only when value transgressions take place. Through a negative set of examples people learn what is and what may not be acceptable behavior. These examples help members of a profession understand what it means to belong because human beings often have problems with abstractions and it is in the storytelling that we seek answers to our questions.

At MindTree, no one is called an employee. We do not like the term. We call ourselves MindTree Minds. When the company started, we asked each MindTree Mind what kind of an organization they wanted to build. From the many attributes that came up, we coalesced a set of five values that define a MindTree professional: Caring, Learning, Achieving, Sharing and Social Responsibility (CLASS). Caring meant caring for customers, people, financial stakeholders and the organization. In that order. Learning meant personal development and emphasis on building a spirit of innovation. Achieving was about aspiration, accountability and action orientation. Sharing meant teamwork, building knowledge and encouragement and, finally, social responsibility stood for being socially minded, imbibing the spirit of integrity and displaying commitment.

Even today, more than ten years later, every MindTree Mind attends a value session on joining. These sessions are usually run by Ashok Soota, the chairman. When the chairman of an enterprise explains why values are central to the people who have just joined the organization, most people get the message.

Not stopping there, we integrate the values at three additional levels—at the level of recruitment, at the level of performance appraisal and at the level of the annual 360-degree feedback. Despite all that, we realize that good intentions are not enough. In any growing organization, value transgressions are bound to happen. In such moments, how the organization behaves determines subsequent value enculturation.

In India, tax laws permit deduction at source for leave travel assistance twice every four years. By law a claimant must furnish actual proof of travel to claim the money. It is a widespread practice in government offices and in the private sector alike for people to furnish false documents that are easy to procure and claim the money without undertaking any travel. In most organizations, it is not looked down upon because the individual is only defrauding the tax authorities and not the organization itself. But regardless of who is defrauded, the fact is, submitting a false claim is a forgery. At MindTree, whether the taxman comes or not, you claim a tax exemption only if you have indeed incurred an expense. No ifs, ands or buts here. People who are not comfortable with such a position do not join MindTree.

One year, MindTree hired a thousand people, and to our utter dismay we found that as many as eighty of them had faked employment information in their résumés. One had submitted inflated salary information which, when verified at the source, turned out to be untrue. Another claimed to have worked in a company that did not exist or was just an employment certificate issuing scam. We asked them all to leave, even though from a technical standpoint these people had cleared our tests. Their social belief that you can bend the rules, fake information to get ahead in life and do what others are doing even if you know it's wrong did not fit in with our core values.

In yet another case, a MindTree Mind working at a client location was shown a competitor's proposal. That evening, he returned to the client's office, and using his access rights copied the proposal and submitted a competitive bid that we ended up winning. The story came out after his subordinates exposed it and questioned the ethics of the act. By then, work with the client had commenced. We removed the MindTree Mind from service and went to the client and informed him about the breach immediately. The client was shocked and saddened. After all, the individual was very valuable to the client, as he was to us. Removing him meant serious dislocation, and such abrupt changes in a project's composition invariably have other implications. But the person had taken a risk that compromised the basic values of the organization, and the issue had to be dealt with without worrying about the inevitable business consequences.

Value transgressions happen for a variety of reasons, and as long as an organization is in business, people will test its boundaries. Most times, the people who breach professional values are the most unlikely candidates, evoking responses ranging from "Are you sure it was him?" to "I never knew she could do such a thing" or "Why did he have to do it?"

At MindTree we have a simple precept. A value transgression is a value transgression—there is no shade of gray here. Irrespective of who is involved and what the impact will be on our bottom line, we part ways. We have been asked about the role of forgiveness in such situations. We believe forgiveness does not really work when it comes to a matter of values. Assume someone has faked a voucher or has sexually harassed someone. Even as you forgive the transgression, people will end up hearing about it and most likely will lose respect for that person. At MindTree we believe it is better for someone who has breached integrity to move out of the organization. That way the individual can make a fresh and hopefully clean start someplace else. Letting such people continue in the organization does not help because both their reputation and their self-respect have already been dented. Years after MindTree was started, we realized that value enculturation remained a challenge, partly because of the social system. We came up with the idea of the Integrity Book, also known within the company as the White Book. This book outlines the integrity policy of the organization, but goes a step further. Unlike organizations that keep their skeletons in the cupboard, we periodically publish them—we reveal the instances of value breach and how the matter was dealt with. This book can be downloaded from the MindTree Web site and is freely distributed among colleges we recruit from and other organizations, including our competitors.

It is the duty of top management to propagate the organization's values through personal behavior and demonstrate the price the organization is willing to pay from time to time to defend what makes up its core. In countries like India, it is extremely critical that there is an organizational framework available so that people can easily relate to the idea. But there are countless self-employed professionals who do not have the advantage of such a framework and are invariably left to their own means. For them, it is a tougher battle when choosing what is right over what is convenient because they lack an effective system that can clarify value questions.

The self-employed professionals in every society have the maximum opportunity to run their vocation without their income being tracked, unlike a salaried individual. The easiest temptation is tax evasion. From that point onward, some people feel bold enough to do worse things. Left to self-regulation, people fail easily because they are also witness to too many of society's wrongdoings and it is very difficult to remain on the straight and narrow. Once they cross the boundary, it becomes a downward spiral for many.

A doctor becomes part of an insurance fraud. A real estate agent withholds critical information from a client. A lawyer bribes a judge. In each instance, the professional breach is justified as the price to be paid to be part of a system. The truth is, every time it is an individual choice.

• • •

Seldom does sticking to your values lead to great glory or high praise. Sometimes a personal price has to be paid in the practice of professional values, but in the end, those who stand by their convictions have a better opinion of themselves, which is far more important than any outside view. Society on the whole may not always put a premium on professional values, and hence most people do not incorporate them into their lives. But practicing professional values is about who you are and what you want to be known as—a professional or merely professionally qualified. And, in the end, even the most corrupt society hails the individuals who choose to be different.

Chapter 37

Commitment to Commitment

As I finished my keynote address at an event in New Delhi, the usual enthusiastic group of young entrepreneurs, press persons and students rushed forward with business cards in hand, their eager faces wanting to snatch a few strands of conversation or the promise of a private meeting later.

A young journalist from a national magazine insisted that we speak for a few minutes. She needed to file a story but I had a flight to catch right after the talk. I asked her to send me her questions by e-mail instead, promising her a rapid turnaround. She accepted, and we agreed that I would respond the same day. On the ride to the airport and while waiting to board a flight, I invariably spend time reading and answering e-mails, so that day, I anticipated receiving her questions and was mentally prepared to respond as quickly as I could.

At the same venue, as I was leaving, a young woman introduced herself as a social entrepreneur. She said that she had come all the way from Chennai to listen to me speak and she wanted to discuss a few things face to face. As I was on my way to her city, I asked her to call me the very next morning when we both would be in Chennai, to arrange a time to meet. She noted my number and gave me hers.

The next morning, I woke up remembering my promise to her but realized I had not yet received a call from her. After I got ready for the day, I decided to call her instead. A sleepy voice answered the phone. It took her a while to realize who was on the line. Then she told me she had stayed back in Delhi to visit her mother. She had obviously not taken my offer to meet the very next day in Chennai seriously.

From Chennai, on my way back home to Bangalore, I was burdened by my promise to reply to the eager journalist's questions. I checked my mail; there was nothing from her. After I reached home, I downloaded mail again. Nothing. Again after dinner I checked my computer and found nothing from her. I even checked my junk mail folder on the off chance her mail had been quarantined as spam. Nothing. She had not sent her questions.

The next morning I decided to call her. When she answered, she was completely taken aback that it was me on the line. It was clear that after our short conversation she had erased the idea of sending me the questions from her mind. I could feel her embarrassment. I had taken her by surprise.

What these two young "professionals" lacked was commitment, or rather commitment to commitment. It is unlikely that I will ever take them seriously if our paths cross in the future.

We tend to think of commitment as a larger-than-life thing, to be reserved for grand gestures like going off to fight a war for the sake of the country. For every soldier, though, his commitment to his country is in the details of his everyday work—in his punctuality, his appearance, his responsiveness and task completion every step of the way. A soldier does not go to war every day, but his commitment to perform every small task decides victory when the big day arrives.

This quality to do what you have said you will do, in the time you have committed to do it, must be applied to the smallest of tasks in your life. Without it, you not only disrespect others, you disrespect yourself.

Without commitment we cannot achieve even small successes, much less large ones. Without commitment, we cannot give our best to our organization. And without commitment, we cannot turn our vision into reality.

Chapter 38

Be Prepared

Great professionals are always prepared—for conversations, meetings, presentations. Prepared individuals project a good image of their company and of themselves, which is the first step toward making a client feel important. Imagine going to a doctor, a lawyer or an accountant and having to repeat everything you had discussed in your previous meetings. On the other hand, when your doctor begins by saying, "The last time we spoke about a change in diet, your exercise regimen and a shift from a high dosage of your medicine down to a maintenance level . . ." you already know that you are in good hands, and have the confidence that your doctor is the best.

In contrast, think about all the times you've had to explain your problem to a customer service representative when you called the toll-free help line number of any service provider. You would feel far less frustrated, better served and more satisfied in the end if you didn't have to repeat your problem each time.

A professional architect will bring along sketches and concepts from the last meeting, and a summary of the points discussed.

A professional salesperson will go over his meeting notes from a previous call before visiting a client, which is why it is also important to take notes in a meeting and have them handy for the next one. And this is not just for the external world. It is equally important to circulate minutes of an internal meeting, if you are responsible for the meeting.

I would never meet someone without having researched him, especially considering how easy it is these days. Go to Wikipedia or LinkedIn, Google him, read the last annual report of the company, visit the company's Web site. If you do not have the time, subcontract the research to a young intern or a junior colleague who has a spare cycle. In the process, you not only become prepared, you spread the culture.

I always tell salespeople that you must "arrive before you arrive." And sometimes, in some miraculous way, when you work really hard and spend time thinking in preparation for a meeting, the message reaches the other party even before you actually arrive for the meeting.

Like the time MindTree had been trying to break into the preferred list of suppliers of a Fortune 500 company. We were doing some work for that company at the fringes but were not one of the five partners the company engaged with for big deals. And I was to call on the head of corporate procurement to exhort him to make us a partner of preferred choice. For days I simulated the meeting. I asked myself why on earth should he meet me? Why should he be sympathetic? What would make our meeting memorable when he is inundated by endless sales pitches?

The night before the meeting, the Oscar awards were announced. *Slumdog Millionaire* walked away with eight Oscars. I got my entry line for the next day's meeting. I decided to tell the executive two things: Who ever thought that one day the United States could have an African American as president, and who ever thought "Jai Ho"—the signature song of the movie—would be on the lips of everyone in the world? My subsequent pitch was based on the idea of change and why it was a good idea for him to let the underdog in.

You will be amazed at what happened the next morning. After I entered the executive's room and we exchanged cards, he opened the conversation with the following line: "What an award ceremony that was last night! Did you see the Oscars?"

I had arrived before arriving!

Chapter 39

Ask Pertinent Questions

A court jester in ancient India once asked the king why he was not treated the same as the prime minister. The king devised a very clever ploy to let the man discover the answer for himself. The king asked the jester to get information about a merchant ship that had just set anchor. The jester rushed out and rushed back to report that a foreign merchant vessel had indeed arrived. The king asked him what merchandise the crew had brought. The jester ran back to the ship and returned with the answer. The king now asked him what did the captain of the ship want to do with the merchandise and again the jester went in search of the answer and came panting back with it. After a few more supplemental questions, the man collapsed. Then the king asked the prime minister to go to the port and find out about the ship, and the man returned with the complete information in one go. The jester learned why the prime minister had precedence in the court. Many variations of this story exist and they all teach the importance of asking pertinent questions.

Ancient India was known for its tradition of questioning and dialogue. We have rich evidence of how these ideas were used in order to seek information, knowledge and even the ultimate truth. The concept of asking a question to get to the bottom of things was often propagated through storytelling. Unfortunately, this tradition seems to have lost its way in the modern education system we have designed for ourselves. Our education system, a product of the society we live in, discourages questioning and dialogue in an effort to mass-produce thousands of engineers, doctors, lawyers, management graduates and others with so-called professional qualifications. Doctors and lawyers are required to use questioning as part of their daily work, yet the art and science of questioning is really found only in a few among them.

The ability to ask questions, no matter what field you work in, is essential for success. When I went to learn Total Quality Management (TQM) in Japan in the 1990s, I came across the concept of the Seven Quality Control (QC) Tools, which included Japanese management guru Kaoru Ishikawa's fishbone diagram. This is a tool for root cause analysis (RCA) among management professionals and is so called because when fully drawn, the diagram looks like the skeleton of a fish.

The idea is profoundly simple yet powerful. It suggests that managers tend to look at everything in a cause-effect manner and jump to conclusions based on what may be an apparent cause, and not the root cause, behind an event or a phenomenon. When you react to an apparent cause of a problem, your efforts do not yield sustainable results. So the trick is to go deep enough until you find the root cause and then consider a solution that truly removes the problem at the core.

To do this, managers need to keep digging for facts or data until the root cause is unearthed and then apply their minds to the solution—only then is the solution sustainable. Quality experts maintain that when a problem occurs, by asking five consecutive "whys" you can usually uncover the root cause.

Here's a simple problem. An employee has been coming late to work for the last couple of days and the manager has to resolve the issue. A conversation ensues between the manager and the concerned employee.

Manager: Why were you late today?

Employee: I missed the bus that normally brings me to work.

Manager: Why did you miss the bus?

Employee: I woke up late.

Manager: Why did you wake up late?

Employee: I went to bed late last night.

Manager: Why did you go to bed late last night?

Employee: My wife has been suffering with viral fever for the last few days and I've had to attend to her.

The apparent reason here for the employee coming late is his missing the bus, but the root cause is that his wife has viral fever. Now consider if the manager had stopped at the first, second or even third "why." The potential conclusion and probable corrective action would have been vastly different! Though we need not stop at the fifth question to find the root cause and can go even deeper, the Japanese believe that most problems can be driven down to the root cause by asking "why" five times and then asking "how" just once.

Thus if the manager had asked the employee how his wife had contracted viral fever, the employee might have told him that they had gone out for a meal, after which she had come down with food poisoning, and the doctor had advised rest and a week's course of antibiotics. Now instead of forming a poor opinion of the worker's punctuality, the manager can adjust the employee's shift timings or give him some time off or simply wish his spouse a speedy recovery and not take any action against the employee.

A true professional, faced with a problem, will always question and find the root cause. Most people do not bother to get to that level. As a result they end up fixing the wrong set of problems.

The idea of professionalism dawned on me in 1990 when I visited the United States for the first time. I was in for a surprise when the hotel plumber came to my room. He was a big man in clean clothes. He wore a special belt around his waist from which hung at least two dozen tools of his trade. When he came, he did not attempt to fix the problem right away. He started by asking me questions, all kinds of questions, and then he asked me to simulate the problem. In many parts of the world, the plumber arrives without a tool kit. He jumps to fix the problem even before you have fully explained it, then tries to fix things with his bare hands, in the process floods the bathroom floor, and only then goes away to get the right tools and parts. And this is not only true of plumbers. It is the same with the cable TV installer, the auto mechanic, the carpenter and countless others up the value chain.

Consider this situation. Your washing machine is not working, and the repairman has come. First he opens the back of the machine, and black liquid pours out onto the bathroom floor. He looks lost. You have an uncomfortable feeling that the man does not know what he is doing.

But what if instead of jumping in to solve the problem, the

repairman began by asking pertinent questions: How old is the equipment? What is the exact nature of the problem? Has it ever happened before? He has come prepared with the right diagnostic tools, and explains to you the subsequent options, costs, parts needed and so on. Finally he asks you how long you can live without a functional washing machine, and then gives you a timeline for it to be fixed. Only then does he go about fixing it. And he closes by assuring you that if anything should go wrong, you can always contact him. He gives you a number to call. You do not call that number, but he calls you a week later to ask if all is well.

The professional competence of a nation is reflected in the way its plumbers, carpenters, auto mechanics and others engage with a problem and a customer. Without this no nation can become world-class even if it churns out neurosurgeons and software engineers.

While there is a clear, direct linkage between professional competence and the ability to ask pertinent questions, true professionals must also ask intuitive questions.

On one occasion I was asked to talk with someone in my organization who was going through a difficult professional patch. We knew each other well as colleagues and I assumed that we were comfortable enough with each other to discuss the details of his problems from the very outset. As we were sitting down for the meeting, I surprised myself by spontaneously asking, "Are you comfortable with me?" No sooner had I finished than my colleague blurted out that he was actually not comfortable. It was a crucial meeting, as the implications were huge, both for him and for MindTree. Without that one intuitive question, a polite conversation would have ensued and I would never have been able to get to the root of the problem or help him. What did I do when he expressed his discomfort? For a large part of the meeting, my job was not to discuss the core issue but to make him comfortable. The core issue could always be discussed, with greater clarity, but actually when he was comfortable.

While learning to ask pertinent questions is a basic skill, as we grow into positions of greater responsibility, it becomes necessary to refine that ability to the next higher level of critical questioning. I discuss critical questioning in Part Five and consider it a quality all professionals must strive to acquire. In the meantime, I'd like to address the corresponding ability to asking the right questions: the ability to listen intently.

Chapter 40

Intent Listening

The pertinent, sometimes intuitive, questions need to be complemented by intent listening. Listen to the answers to the questions you have asked not only with your ears but also with your eyes. Our eyes are trained to notice the smallest sign of pain, impatience, lack of confidence, worry or detachment. Listen with your whole body. Show engagement and signal through your eyes that at this moment the only thing you care about is the person in front of you. When you shift your eyes away from a person while listening to her, it signals disengagement, distraction and lack of interest.

Listening is a physical function as much as it is an art. Some people can see things others cannot. Much the same way, some people can hear things others cannot. Good professionals are invariably great listeners. The interesting thing about listening is that the other person gets to "feel" instantaneously that he is being listened to. As a result, a very positive energy starts flowing between the two people who are in conversation. This positive energy is often a prerequisite for diagnosis, problem solving and collaboration.

The capacity to listen intently can be built with conscious practice.

A colleague sends an e-mail asking a question or raising an important matter. Immediately show interest by responding, not by mail but by calling the individual. During the call, suggest a personal meeting and fix the time on the spot rather than seeking an outlook meeting request. The listening process has already begun.

Now choose the right venue. I would not discuss a serious career issue in a restaurant because I do not want distractions of food or other people's conversations. We often pay little attention to choosing a venue appropriate to the mood of the other person, the intended subject of discussion and the desired outcome. But the right venue enhances listening.

Shut out distractions. In today's world we all pride ourselves on being multitaskers. But it is important not to have distractions like a television on in the background, a cell phone ringing or your BlackBerry in your hand and eyes darting toward it every once in a while.

Ask further questions as the conversation progresses, seek confirmation of your understanding, take notes if required, look for patterns in the information the other person is giving and relay the conversation back to him.

Intent listening reduces the time required for communication, makes the other person feel at ease and builds collaboration within and outside the organization. When we listen well, we signal empathy and engagement. This in turn helps build sound judgment. The world is starved for that one thing—each time you visit a dentist, a lawyer, an accountant, a banker or for that matter a mechanic, you wish they listened a little more!

Chapter 41

Human Beings First and Foremost

My mother had been lying motionless for forty-eight hours.

It all started the evening before when I returned from work and sat down to share my customary cup of tea with her. Susmita, my wife, called out to her to join us. Her frail frame refused to respond. She was breathing, softly, but she was not moving. It was clear this was an emergency. Luckily our family doctor came rushing over and asked us to move her to a hospital immediately.

It looked like she had had a stroke. The next forty-eight hours were critical, and all we could do was wait. At her advanced age, there was no question of an invasive procedure. As there was no internal bleeding, there was hope; the brain can recover magically.

My mother simply lay motionless on the hospital bed as we kept vigil. What was going on in her mind? Was there anything going on there at all? Could she hear us? Was she aware of her surroundings but could not speak? There was no knowing. Time was suspended in that hospital room.

The door opened and a young nurse walked in. She looked

about eighteen or nineteen. She probably got paid a tiny fraction of what professionals in the software industry get paid.

"*Ajji*," she said softly. *Ajji* in Kannada, the local language, means grandmother. "*Ajji*," she called again. I got up from the attendant's bed, startled and surprised that she was speaking to a person who had not spoken to anyone or moved in the last two days. What was she doing?

"*Ajji*, I am going to sponge you," she said.

Mother's body remained motionless.

"*Ajji*, I am going to make you sit up now." She changed my mother's position on the bed while gently supporting her with one hand.

Then she squeezed a wet towel and wiped my mother's face, her hair, her body, covering her at each step to maintain her modesty. All the time, she kept talking to her softly. Then she looked at me and asked me to leave the room because she had to change her. I went out in a daze. After ten minutes, the nurse came out and told me that I could go in. My mother lay motionless as before, but she looked like a freshly bathed child. The nurse had combed her matted hair, powdered her and left her looking content.

The day was long and monotonous. The only interruptions were when the same nurse returned, each time calling out, "*Ajji*, I need your temperature, I am going to open your mouth," "*Ajji*, I am changing the drip," "*Ajji*, I need to take your blood pressure."

That day it occurred to me that each time the nurse said "*Ajji*," she was placing my mother ahead of her work. For her my mother was a human being, not just a motionless patient in the hospital. My mother was a real person who needed her care.

• • •

Seeing the way that young nurse was treating my mother, my memory went back three decades to a hospital bed at a leading national medical research institution where I had been admitted with an undiagnosed high fever. While I was first treated for a viral infection and then suspected of typhoid, the fever raged, my body reduced to bones, hurting from the inside, burning outside. Susmita and I had just been married. My young wife was missing her university classes to be by my side. At times the pain was so intense I would bite Susmita's hand.

We were told that when the head of medicine made his rounds he would attend to my "case." It seemed we waited forever. Then, one morning, I realized there he was, looking down at me, with a group of interns and nurses surrounding him. Wearing a suit and a tie, he was the man. Right now, in his professionally qualified eyes, I was the "case." Someone showed him a sheaf of papers that hung from a metal clip tied to my bed. As I lay helpless, it seemed to me the case history clinically documented in those papers was a mysterious collusion between the doctors, the nurses and the gods. He briskly looked through the papers and then he spoke, in a clipped accent. "Young man," I heard him say, "you have tuberculosis. The good news is that it can be treated and you should become all right."

And then, in the very next moment, he and his convoy were moving on to the next bed. The man had just told me I had tuberculosis and simply walked away! I was completely devastated for the days that followed, until a diagnostic test established that I did not have tuberculosis and saved me from being put on medication.

• • •

My mother recovered gradually and returned home. Years after, she passsed on. But the young girl who called her *Ajji* each time she came to attend to her needs has become an unforgettable part of me, a beacon of professionalism. The doctor who informed me I had tuberculosis is also unforgettable, but in contrast, as an enduring nightmare.

A professional must be empathetic, able to look at another person as a human being, to respect the other person no matter the state in which he or she appears—as a motionless patient, a demanding client or simply a harried customer. It is not enough to have the best skilled people on the job; they must respect life and living things.

Chapter 42

The Rewards of Transparency

In 2008, MindTree won its single largest deal at the time against formidable global competition. The L. N. Mittal–led Mittal Steel had just bought out Arcelor, the largest steel conglomerate in Europe. Postacquisition, one of the tasks on hand for the new management was to improve cost, performance and profitability. As part of that overall strategy, they had decided to use offshore IT capability. At that time, MindTree had fewer than six thousand employees and did not really consider itself to be large enough and deep enough in terms of specific domain experience to meet the needs of ArcelorMittal. A steel company that size demands not only a large IT services company as a partner but also a very diverse set of capabilities, requiring both understanding of the steel industry as well as a multitude of software platforms.

When the bid papers did come to us, our European operations chief, Vishaal Gupta, studied them in detail. He realized that we did not have the full-spectrum capability to meet ArcelorMittal's needs. So he told ArcelorMittal that we were probably not their best choice based on our skill assessment and that we would participate only if we were allowed to make a partial bid. Their response surprised all of us.

ArcelorMittal said that we were the only company that had been honest enough to admit our limitations. They asked us to go ahead and put in our bid.

The next perceived hurdle was the future role of an existing vendor who had on-site presence, longtime entrenchment and knowledge of local conditions in ArcelorMittal's multiple European locations. We offered to take on the work but not as the prime vendor. We did not see it as being in the client's best interest to dislodge their existing relationship. We had value to add but not the capacity to replace the existing partner. ArcelorMittal did not reply. Instead, they sent a high-powered team to India for a detailed site visit. During a dinner we hosted for them, a call came from their corporate office informing us that, based on all the inputs they had, ArcelorMittal believed we could be the prime vendor with the existing organization as a subvendor, and if we were willing to take on that role, the business was ours.

Next thing we knew, we had a multiyear, multimillion-dollar relationship with ArcelorMittal. The genesis of that relationship was in Vishaal Gupta's ability to state clearly and up front what we were not capable of doing. Good clients value that because they know trust cannot be replaced by competence. Competence can be acquired, but trust either exists from day one or it does not. Knowing your individual limitations and the limitations of your company, telling a client, "No, we do not have the capability you require," is the professional thing to do.

Chapter 43

The Responsibility of Dissent

The word *satyam* in Sanskrit means truth.

On December 16, 2008, the sky above the Indian IT industry, which prided itself on its corporate governance, went dark. On this day, Satyam Computers, the country's fourth largest software exporter, had declared it would buy two companies owned by the chairman's sons for an astounding $1.6 billion. The chairman, Ramalinga Raju, and his family owned a meager 8.6 percent of the listed company. The two companies Satyam intended to buy were not only managed by Raju's own family members and largely owned by them, they were in the real estate business—an area that had absolutely nothing to do with the software business. On top of all this, the astounding valuation at which they were to be purchased was fictional. Satyam had decided that the land bank in the hands of the chairman's two sons, if constructed upon, was worth $1.6 billion. The entire board of directors, which consisted of several independent directors of international standing, simply agreed to a completely unrelated business acquisition with an interested party. Who were these men and women around the table who consented to one of the least transparent deals in Indian corporate history? What was on their minds when they raised their hands and said, "Aye"?

To answer this question, we must first look at the members of the board.

Dr. Mendu Rammohan Rao was dean and director of the Indian Institute of Management, Bangalore, prior to joining the Indian School of Business at Hyderabad as dean. He had also been a tenured professor in operations research at the famed Stern School of Business at New York University, served on several government boards and committees and won innumerable academic awards.

Dr. Krishna Palepu is a professor at Harvard University—specializing in strategy and governance! He teaches courses on making corporate boards and audit committees more effective and even co-led Harvard's Corporate Governance, Leadership and Values initiative.

Mangalam Srinivasan, arguably the lone dissenter since she was the first to take moral responsibility for the acquisition even though she did not really oppose the deal when it happened, is an adviser to the John F. Kennedy School of Government at Harvard.

Vinod Dham is a well-regarded Silicon Valley serial entrepreneur and investor who program-managed the Pentium chip at Intel and is considered a coinventor of flash memory. In addition, there was V. S. Raju, an ex-dean of IIT, Delhi, and T. R. Prasad, who served as cabinet secretary, India's senior-most bureaucrat, at one time.

Eminent people all, but they individually and collectively failed their shareholders, fifty-three thousand employees, families and their entire industry by their bizarre approval of the deal. While Mangalam Srinivasan was the first to step down in the wake of the scandal and admit to not having been vigilant enough or asking critical questions, Rammohan Rao defended himself in public, saying that the board had done no wrong in approving the deal.

In the following weeks, the company and the country went into convulsions as the Central Bureau of Investigation started unearthing murk in Satyam's financial dealings and revealed what it claimed to be a potential $3 billion fraud. Ramalinga Raju, his brother and the company's chief financial officer were put in jail and the board dissolved. *Business Line*, a leading financial newspaper, wrote on December 18, 2008, that the board did not have "a fig leaf of a cover."

As they sat down to consider the $1.6 billion deal, the Satyam board began their trip to Abilene.

The Abilene paradox is a management story in which a group of people agree to a course of action that is contrary to their individual preferences. The concept is attributed to Jerry B. Harvey, a management expert, and has since been captured in a film of the same name. The movie powerfully illustrates how people will agree to do strange things if they suppress their own voice and simply go along with what everyone else is saying. *The Abilene Paradox* should be seen by any professional who is part of any group that deliberates on issues of consequence.

The story begins one hot afternoon in Coleman, Texas, as a family is playing a game of dominoes. The father-in-law casually suggests that the family go to Abilene, a good fifty-three miles away, for dinner. The group concurs passively. The car ride is hell.

The food and service in the restaurant in Abilene are bad. On the way back, no one speaks. After they return, emotions erupt. Everyone points fingers at everyone else and disowns the idea of going to Abilene. Finally, when all fingers point to the father-in-law, he shrugs off responsibility by saying he never really wanted to go—he was simply testing the waters.

Every day, groups of professionals who work together take a trip to Abilene. That is how noncommittal agreements emerge and projects get kick-started. When things begin to fall apart, everyone blames everyone else for the decision and claims that he or she had simply gone along to avoid conflict.

The Satyam fraud was not limited to that one decision to buy out two family-run companies in a completely unrelated line of business for all the cash the company had. Nor was it one man's fault. No one—no matter how smart or cunning he is—can pull off something as gigantic as this, and all in one day.

I visited Satyam for the first time in 1996 when I was asked by Wipro chairman Azim Premji to go and see their campus under construction in Hyderabad. I was given a warm welcome. At the campus, I was served lunch, which turned out to be a dish of wild rabbit. That was rather unusual for an official lunch for a visitor, I thought. But I was told it was in my honor, as a special gesture. The rabbit had been shot on campus. Inside, I flinched, hoping the honor was not extended to visiting customers, some of whom could be animal rights activists.

Sometime afterward, I came across Satyam's annual report and balance sheet. It included the names of board members as is customary for any listed company. I was perplexed to see among them the name of a retired income tax official. What, I asked myself, was an ex-bureaucrat from the income tax department doing on the board of a software company?

Years later, I interviewed a Satyam consultant in Silicon Valley. He seemed like a good candidate for a job with us. After the interview, he surprised me by saying there was a component of his salary that was "unofficial."

It seemed that when Satyam had issued American Depository Receipts (ADRs) in the United States, a certain amount was reserved for employees as stock options. The candidate I was interviewing was unable to apply for them on time because of some official lapse. When the allocations were made, he lost out. He agitated about it, and the matter finally went to Ramalinga Raju. The chairman could not issue the ADRs to this employee for regulatory reasons. Instead, he said that each time the employee came to India for his annual vacation, the company would "unofficially" buy his family's tickets and that the arrangement need not be formally recorded.

While Raju was the entrepreneur behind Satyam, he was not its owner. In making many such decisions, he did not respect the difference between entrepreneurship and ownership. Mixing up the two ideas, he ran the company by fiat—sometimes malevolently, as in the case of the rabbit they shot for lunch, and sometimes benevolently, as with this employee enjoying the unaccounted perk.

But our focus in the Satyam fraud should not be on Ramalinga Raju. Our focus should be on the professionals around him who knew all along the company was taking one wrong step after another but chose to go along and earn their fat salaries and hefty bonuses. These people were failing the unwritten code of conduct for any professional—the responsibility of dissent.

Satyam is not the only organization where professionals fail to discharge their responsibility of dissent. It happens all the time and all around us wherever groups work together. When professionals meet, they often assume that the purpose of every meeting is to get consensus. But consensus is not always beneficial and can sometimes lead to disaster. This result can be avoided if each professional in a group exercises his responsibility of dissent and the purpose of the group's decision-making process is shifted from the urge to agree to the need to do the right thing.

From the power of vision to doing the right thing, individual decisions by professionals build the reputation an organization ultimately enjoys. The imperatives discussed in this section not only enable a professional to manage high volume, they also make good business sense. Outstanding companies recognize and actively incorporate them into their way of doing things, which then becomes corporate culture.

From managing volume, let us now shift to managing complexity. As a professional reaches the next higher plane, it is this ability that separates the best from the rest.

Part Five

Managing Complexity

Chapter 44

Three Disasters, Three Great Professionals

At eBay, an uninterrupted power supply is literally worth billions of dollars. The business of this California-based Internet auction site totally depends on its computers running 24/7 and, therefore, on the availability of power. And yet, ironically, there was a time when companies in California were more prepared for major disasters like earthquakes than for power outages. One day in 1999, all that changed. California suffered one of its worst power outages. The first outage lasted twenty-two hours. Power was restored briefly for eight hours, and then the second outage occurred. The power failure on that fateful day impacted every business in the region. When the servers unexpectedly went down, databases crashed, software programs went haywire and critical customer transactions were lost. All this meant lost money and reputation, and potential claims for damages.

At 5 P.M. on the day of the first outage, eBay CEO Meg Whitman got a call from her team estimating they could bring the site back up in a couple of hours once power was restored. Yet hours later it was still not up. Around midnight Whitman decided to go to her office and take charge of the crisis. For thirteen consecutive days she stayed right there, sleeping on a bed in the conference room, until business returned to normal. The fact that the CEO of eBay was there on-site twenty-four hours a day brought not only her entire organization together but also the best people from all the IT vendors whose collaboration was critical to repair the servers. She attended every single meeting—with the technical team and the marketing and finance teams—and it made a huge difference in their ability not only to get the site back up but to get it stabilized and confirm the integrity of the data. eBay was one of the first companies to recover from the outage and get its customers and business back.

At FedEx, time is money. How much money? Every minute's delay in turning around the planes that bring parcels to the company's hub in Memphis, Tennessee, costs FedEx a million dollars. In 1982, thirty-year-old Ken May was the managing director of what the company called Global Operations Control and Coordination (GOCC). One of the many tasks before his team was the deicing of aircraft in winter. Deicing preparedness obviously depends on the weather forecast. One night, Memphis was battered by an ice storm that moved in without warning and took Ken's entire setup by surprise—causing hours of delays and many millions of dollars in losses. Fred Smith, the CEO, asked for a postmortem. Explanations were sought and the place was abuzz with finger-pointing; the blame game had begun. Management author and Dartmouth professor Vijay Govindarajan recalls that when May walked into the room for the final conversation, there were signs of stress on everyone's face. And then Ken said that since he was the managing director of GOCC, the coordination of the deicing program was his responsibility. He said that he and his team had done a poor job that night and it was his fault. He took total responsibility for failure to execute. Ken had taken everyone by surprise, and the tone of the meeting shifted from witch-hunting and finger-pointing to how they could be prepared for such an event in the future.

The year 2001 will go down as the worst year in passenger airline history for a very obvious reason—the September 11th terrorist attacks in New York and Washington, D.C. The already ailing airline industry went into a tailspin after this. All airlines mothballed 20 percent of their planes, laid off 20 percent of their workforce and introduced across-the-board pay cuts. The only exception was Southwest Airlines, the low-cost leader of the U.S. airline industry.

Three days after 9/11, CEO Jim Parker and his leadership team had one monumental decision to make. The company was supposed to deposit $179 million into an employee profit sharing account, and everyone expected that Southwest would not make the payment. After all, who could tell what the future was going to be?

In what would appear to be a counterintuitive move, Parker decided "it was the right thing to do" and went ahead and made the payment. In addition, he decided Southwest would not ground planes or lay off any employees. As some flights ran below capacity, the company decided to let the excess staff handle customer calls for refunds as many passengers were wary of flying. To restore confidence, the airline announced a "no questions asked" refund for every cancellation. It went one step further by announcing more lower-fare deals in order to encourage people to fly again. Southwest made no salary cuts, except for board members and executive officers who volunteered to take them. The result? The company made a profit in the last quarter of 2001—a feat unsurpassed in aviation history—and its market cap was ahead of all other airlines combined.

When eBay CEO Meg Whitman decided to leave for her office on the night of the power outage and then stayed there for the next thirteen days, or when Ken May decided to own responsibility for the Memphis debacle, or when Jim Parker decided to do the right thing, what was going on in their minds? Were these decisions born of logical, rational thinking? What makes a professional think the way he or she does in a make-or-break situation? What is the source of the intelligence that guides such pivotal decisions? How do their minds actually "read" the big picture and yet focus on the small details at the same time? And more important, as we live in a world of unfolding complexity and unscripted events, what do professionals need to succeed in the future?

In the past, professionals like Whitman, May and Parker were told to make decisions based on data, facts and precedence. Other professionals who followed the same principle succeeded, and some did not. The future will be quite another story. As we look ahead, we need to get under the hood, to understand how the mind that makes critical choices in difficult moments works.

Chapter 45

Logic or Emotion?

One of the two oldest-known skeletons of an upright human is estimated at 4 million years old. It is what remains of a woman who is affectionately called Lucy by her keepers at New York's Museum of Natural History. Looking at her, we can safely conclude that we have been around in our current form for millions of years. And for the most part during our existence on the planet, the most prevalent human profession has been that of hunting. Radiocarbon tests indicate that only in the last 3,600 years have we known organized agriculture. So for millions of years we were hunters.

During the period when we were predominantly hunters, supporting professions rose up—the medicine man, the potter, the flint maker, the ironsmith. During the predominantly agricultural period, as civilizations evolved, newer professions came into being. In the last 400 years of the industrial and then the information economy, there has been an explosive growth in that number. But compared to a Wall Street dealer or a software engineer from Bangalore, professions during the hunting and farming phases of man's evolution had a longer sense of history; knowledge and skills were passed on from one generation to the next on a vastly large canvas of time.

In contrast, most modern-day professions have developed all too suddenly—in the course of a century at best. There is a far lesser sense of history to being a psychologist, a patent lawyer, a neurosurgeon, a dietitian, a television reporter or a software engineer. In that sense, we are still evolving and in some sense the so-called modern professional is really at the bottom of an evolutionary phase.

One significant difference between the modern professional and the hunter or farmer is probably the extent to which the former must use her brain. The brain as a tool has always been used, but its usage has varied dramatically among Lucy, her farmer descendants and her current-day service-sector progeny. This dramatic shift requires us to ask a very fundamental question: How well do we understand this tool which we use and rely upon as professionals? Do we know it as well as the hunter knew his spear and the farmer knows his plow?

The brain is our thought factory, and thought is the mother of all voluntary action. In the brain, billions of brain cells interact with one another to generate the thoughts, ideas and decisions that shape our activities and make us who we are. Until the 1980s, when magnetic resonance imaging (MRI) technology became available, there was little to no scientific understanding of how the brain works. In the last three decades, we have made tremendous progress; yet our current knowledge is probably just the starting point of an exploration that may possibly have no end.

Broadly speaking, the brain is divided into the left hemisphere and the right hemisphere. The left side controls the right side of the body and the right side controls the left. The left side is driven by logic, responsible for all analytical thinking; the right side thinks in terms of feelings, relationships and the big picture in any given situation.

Professions that emerged in the last hundred years emphasized analytical capability that largely requires left-brain thinking. This is the part of the brain that looks at everything in a causal sense; that is, for every cause there is an effect and for every effect there is a cause. It aids logic and problem solving. As we look back, we can see why the industrial economy that predominantly required left-brain thinking also saw the ascendance of accountants and engineers in the last century. During this time, modern science flourished. Modern management, as we know it, is an extension of modern science. As a result, logic has been the driving theme for most professionals over the past century.

Emotions, feelings, big-picture thinking and relationships were considered to be touchy-feely, not given easily to measurement and hence deemed outside mainstream management thinking. That paradigm is going to change in the decades to come. We are actually witnessing the emergence of a more balanced view in which emotions and feelings—attributes that determine creativity—are becoming as valued as logical thinking and analysis.

Daniel Pink, the author of *A Whole New Mind*, argues forcefully that right-brain thinking will actually dominate the future. It will help build differentiation in products and services we offer while factors like abundance, automation and Asia make everything into a commodity in every sphere of activity. He cites several powerful examples. Students of medicine at Columbia University are learning "narrative medicine," which goes beyond computer-based diagnostics and urges students to find the clue to the cure in the patient's story. Beyond the apparent illness, every patient has a story. Remember my lesson from Part Four. Often the difference between life and death is not technology, but how well the doctor "listens" to the story.

At the Yale School of Medicine students are honing their power of observation through courses at the Yale Center for British Art. When you look at a painting, you see the whole picture; you also see the details. A patient's face is no different from a painting. When a doctor examines a patient, the face can reveal a lot that could change the diagnosis. At UCLA Medical School, students are required to enact role plays pretending to be patients; they get admitted to hospitals so that these future doctors understand what it feels like to be at the receiving end. Ideas such as these originate in the right brain, which was largely allowed to hibernate in the twentieth century.

Remember my story about being diagnosed incorrectly with tuberculosis? While I lay listlessly on the hospital bed, waiting for the results from the tuberculosis test to arrive, Dr. Nabin Patnaik, an ophthalmologist, dropped by for a social visit. While talking to me, he noticed that my left eye was abnormally red. He "sensed" something was wrong and "intuitively" sought permission from the head of medicine to move me to the ophthalmic unit for a checkup. There, the tests revealed I had iritis, a condition in which the iris gets inflamed. Unchecked it could lead to a total loss of sight. My vision had actually dropped to 20/200 from a perfect 20/20. I was on my way to blindness and did not know it. The raging fever had taken my attention away from my failing eyesight, and I had not noticed I was steadily losing sight in one eye because

I was lying down all the time. What followed was a series of steroid injections on the eyeball and eventually I regained sight.

Why did the visitor notice something that scores of attending doctors and nurses had missed? What would have happened if I had recovered from the fever but lost my vision? What makes one person a better observer than another? Developing the right brain helps you look at the big picture, the context so to speak. When you see the big picture, you realize the interconnectedness of various facets of the problem. That is when you realize the problem is not what you thought it was, and hence the solution too cannot be what it has traditionally been. Unfortunately, in most countries, the educational system is designed to overfocus on the left brain at the cost of the right brain. As children, we are not encouraged to use the right brain; it then simply takes a backseat.

The professional in the twenty-first century must learn to use both sides of the brain, to harmonize both fact and feeling when making decisions. *A Whole New Mind* has a powerful illustration to drive the point home. In the book, there is a picture that shows the evolution of man—from ape to farmer, from farmer to blue-collar factory worker, to white-collar worker, and finally to a physician with a stethoscope around his neck, and, wait a minute, a paintbrush and palette in his hands.

This is the professional of the future—the master of facts and feelings.

Chapter 46

Multiple Intelligences

The brain's biggest gift to us is intelligence—it is our tool to learn, understand, explore and create. Just as we are only beginning to understand the brain itself, we are also only beginning to scratch the surface of "intelligence," which is the most important tool for any professional. What is the nature of intelligence? From the earlier discussion on what constitutes the left brain and the right brain, we looked at how the twentieth century saw the emergence and the somewhat misplaced priority accorded to the left brain. Not surprisingly, the idea of intelligence was also equated with left-brain activity, which is mathematical-logical in nature. In effect, this became the basis of intelligence and its measure is known as intelligence quotient or IQ.

The real story behind IQ is recounted in Harvard professor Howard Gardner's pathbreaking work *Multiple Intelligences*. It was the year 1900. The city of Paris was witnessing a huge influx of villagers who migrated along with their children. In the preceding year, France had made it mandatory for all children between the ages of six and fourteen to attend school. So the law required that no child in that age group be left behind. Suddenly, the pressure on the Parisian school system became huge. In order to cope with the influx, the authorities needed a way to assess the intelligence of children so that the more intelligent among them got priority. They turned to a self-taught psychologist named Alfred Binet, who devised a method that has come to be known famously as the IQ test. At the time, it was largely a test of mathematical-logical capability in an individual. The idea of the IQ test as we understand it today has its roots here. Later, the U.S. Army became its largest user during drafts for the world wars.

Howard Gardner's work brought the limited nature of mathematical-logical capability—and the mistake we make when we equate it with intelligence—to the attention of people the world over. His book lists eight other forms of intelligence to illustrate the point. These are: literary, musical, kinesthetic, interpersonal, intrapersonal, spatial, naturalistic and, finally, spiritual. Professionals must be aware of each of these intelligences to harness the brain, to think, to ideate, to innovate and solve problems. It is important to understand them and how each one works, because in varying degrees we are all gifted with all nine intelligences to deliver better results and enjoy life more.

Literary intelligence is about the capability to read and write as much as imagine. Today, between a chief executive who can do the math and another who also reads extensively and writes some, the latter would succeed many times over. In a world in which we have to network and collaborate with dispersed teams, some in faraway lands and time zones, the capacity to build leadership proximity is determined by how well you express yourself through language. Developing excellent language skills for any professional then is extremely important.

When I was growing up, music or music appreciation was not a usual part of school or college curricula. It was considered "feminine" to pursue such activities; boys were encouraged to play sports, not learn music and dance. Music is about harmony. Studies indicate that children who play an instrument learn to listen much better—they grow up to be empathetic professionals. Not everyone is endowed with the intelligence to write lyrics, compose, perform or play an instrument, but the capacity to build appreciation can be cultivated by anyone, and it has several benefits beyond just entertainment.

Great athletes have what is called kinesthetic intelligence. In certain kinds of sports, you do not have time to think. In fact, thinking actually sets you up for failure. Consider boxing, fencing, soccer or even baseball. When you see a baseball player like the Yankees' Alex Rodriguez at bat, you realize that he has intelligence in his muscles and learning ability in his hands and legs. Even when a ball is coming at him at ninety miles per hour, he can decide whether to swing and how based on the spin and arc of the ball. Such people are able to make critical decisions involving the body and the mind on the fly, where each move is a new move. For some, life depends on this ability. How is a firefighter, a policeman or a trauma surgeon's job any different? How much time do you think they have to process information and take action in an emergency? Increasingly, professions of the future will need leaders with greater kinesthetic capability to reduce the gap between sense and response.

You must have come across many people who are naturally able to converse, make friends quickly, read other people's emotions or solve difficult problems collaboratively. These people have a higher degree of interpersonal intelligence. They build more empathy and can work better as part of a group. This is hugely important for people who want to lead. In the future, the ideas of leadership, followership and situational collaboration will be extremely important. Sometimes you lead, at other times you follow, but at all times you have to collaborate—and a professional of the future has to be able to collaborate effectively.

Given a position of power and authority, why does one person behave in a despotic manner and another in a humane way? The former is not self-aware, the latter is. Some people are more self-aware than others. Yet everyone has the capacity to build self-awareness. In my work with young leaders, I find that those who are more self-aware have a realistic understanding of who they are and what their true needs are. As a result, they are emotionally more stable and can deal with the ups and downs in life and work far better than others. These people have a higher degree of what Gardner calls intrapersonal intelligence. This is extremely important in professionals whose decisions affect a large number of other people.

Spatial intelligence gives us navigational capacity. In the days of yore, some seafarers were specially chosen for their navigational skills. On an overcast night, they could steer a ship and instinctively know true north. In today's world, in a profession like sales, some salespeople can map and chart a client organization much better than others. Some people can land in an alien town and map it so well that they become demonstrably more productive than others. I suspect these are more spatially intelligent people.

In our hunting and farming days, every tribe had someone who was better at recognizing animal calls or could tell one animal from another by looking at hoof prints or pugmarks. That individual was very important for the safety of the hunters and the success of their mission. When farming dominated human activity, this person was the one who knew one herb from another, could tell you which wild mushroom to eat and which one would kill you. He was the one gifted with naturalistic intelligence. As we all become more environmentally aware and sensitive to the planet, naturalistic intelligence, that is, the ability to relate to the natural world, will be much more in demand across professions.

Finally there is what Gardner calls spiritual or existential intelligence. We are all hardwired to appreciate the presence of a higher power. Some people can connect more easily with the spiritual world and as a result they can better deflect anxiety, can better deal with loss, and can better balance decisions with yardsticks of morality and self-governance. Such professionals, who know and can do the right thing the right way, will increasingly become relevant in the world of business.

Chapter 47

Three Levels of Knowledge

All professionals work with bodies of knowledge within a certain domain. And all things being equal, different individuals relate to the same body of knowledge within that given domain very differently and with astonishingly different outcome possibilities. According to Yves Doz, a professor at INSEAD, France, people relate to knowledge at three different levels.

At the lowest level, we relate to knowledge in a technical context. At this level, knowledge is all about specifications handed down to a professional who needs to create something. It is the kind of knowledge required for reverse engineering. Someone shows me an object or a model and I build a similar artifact with the desired form and functionality. I do not question, I do not add substantial new value. This is also called the adaptive layer. There is no innovation involved at this level. A professional who works at this level is unlikely to be differentiated and, in reality, as the world changes, could become extinct one day.

At the next higher level is the experiential layer of knowledge. This level is not about technical specifications and the functionality of the product or service we make or offer. Here the professional gets into the shoes of the end user, building a product or creating a service based on the experience of the target customer. When Nissan wanted to design a car for the European market, their engineers did not merely take a Japanese engine and create a body with European styling. Instead, Nissan sent a delegation of auto designers to Europe. When they got off the plane, each rented different makes of cars and drove thousands of miles all over Europe to understand what it meant to be a motorist there. They imbibed the motoring experience in the French Alps, the Italian countryside, the German autobahn—all vastly different from one another. Then they came back and designed a car that was right for the European driver. They created a product after they understood the consumer experience.

Beyond these two levels is the existential layer of knowledge where it is not about getting into the shoes of the customer but about "creeping into the minds" of customers. When Sony designed the Walkman, it was operating at this level. Sony is Sony not because it manages five thousand different products, or because it knows ferrite magnet technology, DSP chips and liquid crystal displays better, or because it builds more competitive and more optimal supply chains. What differentiates Sony is the fact that its designers understand what goes on inside the heads of its customers. For example, they have a feel for what happens to a kid in the Bronx when he dons his headphones, hits the play button and begins to wriggle in tune with his favorite music. The professionals at Sony work backward from that feeling to design innovative products and services.

Individual professional progression charts a very similar path. The best professionals operate at the existential level where it is about building the capacity to engage with people, problems, processes and opportunities by creeping into the minds of customers. This requires empathy, inclusion, 360-degree thinking, recognizing the interconnected nature of things, looking for solutions outside the box, learning from unusual sources and finally cutting through complexity and doing so with empathy.

A professional I have deep admiration for is the internationally renowned cardiac surgeon Dr. Devi Shetty. When you visit his famous hospital, Narayana Hrudayalaya in Bangalore, you first encounter a small shrine in front of the portico. It is a small, four-cornered structure. One side is a temple, the other a church, the third side is a mosque and the fourth a *gurdwara* (a place of worship for Sikhs). The doctors at Narayana Hrudayalaya will tell you this shrine is where all healing begins—from faith and then on through the hands of the doctor. Beyond the portico, as you step inside the building, is a well-organized reception area, which has a sign in Bengali that says "Bengali Spoken Here." In Bangalore, almost everyone speaks either English or Hindi or both. Why does the hospital require a separate area within reception for patients who are more comfortable speaking Bengali? Because demographic analysis of patients has shown that most heart patients who come to the hospital are from West Bengal in India and neighboring Bangladesh and are often escorted by a fellow villager, spouse or family member for whom an alien city, a hospital and the burden of an unknown language can be daunting. Dr. Shetty has created his institution based on existential knowledge, on understanding how a patient who visits his hospital feels.

At the end of a long day, Dr. Shetty sits in his office and sees outpatients. He always tries to say a few words to each patient in their mother tongue. Sometimes his accent is clumsy, but his gesture instantly endears him to them and makes them feel at ease. He uses his stethoscope to examine them, something that is probably meaningless in a clinical sense because he has their entire case history, including CT scans, in front of him. But he does it because most patients from rural areas feel a doctor has not paid attention to them or examined them thoroughly unless he has listened to their heartbeat with a stethoscope. And, invariably, he will touch them at least once during the conversation. When I asked him why he did that, he told me that these days doctors do not understand the power of the human touch. There can be no healing without touch. After he has examined the patient, a completely unhurried Dr. Shetty ends the consultation by asking, "Do you want to ask me anything?" For a man strapped for time this can be risky, because patients may engage him in a long conversation. But for Dr. Shetty this is the most important thing to do.

Dr. Shetty relates to the world of medicine at the existential level. And this is what will be required of every world-class professional of tomorrow—be it a dress designer, product manager at Procter & Gamble, a physician, a family attorney or an air traffic controller.

Chapter 48

The Five Minds of the Future

Our brain is the repository of nine intelligences. The mind, which is the abstract layer of the brain, works with these nine intelligences. Howard Gardner tells us we have more than one form of intelligence and in reality more than one mind. We can be masters of many minds by using our multiple intelligences. It is in this context that he writes about the "five minds of the future." These are: the mind of discipline, the mind of synthesis, the mind of creativity, the respectful mind and the ethical mind. To illustrate the five, I often ask audiences to simulate a situation.

Imagine you are the police chief of your city. You get a phone call from the governor of your state informing you that in an hour's time a hijacked plane will land at the city airport. You have to take charge. What are you going to do? What decisions are you going to make? And how are you going to make those decisions?

It is amazing how groups can spend hours discussing what needs to get done even though in a real hijacking situation, there is generally no time to discuss.

You need to rush to the airport, of course. You need to gather all intelligence reports and get a well-briefed police force in there. But you also need commandos who are better trained to handle terrorists. While the police can handle the outer perimeter of the airport, it is the trained commando who must handle the plane and the hijackers. You need a trained hostage negotiator. You probably need a language interpreter (the terrorists may not be conversant in your language). You need aviation experts. You need doctors—first-aid experts, specialists who deal with trauma victims, a cardiac specialist and a burn injury specialist just in case. You need paramedics. You need food and water. You need fire engines and fire experts. You need the bomb disposal squad. You need forensic experts and body bags and DNA fingerprinting capability. You need trained communicators because you will need to talk to many different stakeholders and not just the media. Above all, as the police chief who has an hour to put preparedness in place, you need the five minds of the future.

First and foremost you need the mind of discipline, which means your own discipline as a police officer. You must be a master of all the things needed to run the police force. Studies done at Harvard and Stanford universities indicate that mastering any discipline takes as long as ten years. To be a good surgeon, a good photographer, a good teacher or a good software engineer, it is not enough to be technically qualified or professionally educated; you need sustained, devoted practice over years to truly know the nuances of your discipline, understand the big picture, the dependencies and the domain.

But knowing how the police force works is just one small part. Mission success now depends on knowing how other disciplines work as well. All professional problems of the future will require an interdisciplinary approach. Thus the police chief needs to know the workings of aviation experts, the antihijack squad, doctors, psychologists, etc. Should he make the decision to storm the plane or negotiate and psychologically wear down the hijackers? This calls for the capacity to synthesize information from many different disciplines. This is where you will need the mind of synthesis—the capacity to look at any issue, any solution from a multidisciplinary viewpoint.

Because life is going to throw at us increasingly unscripted problems—like the hijack drama unfolding before us—we need to build the creative mind. Solutions in life cannot always be resource intensive; sometimes the best solutions are found on the fly, using available resources. Until the trained interpreter arrives, the air traffic controller may be a better bet to keep the hijacker engaged. Persuading the hijacker to let a multilingual nurse on board help with a woman in labor while waiting for a gynecologist may be the smarter thing to do; asking the local army hospital to send a gynecologist may be even better than sending for a civilian doctor. Now you begin to see how almost all steps, thousands of them, through the hijack crisis need to be extremely creatively thought out and how each decision has downstream consequences, like a game of chess.

The police chief cannot handle the problem on his own. He requires the fire chief, the chief medical officer, the head of public relations, army and aviation experts, to name a few collaborators he must rely on. Now we have a different dimension to the problem. These people are experts (not amateurs) in their own disciplines. There are potential problems dealing with experts—they are all extremely knowledgeable in their respective fields, and they can be egotistical people who do not necessarily get along with other experts. So in a situation such as we have, we need a police chief who has a respectful mind. Only a person who knows how to iron out differences, bring out the best in other people, get consensus and buy-in for difficult options—all in the shortest possible time with the fewest possible unwanted consequences—is a true professional of the future. Leaders in every profession must learn to deal with others with respect. In the situation we are discussing, that one capacity may make the difference between large-scale destruction and a negotiated settlement that saves innocent lives.

Finally, every great professional requires what Gardner calls the ethical mind. To me, the ethical mind is not just a morally sound mind. Should the police chief pay a ransom or not? Is it better to concede to the hijackers' demands to save the babies on board? Professionals must begin with the basics. Being a professional will also mean understanding the larger implications of our decisions and our processes, and deriving our actions from a higher code of conduct—that of self-regulation and belief in the practice of fair dealing in every transaction.

Chapter 49

Critical Questioning

The true professional has the capacity to raise incisive, difficult and sometimes uncomfortable questions that become potential game changers. Such questions can be inherently counterintuitive and often differ from collective wisdom. Critical questioning is a powerful tool available to every professional. However, it requires understanding and then cultivation. As you sharpen the power of critical questioning, you begin to ask the right questions. This opens up many possible solutions beyond what is apparent on the surface.

Michael Marquardt, author of *Leading with Questions*, talks about how people refrain from asking questions because they do not want to rock the boat, risk looking foolish or interrupt the "harmony" of an organization.

The *Titanic* sank on her maiden voyage from England to the United States in April 1912. More than 1,500 people died. Some of the people who had designed or been involved in building the ship had serious doubts about her design, but they kept quiet as she sailed because they did not want to look foolish in front of the majority of their colleagues who did not seem to share their doubts. Even as she was undertaking the crossing, it seems other ships in the vicinity had radioed the *Titanic* warning her of icebergs. But those on the ship ignored the warnings. Read: Someone did not exercise the professional requirement to raise critical questions about how the presence of icebergs might affect the ship. Someone did not agitate enough. What followed was one of the worst maritime disasters in history.

In 1961, President John F. Kennedy gave clearance for a covert attack on Cuba by 1,400 Cuban exiles in what has been termed the Bay of Pigs disaster. Intelligence reports indicated that such an attack would unleash a popular uprising in Cuba resulting ultimately in the overthrow of Fidel Castro. Contrary to expectations, when the 1,400 exiles landed, they were far outnumbered by the Cuban military, which was forewarned of the assault. More than a thousand were taken prisoner and hundreds died. One key assumption when making this decision was that the exiles would scatter into Cuba—how they would do this when between the Bay of Pigs and the mainland lay eighty miles of swampland remains unclear. Was the question ever asked? When the mission failed, it was apparent that the president had been misled and it was even clearer that people who had doubted the success of the strategy had suppressed their urge to question it because everyone else seemed to be so gung-ho on the idea.

After the Bay of Pigs fiasco, "groupthink" as a concept emerged and was researched extensively by celebrated Yale psychologist Irving Janis in the 1970s. Janis brought to light how groups of decision makers become blind to impending warnings when they are bound together on a mission and how group members fail to raise critical questions because of the need for false harmony.

When we ask critical questions, we begin to understand the interconnected nature of things. In other words, we are able to build a larger, systemic view.

Peter Senge at MIT uses an interesting example—that of highway congestion—to explain this. In Senge's scenario, traffic planners decided that the best way to ease congestion on a busy road in a city was to add another lane. After the lane was created, the congestion worsened despite the fact that traffic volume had not increased. What was the reason? The availability of the additional lane had actually encouraged drivers to change lanes more frequently, hoping that they could get to their destinations faster. When everyone tried to do that, the result was chaotic driving and worse congestion.

The inadequacy of the well-intended measure was caused by nonexploration of systemic issues—that is, not exploring all aspects that cause traffic snarls, including bad driving, an effect of human psychology that involves opportunity seeking, risk taking and a host of other factors that make drivers as much a problem as the width of the road.

This is not to say that the road widening was not needed. It is to suggest that other factors needed to be unearthed by asking the right set of questions. These could have been remedial measures like teaching drivers to follow lane discipline, allocating the new lane for only certain categories of vehicles at certain times or simply allowing one lane on the reverse side of the road for traffic to move in the opposite direction during peak hours.

• • •

There are stories galore in management literature that tell us how the predominant role of a CEO is to ask questions, not give the answers. Jack Welch, one of the most admired CEOs of the twentieth century, was famous for asking questions and using them to build a high-performance culture among his leaders. In the process, he helped build one of the most valuable companies in the world: General Electric. GE is an extremely diversified company, with activities ranging from financial services to aircraft engines, medical equipment to plastics and domestic appliances. The company could well be a CEO's nightmare to manage.

Jack Welch knew the power of questions. In managing and uniting GE's very diverse businesses, he would invariably ask himself a set of five specific questions about the leader of each of the various businesses.

- Is he real?
- Does he see around the corners?
- Who's around him?
- Does he get back on the horse?
- Is he pro-business?

Welch's penchant for asking questions as a way of simplifying complex issues extended to areas beyond business as well. In a much talked-about article he wrote in *The New York Times* on October 30, 2004, to help Americans choose a new president, he asked voters to ask themselves these same five questions when evaluating the candidates.

Through these five questions, Welch brought out five important leadership imperatives of the U.S. president: authenticity, capacity to see the big picture, ability to attract and surround oneself with competent people, capability to recover from mistakes and the ability to see business as a driver for nation building.

Welch was not the first to understand the power of questions as a professional requirement. In the 1970s, Motorola Corporation embraced the idea of Total Quality Management (TQM). The company not only won the Malcolm Baldrige Award, considered the Nobel Prize of quality management in the United States, but used the idea of TQM to completely reinvent itself. Motorola used Six Sigma—a TQM adaptation—and following Motorola, companies like Allied Signal, GE and many more began to apply Six Sigma principles. The concept of Six Sigma quality can be very complex and sometimes difficult to understand for people without some knowledge of statistics. To ensure that everyone in the organization embraced Six Sigma, Motorola used a framework of building questions that could help even the mathematically challenged understand the idea of quality at some basic level and build an action plan for continuous improvement. Toward this end, everyone in the company was required to answer the following questions:

- What is my product or service?
- Who is the customer for my product or service?
- What do I need to do to deliver my product or service?
- What are the Critical to Quality (CTQ) requirements of my customers?
- What is my "As-Is" condition?

- How do I Define, Measure, Analyze, Improve and Control?

When people start seeing their work as an output on which someone else depends, they start recognizing the other person as a customer. They begin to realize their dependence on others to create their product or service. Once they see their recipient of the output as a customer, they learn to go and seek out the voice of the customer. This becomes the CTQ requirements without which the product or service falls short of gaining the customer's approval. They learn to measure the current level of quality vis-à-vis what the customer needs and thereby create a process for continuous improvement.

But it all starts with the idea of critical questioning. Critical questioning is the starting point for every quality journey. And that includes the journey toward personal quality as well.

Chapter 50

Dealing with Personal Pain

In our jobs, particularly at senior positions, each of us deals with a host of complex situations and difficult problems every day. These help all of us to grow as managers and leaders. At the same time they also build pressure, and before you know it, you are in your forties, and have to face many challenges on the personal front as well. It could be your own health, an aging parent diagnosed with a terminal illness, perhaps marital discord that ends in separation, or problems associated with raising teenage children. These are life-changing events that are bound to affect anyone. How you manage yourself and sometimes deal with personal tragedy determines how you will grow as a professional in the years ahead.

One of the most uplifting stories of someone dealing with personal pain while in the line of fire is that of Anu Aga, former chairman of Thermax. Anu took on the mantle of chairmanship after her husband died suddenly in 1996. She had no time to mourn. The board asked her to step in and take charge of the company her father had started and her husband had built as a nationally reputed engineering organization. She took over the reins within forty-eight hours of her husband's death. Barely a year later and within a fortnight of one another, her mother-in-law, her pet and her son died. Her son died in a road accident while driving back from Bangalore to Pune after personally fixing the problem of a dissatisfied customer. Anu balanced herself through all this and remained steadfast. Under her leadership, the company grew, and she reconstituted the board, divested unrelated businesses and improved performance several fold. In 2004, she handed over the reins of the company to her daughter Meher Pudumjee.

Dealing with personal pain can have a debilitating impact on one's work and that is understandable. Yet somewhere we must draw the line so that the deeply personal does not override duty. Surprisingly, a personal crisis can also bring focus and attention to one's job, helping us achieve higher and better standards. In many ways, it sometimes transforms us as professionals.

Faced with a personal setback, the first thing a professional must do is to secure his or her professional position. Your commitment to your job and to performing to the best of your capabilities cannot be compromised. It makes no sense to let a personal tragedy submerge the platform that provides us with the power of continuance. Allowing personal issues to drown professional obligations is like abandoning the shelter before the storm has passed.

If a personal problem is beginning to overwhelm you, inform your organization. Seek help in reprioritizing your work on hand so that deadlines and commitments do not fall between the cracks. If required, take time off. The period of adjustment varies depending on what you are up against. Renegotiate your contract, work hours and compensation until you feel stronger. The key is to plan, strategize, know the trade-offs, build backups and keep people informed. Most professional and progressive workplaces understand this and find creative ways to help a valuable individual cope with a period of stress. Return to full-time responsibilities only when you feel you can once again give your best to your organization. And come back with full force.

Part Six

New World Imperatives

Chapter 51

Rainmaking

I have spent the last several chapters discussing the importance of managing volume and managing complexity. But in today's increasingly fast-paced and competitive world, it is not enough to simply manage what you have been given; you must also seek to create new opportunities for yourself and your business. You must see yourself as a rainmaker.

The term "rainmaker" was first used by Native Americans to connote a medicine man who by various rituals and incantations sought to literally make it rain. Today, the term implies an organizational man (or woman) with an exceptional ability to attract clients, leverage the network, amplify the brand values and increase profits. In short, a rainmaker is a professional with disproportionate capacity to add value to the organization. This ability has become a fundamental requirement for all CXO positions in any organization.

Rainmaking is often associated with the act of selling. We are all familiar with stories of the sales guy who, upon finding no leads in a territory or an account, consciously skips it and begs to be shifted to a better ground. Then, when another salesperson is assigned to the same territory, she is miraculously able to win business and make deals on the same barren land! Managers love her, and the entire organization looks up to her, but no one really knows how she does it. She is a rainmaker.

But rainmaking is not about selling alone. To be a true rainmaker, one must rise above anonymity in every function and become a critical asset in every organization.

Imagine an editor of a publishing house who is not just good at her job but has a knack for picking up best-selling authors. That individual is a rainmaker without being a salesman. Or for that matter, take a chief technology officer at a high-tech Silicon Valley company who is an inveterate geek but has the special capability to anticipate trends in technology, which allows his company to gain a competitive edge and raises the bar for the entire industry. Think of a surgeon who actively cultivates industry contacts and is visible at various platforms. As a result, she gets favorable press that helps her hospital stay ahead of the competition. In the process, the hospital is chosen to receive government funding for research in a new area, leaving its competition behind. In each case, the professional involved is not just doing a first-rate job in his or her own specialized area but also actively seeking opportunities to expand the organization's revenue, profitability, reputation and overall impact. The professional is personally enhancing the organizational brand value. She is a living logo.

It is important to remember that rainmakers were not born with any sort of special ability. Rather, they cultivated their skills with conscious effort and over a long period of time. They recognized rainmaking as a professional requirement that would allow them to make a larger impact. When you observe such people closely, you can discern a set of practices they do almost effortlessly. But as I said, developing this talent begins with recognizing that it is a matter of priority, an important deliverable for anyone in a position of consequence in the organization. At the core of that recognition is respect for the larger goals of the organization. You cannot be head of R&D or production or supply chain or HR and not recognize the value of improved sales, profitability or other things that affect the competitiveness and leadership of the overall system. This means, in some sense, we are all salesmen and -women. In today's hypercompetitive world, selling is not an option at a senior level. Someone sells the product or service to an end customer; others sell the brand, the image, the technology prowess, employment and the financial transparency of the organization to stakeholders. That said, people are seldom trained for it.

I have met Bill Gates only once. He was visiting India in the year 2000 and my boss was invited to be part of a small group of industry leaders to listen to him. My boss could not attend and so sent me. I was very flattered and flew in all the way to Delhi from Bangalore hoping that the man would enlighten me about the future of future. When Gates appeared and took the podium, he spoke evangelically about something I felt to be quite devastating. He spoke about how Microsoft Windows would change the world. My first reaction was disbelief, replaced by disappointment. As I returned to Bangalore, suddenly it dawned on me why Bill Gates was selling Windows. After all, what else did he have? And if he did not sell Windows, how on earth could he expect his field force to believe in it? Bill Gates believed his number-one job was to sell Microsoft and its products. Anything esoteric fills his ego but not his coffers. That is the reason why Microsoft is where it is today, despite what detractors may say of it. Whether it is Jeff Immelt of GE, Larry Ellison of Oracle, Richard Branson of Virgin or Indra Nooyi of PepsiCo, today's successful CEO is focused on his or her company's revenue and reputation. He uses important platforms and connections, goes along to make a sales call or join in for a closing. These initiatives give him a firsthand customer perspective as well as providing the sales organization with the motivation to do a better job. Today's CEO is a rainmaker. People who aspire to larger responsibilities better master the art of rainmaking well in advance and certainly before the organizational line-up to pick the chosen one.

Rainmakers are constantly searching for opportunities to pass on to their sales organizations any knowledge they gain through contacts, competitor information, reading, trade shows, meetings and networking. "Who is doing what?" and "Who is saying what?" are often critical pieces of organizational knowledge that can only be acquired through the eyes and ears of a seasoned professional. Thus attending industry events and trade shows results in important learning points for any organization, and it is not just the sales personnel who need to go. Sometimes, the most valuable leads are generated not by salespeople, but by employees in other departments who must train to spot and seize opportunities.

There are a few other aspects of rainmaking that all of us should know about.

A rainmaker is invariably a good spokesperson. He is able to articulate statements on behalf of his organization to the outside world. Sometimes all of us need to present a point of view to the trade press; sometimes we are called upon to make a presentation to a trade body or a financial analyst. In a world of global business, we all have to play host to an overseas delegation every now and then. Chances are, in addition to making a presentation on your organization, you will have to answer a barrage of questions: on your organization, on the industry and, quite often, on your country's policy issues. A lot of people fumble through such occasions. Few organizations train their senior professionals to be spokespeople, but it is not that hard to learn. It is different from public speaking and simply making good presentations. It involves articulating a point of view on a subject of importance, and sometimes handling questions on difficult issues, particularly in difficult times. Public relations agencies are usually a good resource to get trained in how to become an effective spokesperson. Effective spokespersonship requires understanding of how different media work. I always considered myself to be a reasonably good presenter but was not aware of how to handle television interviews. I used to be long-winded whenever a question was thrown at me. My trainer taught me a simple thing. "Publish the headline first," he would say. When we read the newspaper, we do not begin with the body copy first. Imagine having to read through an entire article just to find out what the story is all about. Instead, I was told, one must speak in short sentences, provide the position on the subject as a headline of sorts and then allow follow-up questions to get deeper into the subject. During the conversation, one needs to use data, illustrations and examples to leave memorable key points. But the hook is always the headline.

In today's world, every professional has to frequently attend video- and teleconferences, which present their own nuances. Speaking to a live audience is vastly different from speaking to someone halfway across the world, even if you can see them, and it is much more difficult to engage people this way. The ability to have a conversation in a remote mode, sometimes with people you do not know, can be very daunting. All of us must learn phone, video and Web skills; these are no longer a nice-to-have capability.

Rainmakers understand and stay current with their organization's key business parameters and the issues concerning them in the environment; they scout for opportunities for new business, sometimes an unusual source of supply, an alliance that expands operations to a hitherto new area, and they attract high-quality talent. These require a professional to learn how to speak the language of the business he or she belongs to, and it begins with the ability to make an engaging presentation on what their organization does. It is amazing how few people realize the importance of something as basic as this. If you cannot make a good presentation of your organization, you will never be able to speak confidently and clearly about issues at an industry level.

Last but not least, as we all progress to positions of seniority in any field, we need to understand that we can no longer live off the organization's brand. In fact, the brand of any organization is an aggregation of the personal brand of its top team. Thus building a personal brand and making it visible are important professional requirements for everyone. My personal brand is all about what I stand for. What are my key beliefs and positions on important issues? This is something that gets built over time through what you say in public forums, what views you propagate, what organizations you endorse and what you write in newspapers, magazines and blogs. My cofounder at MindTree, Kalyan Banerjee, tells young leaders to frequently check their "Google Quotient." If someone makes a Google search on your name, what is likely to show up? If the answer is nothing or nothing substantive from an organizational point of view, then you are probably not a rainmaker yet.

Rainmakers are invariably visible in what I term "alternative spaces." These are your organization's Web site, your personal blog, YouTube, Facebook and such other places on the Internet where people may look for you before contacting your organization. What they find about you will help shape their first impressions about your organization. Take for instance Steve Jobs of Apple. The commencement speech he delivered at Stanford University is on YouTube and it has been viewed more than 600,000 times! In it, you not only learn about life lessons from the man who founded Apple, but by listening to him speak, you build an opinion about Apple itself. How much did Apple have to spend on creating the video, posting it on YouTube and keeping it there? Not a dime. The value of that one video is worth millions of dollars of paid advertising. That is the power of rainmaking.

Chapter 52

Inclusion and Gender

In 1998, Susmita and I decided to build our own house. We bought a piece of land, but we needed to take out a home loan so we could start construction. We went to a leading private sector bank in India to discuss the loan. Sitting in front of the manager, I told him off the bat that although I would be the borrower on record, it was really my wife who managed all our finances and made all the financial decisions, and she had some questions to ask before we made up our minds. But as Susmita began to ask the relevant questions, the manager would steadfastly ignore her and look at me while answering. It was clear that he did not consider her important. In the end, despite the bank's marginally lower rate of interest, we decided to take a loan from another bank. It was clear that the first bank would not treat Susmita on par with a man. This was a practical problem since I would not have been able to accompany Susmita to all meetings with the banker. She would be dealing with him on an ongoing basis, and she would obviously not have been taken seriously.

Countless people unconsciously discriminate against women

in everyday work life. A man is assumed to be more competent, more technically savvy than a woman, and some men still believe that a female colleague is someone to be accommodated, not accepted or treated as an equal. Despite an increasing number of women at the workplace and in positions of power, in the average workplace transaction we remain very male-centric. Going forward, professionals have to be more aware of gender issues while dealing with colleagues, customers and suppliers.

When a group of fellow workers go out for lunch and there is a lone woman in the group, without realizing it, the men begin to engage with one another and usually the woman is expected to blend in. Some people believe that it is the woman's responsibility to do so; men do not realize how quickly they actually create a fortress of sorts that shuts women out. By virtue of being a woman it becomes difficult to be part of the group, even though the person to be included is a coworker, not an outsider. It is a silent problem in every workplace.

Many may think that in a developed economy, gender discrimination is not a big issue. That with nearly half the workplace already populated by women, we have come a long way from the day the first American woman joined a factory. With so many women CEOs and heads of government departments, do we really need to discuss gender sensitivity any longer? The truth is, while in developing countries sexism is widely prevalent, in developed countries it may not be endemic but still requires a conversation.

In every profession around the world, more women are joining the workforce. In addition to the issue of gender discrimination, professionals need to understand the issue of sexual harassment at work. Consider the case of the head of sales in a leading Indian company with a great professional reputation. He was the ultimate rainmaker. Based out of the company's U.S. office in California, the man felt an attraction to his secretary and, going by her account of the events that followed, used his positional power and influence to have sex with her. After going through the awkward relationship for some time, she left the job. But he began to stalk her. She registered a case, got a restraining order against him and gathered the courage to sue her former employer and the man for sexual harassment. The company made an out-of-court settlement, as did the man, who also "resigned" from his job.

In another instance, a project manager at MindTree started showering extra attention toward a young woman developer. He called her into his office every now and then under the pretext of giving her feedback, but instead would get personal with his questions in one-on-one meetings, make statements about her dress and started sending her romantic text messages. She asked him to stop, but he persisted. She reported the matter to the organization. The text message trail was enough to show that the man had indeed crossed the line. He was a star performer but was asked to leave.

Research in the United States has shown that 95 percent of women have received sexual material such as letters and phone calls at work. In roughly one out of two cases, the sexual harasser is the victim's supervisor; and nine out of ten sexually harassed women suffer from debilitating stress reactions, including depression and headaches. According to Linda Gordon Howard, author of *The Sexual Harassment Handbook*, most ugly situations can be avoided if people follow the three rules our parents taught us whenever we went to the playground early on in life: do not force someone to play, do not be mean and do not pick on little kids. What this conveys is, first, that you must not coerce anyone to have a relationship. Second, you cannot be vindictive if the person rejects your advances. Finally, those who are in positions of power, authority, rank or hierarchy must not prey on a subordinate in the workplace.

Linda Howard rightly points out that not all sexual harassment involves men harassing women. It can be the other way around, as well as between two persons of the same sex. Despite the fact that sexual harassment at work can occur across genders, it is still largely the female worker who is at the receiving end. In a social culture that believes women must be wooed and "pursued," and a "no" is supposed to mean "yes," many of those accused of sexual harassment look perplexed and innocent and claim that they did not know they were doing something wrong in the first place.

Howard also brings forth the need for articulation of an organization's stand and for educating people at work across genders. The crux of the matter lies in an organization's clearly defining what is and is not acceptable behavior. In short, any "unwelcome" gesture is a potential cause for harassment if it persists. Thus whether it emanates from a man to a woman or otherwise, a "no" must always be taken as a "no."

For the professional organization of the future, addressing gender sensitivity and sexual harassment will be very important. Gender sensitivity, however, cannot just be an organizational priority; it is also for every individual to take the initiative to observe, learn, cultivate and demonstrate. Wherever adults work, there is bound to be mutual attraction, but every professional must know where consent stops and harassment begins.

Chapter 53

Cross-Cultural Sensitivity

The Lufthansa flight from New York to Frankfurt was delayed by seven hours. When the flight finally landed in Frankfurt, there was chaos. People were tired and anxious. Most had missed their connections to other parts of Europe and Asia. As soon as people deplaned, everyone ran to the nearest customer service representative and there were long lines of passengers wanting to know how they would get rerouted, when their baggage would arrive and a dozen other things. The airline representatives were as harried as the passengers themselves.

Standing in line ahead of me were a young American woman and her friend, fuming like most of the others. In times like these, lines seem endless and the person at the service desk always seems to be taking forever. Finally, the young woman's turn came. The German woman at the service desk asked her, "What's your problem?"

All hell broke loose.

The young American woman screamed, "It's *you* who has a problem. Not me." It was very obvious she was furious at the question. The German woman was taken aback by the passenger's fury and soon enough an angry exchange ensued.

In her own country the American passenger was used to being asked something like, "May I help you?" "What's your problem?" was an affront in the American context. To the German, trained for efficiency, "What's your problem?" was just an expression of interest to immediately engage with the issue at hand and offer a solution. The German was expecting a specific reply, such as that the passenger required hotel accommodation, or needed to make the next connection to Prague or wanted to check the status of her baggage. She did not realize that the direct nature of the question could have a different nuance for the American. The encounter would have been very different if the parties were cross-culturally more aware. Then, before climbing up an emotional ladder, they would first ask, Is there something I might be missing here?

In 1991, when I was working for Wipro, I was in a tight spot with a leading technology organization from Silicon Valley. Doing work offshore was not the norm in those days. As a result, few knew the intricacies involved, far less the impact of culture on globally distributed teams. My client was clearly upset. The work delivered was different from what was agreed upon, they insisted. I drew their attention to the agreement. Then they said that the agreement had been revisited after our technical lead showed them the first prototype; at that time they had told him they now needed many other things included as part of the same specifications. Our technical lead seemed to have agreed with everything during that one meeting, they said. It was my turn to be surprised. I had not been updated about this development.

"But right through the long meeting we had in our office when we asked for all the features, he kept nodding his head," the perplexed American customer told me. Then I realized what must have happened. To the American, a nod means "Yes." To an Indian, a nod simply means, "I am listening."

It took a lot of convincing on both sides for me to contain the damage done, but finally, everyone got a better handle on the impact of cultural differences on such projects. Despite two decades of globally distributed software development since that incident and a lot more knowledge on how to bridge cultural gaps, every day I learn about new ones.

At MindTree, we assess emotional bonding in large, long-playing project teams that invariably consist of our own people and the client's, and they are often from many different parts of the world. Recently, one such high-performing team was seen experiencing difficulties, and a joint off-site meeting among the leaders from both sides was conducted in London. I invited Lu Ellen Schafer, a consultant from California who advises leading-edge companies around the world on team dynamics, to join us as a process observer. With help from her, we realized that apart from the hiccups of a complex software project, there were some significant cross-cultural issues that were affecting perceptions in the minds of leaders from both sides. The Indians on the team were overall very comfortable with their British counterparts but felt that the British "escalated" issues far too often. When this came up, the British team was perplexed. They said they had escalated issues only twice in the course of an entire year. It turned out that while reporting day-to-day issues that remained unresolved for some reason or another, the British frequently copied their e-mails to their other colleagues. To the Indians, who were still working on a solution to a given problem, this amounted to "escalation," and over time, they thought the British were showing them in a poor light by their frequent act of putting on the copy list. The extension to this perception was that the British did not trust us and were unappreciative of the very diligent efforts being made by the team back in India to fix problems. The British were quite taken aback. They believed that they were simply "keeping everyone informed" and not escalating when copying a mail to others.

Lu Ellen Schafer sees lack of cross-cultural sensitivity among professionals as a real issue around the world. In a recent case study, she describes the following situation:

> *During a global dinner in Bangalore with his new engineering team, Heinz, a German engineer, complained that he didn't trust Americans. "Really? Why is that?" asked one of the Indians, Ramesh, since he had had mostly positive experiences with American teammates so far. "They pretend they support you, but they don't. They do not tell you the truth," the German replied. Ramesh was startled that Heinz had such negative feelings, but perhaps he was right. Ramesh would listen carefully in this week's meetings to see if he could see what Heinz was talking about.*
>
> *The next day Heinz laid out his ideas for a possible new feature. Susan, from California, exclaimed, "You rock, Heinz. I love it!" Heinz narrowed his gaze on Susan and said rather assertively, "So you are one hundred percent behind me? You will support the idea of this new feature?" Susan laughed and replied, "Well, I don't know yet. There are a lot of features to consider."*

Heinz shot Ramesh a look as if to say, "Ha! I told you that you cannot trust them." He put his pen down and said, "Okay, just forget it."

Ramesh was puzzled. Something was going on but he couldn't quite figure it out. After all, Heinz had just met Susan. How could he distrust her so soon? What she said sounded like a typical American response, full of enthusiasm and encouragement.

Ah, then Ramesh got it! Susan's enthusiasm must have been interpreted by Heinz as a commitment to his idea which she then appeared to take back. But Ramesh knew this was not the right understanding. Rather, it was just an American way of acknowledging someone's idea. Ramesh recalled the first few times Americans had said to him, "Great job, Ramesh!" He had loved the compliment, but then realized that Americans usually say "great job" to everyone all the time. Surely not everyone was doing a great job. And then it occurred to him: "Great" in the US had another meaning. It appeared to mean "thank you." He would quietly tell Heinz about this cultural difference the next time they spoke.

Lu Ellen Schafer tells me that cultural differences are but one of the many factors that affect how people respond to various situations. Such differences seem to matter most when things are not going well. When professionals are communicating openly, when a project is on track and the deadlines are reasonable, most cultural differences are noted as interesting, if they are noted at all. However, when distance, time zone differences, overly aggressive deadlines, highly stressed clients or unclear roles and responsibilities create tension, cultural differences can exacerbate the apprehension on both sides.

Just because most of us speak English does not mean we understand one another while doing business around the world. In an increasingly global world, we have to learn cross-cultural nuances because we deal with customers, suppliers and other collaborators who are as new to us as we are to them. A professional who understands these differences and finds ways to work with them, rather than letting them work against the organization, will be in much demand in the new world.

Chapter 54

Governance

Let me tell you the story of the time MindTree went public and what happened to the career of an exceptionally bright young man. A public listing is the perfect occasion to say thank you to employees of a company through an assured allocation of stocks that is permissible under the law. However, such allocation comes with a lock-in because stock exchange regulations do not allow employees to sell (dump) their shares in the market and make quick money the moment the share price goes up. The reason is simple. While only a few retail shareholders may get an allocation in an oversubscribed public issue, an employee who gets an assured allocation is already getting preferential treatment, and the law requires that such a person return the favor by holding on to the stock and not making short-term gains at the cost of a retail investor. Hence the mandatory lock-in period. Hundreds of MindTree employees benefited under the scheme and were bound by the lock-in period of a year. The young man in question took a large bank loan and purchased shares from the special employee quota. The moment the stock listed and its value went up within days, the employee off-loaded his shares. He thought that in the thousands of transactions performed every day, nobody would notice. The problem for people who contravene the law today is that digital footprints are easy to track. As soon as the matter surfaced, the individual was asked to leave and the stock exchange was notified, for he had broken insider trading rules that employees of every listed company must abide by. It was a clear breach of governance.

In the days to come, the term "governance" will touch professionals in every field in one way or another. Broadly, it is a word that encompasses compliance with the laws of the land but in spirit asks for self-regulation. If you think about it, a company, or any organization for that matter, is not a living thing and is quite incapable of breaking laws. Thus when institutions break laws, it is actually someone working there who has broken the law.

Basic governance issues could arise in numerous situations, as when purchasing goods and services for personal use from vendors to your company or when selecting vendors and awarding contracts. Paying money or other forms of inducement to get your work done is a clear breach of governance.

If you are the purchase manager and if one of the bidders is even distantly related to you, you need to step out of the process voluntarily. If you do not, that is a governance issue.

If you are the head of finance and your brother-in-law is the banker you deal with but you fail to disclose that relationship to your organization, that is a governance issue.

A doctor, a lawyer or a chartered accountant who has advisory capacity cannot take advantage of his position with a client to secure preferential benefits.

This is not a book on governance. The intent here is to take the first steps in sensitizing professionals. They must hereafter actively seek information and guidance on this issue when they are members of any given profession.

Most governance issues can be avoided with a culture of education and a spirit of full disclosure. When there is the slightest possibility of a conflict of interest, prenotify the relationships. Dr. Sridhar Mitta, a pioneer of the Indian IT industry, had a nice way of explaining how to avoid problematic situations in governance that sometimes arise out of the seemingly innocent acts of a professional. "Never drink milk under a palm tree," he would say. For the uninitiated, a palm tree is a source of toddy, an alcoholic beverage of great potency. If I were seen drinking under it, even if it was lemonade and not the real thing, the conclusion people would make is going to be very obvious. Thus it is important to be careful about perceptions we create. The point here is that we must be doing not only the right thing or staying away from the wrong, but we must also be aware of perceptions sometimes created with unintended but avoidable acts

As businesses become global and more complex, there will be an increasing demand for governance that meets requirements across countries. How do you self-regulate to ensure that you do not break the laws laid down by different governments? Is it something that only organizations need to worry about? Or is it something every employee of an organization, or for that matter every self-employed professional, needs to understand?

Chapter 55

Intellectual Property

When he joined MindTree, he looked infallible in every respect. He was an outstanding technologist who would do any organization proud with his great ability to build embedded applications on semiconductor chips. But after joining us, he built an application for a MindTree customer in which he reused portions of software that he had previously written while in the employ of an earlier organization. When the issue surfaced, he claimed innocence. He had no idea he had created a legal exposure for himself as well as for MindTree. The man did not know that the software he had designed while in the pay of his earlier employer was what we call "work for hire" and the rights did not belong to him. Worse, his bringing out that code when moving to MindTree was itself an illegal act. If by chance the code had gone unnoticed and the ownership issue surfaced after the chip went to market, the client for whom we had designed the chip would have been liable as well.

• • •

In another organization, a bunch of designers created a prototype of a piece of equipment and got the customer's approval for the project. When the actual design was in progress, it turned out that the prototype contained what is called "open source software." The problem with using certain kinds of open source software is that by virtue of its being available free, if you build an application based on it, that application may not enjoy any trade protection. For a customer, it is of no use to buy a design that starts off with no competitive advantage because the design has been created in part from open source code. When this fact surfaced, the customer asked the team to rewrite the entire code, and then all hell broke loose. It was only then that the team realized it was a Herculean task and that they did not quite understand the logic and algorithm behind the components taken from the open source software. Though they might have the capability to figure it out, the timeline for doing so was going to be very different from the original understanding with the customer. It became a reverse engineering nightmare that could have been avoided if engineers knew the basics of governance.

An advertising agency showed a prototype Web site design to a client. The client loved it. The Web site went live. After many months, a U.S.-based company that sells stock photographs sued the client because the Web site had used stock shots downloaded from its site. When the client asked the Web site designer, he washed his hands of the issue. He simply drew attention to the fine print in his agreement that required the client to get the necessary approvals for any visual/graphics that might be used. But it was the moral responsibility of the Web site designer to draw specific attention, based on that clause, to the source of the photographs he was using and to clearly inform the advertising agency that the rights to use those photographs had to be obtained from the source before using them. And it was negligence on the part of the advertising agency not to pay attention to this clause in the contract which exposed them and their client to the possibility of being sued.

What was the issue the employees above failed to grasp? That using someone else's creative work comes with the responsibility to seek prior permission; it comes with the requirement that it is duly acknowledged and, where necessary, paid for. And if the work was done while in the employ of another organization, you do not have the right to use it in your current organization: it does not belong to you, unless otherwise stated in writing.

You could be a doctor, engineer, graphic designer, architect, teacher or musician. The new world requires every professional to understand the basics of intellectual property rights. In the old world, we had little everyday interaction and exposure to such concepts. It was okay if we broadly understood the rights of usage and ownership of immovable and movable property like land, building, machinery or cattle. But today, with more value getting generated by the use of knowledge, it is every professional's responsibility to acquire a basic understanding of concepts like copyright, patents and trademarks. And it is every professional's responsibility not to expose an organization or a client to lawsuits that could result from negligence in this area.

Chapter 56

Sustainability

There was a time when I used to agitate a lot about the overflowing dustbins in the neighborhood I live in. I could not stand the apathy of people who simply threw the garbage around and not inside the dustbin. On weekends, I would routinely pick up the garbage strewn around the bins and put it all inside.

This taught me valuable lessons about my own limitations. But it also taught me that the biggest culprits for creating non-biodegradable material are consumer goods companies, those that use shampoo sachets, food wrappers, plastic tubes and containers, and those that make diapers and all other forms of material that will sit in a landfill for the next thousand years and release toxins into the soil and air. This includes the thin plastic bags all of us pick up at the grocery or convenience store. In the developing countries of the world, even before they go to the landfill, municipal workers simply take such garbage and set fire to it just outside the town limits, releasing poisonous gases into the atmosphere.

In almost all consumer products companies, most of them

multinationals, the people in charge are qualified management professionals, and while they may choose to ignore the environmental consequences of their products today, tomorrow is another story. Professionals will have to know how their work could potentially destroy the earth and, conversely, how they can help sustain it.

Today, we think that only corporations and the government have to worry about issues like global warming and carbon footprint. The truth is that every profession and as a result every professional has a much closer relationship to these issues than we realize.

When a doctor opens a clinic, how does he dispose of medical waste?

An architect suggests a design for a house or an office. What about educating the client on the implications of lighting and air conditioning and ways to save energy?

You open a new campus for your organization. Initially, all you worry about is following local laws that regulate zoning and such things. But now, thousands of people begin to commute to work and when they commute from afar, they add to the carbon footprint. When you pay attention to their transportation needs and look at location options from all angles, you contribute to the process of creating sustainability. In places where thousands of workers eat every day, food waste can create serious health hazards. Wanton use of plastic cups and other such nonbiodegradable items will add to the problems of environmental management. And what about reusing waste paper? How about reducing use of water in toilets, encouraging rainwater harvesting and the use of solar-powered lighting?

You are the buyer of raw material in a food products company. You have to pay attention to the practices of your supply chain. You must apply your mind and understand how much water the farmers consume, what pesticides they are using, how they may be subjecting the earth to chemical torture by excessive use of chemical fertilizer. If you source materials from developing countries, you need to understand the social implications of inducing farmers to move away from cultivating rice and wheat to raising other cash crops. Each of these examples expands the scope of a professional's responsibilities vis-à-vis the environment and the people who are affected by our strategies and actions.

The next time you buy coffee at Starbucks, pay attention to the paper coffee cup. It says Starbucks buys beans from growers who abide by standards laid down by the Rainforest Alliance. Starbucks does so because it clearly matters to Starbucks customers what sustainability measures the company follows. In March 2007 the *Financial Times* reported, "As big global companies address sustainability, many are focusing on agricultural water use. Companies such as Nestlé, Starbucks and Kraft have recognized the need to address water conservation in supply chains. To do so, Kraft and Chiquita have turned to the Rainforest Alliance, a non-profit conservation organization that monitors the sustainability of agricultural producers."

Sustainability is no longer a "peacenik" fad. Issues like potable water, waste disposal, greenhouse gases and carbon emissions can be directly linked to every profession, and more significantly to how every professional makes a living.

But even before people take responsibility for the larger cause, the first step is to get educated on the basics of sustainability and become aware of all subsequent individual actions. Professionals in the new world will need to develop a personal context for the phrase "doing well by doing good." Because tomorrow's customers will seek out those who are proactive in their practices to make the planet a better place to live in.

Part Seven

The Professional's Professional

Chapter 57

A Yen for Professionalism

One of the countries I deeply admire for its professionalism is Japan. I have visited Japan many times but learned about the Japanese people closely during my training in TQM in 1996. Japan has become what it is today due in some part to its great corporations and the people behind them.

I tasted the professionalism of the country the very first time I visited. The taxi driver who was to take me to my hotel did not know the shortest route (these were the days before navigation systems in cars). Though he brought me almost to the road on which the hotel was located, he took a wrong turn, and as a result of the one-way system, had to go around a couple of blocks before arriving at my destination. I did not mind because I was not in a hurry and it wasn't his fault at all. But when I paid him the metered fare, the driver returned some money. I was very surprised and after a labored conversation—the driver did not know English and I did not understand Japanese—I realized that he was refunding the money for the last leg of the trip when he had gone around a couple of blocks. It was his job to know the route and he could not charge me for what he felt was his mistake!

The second immediate lesson was waiting for me as soon as I entered the hotel. I was still recovering from the shock of the taxi driver's honesty. The porter from the hotel brought in my suitcase. I offered him a tip. The man went red in the face. He backed off a few steps and kept mumbling, "No, sir, no tip. This is Japan."

During subsequent years as an occasional visitor to Japan, I learned how the idea of professionalism permeates the nation and to what lengths Japanese go to live the idea. Professionalism in Japan blends with three other things: national pride, precedence of the group over the individual and, finally, the spiritual identity of the culture.

Years after my visits to Japan, I was exchanging notes on professionalism with Infosys chairman Narayana Murthy and he told me the story about the time he visited the Japanese department store Matsuzakaya, which has a ten-story outlet in the Ginza district of Tokyo. Murthy was being shown around the store by his Japanese host. As the two got out of an elevator, a young Japanese woman in a kimono welcomed them with a smile and ushered them in, as is customary at Matsuzakaya. Murthy blurted out in English that this must be a boring job. His host immediately translated what he had said. The girl listened attentively, smiled, bowed and replied in Japanese, which was dutifully interpreted by the host. She said, "I have been paid by my company to make my customers feel welcome. I try hard to do a good job. And I am very proud of it."

Years after the incident, Murthy still shrinks with embarrassment at what he calls his "silly, thoughtless remark."

• • •

A Japanese passerby stopped to watch a monk at work. He was picking up dry twigs from a garden patiently, dutifully, one by one, the whole day. The man finally asked, "O Holy One, when will your work stop?" The monk replied, without looking up, "When the last dry twig has been removed."

To the Japanese, work is a living experience. How can we all try to improve things around us? Where does the human endeavor to be a better professional stop? In the garden of work, as long as the garden is alive, there is no last dry twig.

I stated at the beginning of this book that I have drawn on the experience of many men and women whom I have met in the course of my work over the years and whom I consider to be benchmarks of professionalism in their respective fields. They come from various backgrounds: banker, lawyer, advertising head, salesperson, software developer, program manager, doctor, journalist, accountant, photographer. I asked these eminent men and women what qualities they admired in a professional and what they considered unprofessional conduct. I collated all their feedback to rank the top ten attributes of a professional:

1. Integrity
2. Commitment and ownership
3. Action orientation and goal seeking
4. Continuous learning
5. Professional knowledge / skills
6. Communication

7. Planning, organizing and punctuality
8. Quality of work
9. A positive attitude, approachability, responsiveness
10. Being an inspiring reference to others

It is heartening that integrity came out right on top, and that is how this book opens, with an expansive discussion on a subject that is often considered prudish but in reality is the keystone of the arch. Commitment and ownership come second. What is interesting to note is that action orientation precedes continuous learning. I believe this implies that sometimes it is learning on the job that saves the day and that requires agility and networking. We live in an age of just-in-time learning. So it is not about how much you already know but about how rapidly you can tap in to other experts and utilize their help and their knowledge required to solve a problem. And that requires commitment, ownership and action orientation. Planning, organizing, punctuality and quality of work have to be delivered with a great attitude; otherwise a client will in all probability not like to do business with you a second time.

But the last attribute really opened up my eyes: being an inspiring reference to others. Tomorrow's professional must have a beaconlike presence in a world that will ask for memorability. Because being ordinary will no longer be considered professional.

Chapter 58

The Unprofessional

The preceding pages have discussed what it means to be a professional, how one can become a better professional and what it will take to be a professional in the twenty-first century. Unfortunately, the discussion would not be complete without mentioning how equally important it is for any professional to be aware of unprofessional conduct, because sometimes it is very easy to slip into this behavior without realizing it.

Here are my top ten markers of unprofessional conduct:

1. Missing a Deadline

It happens to all of us all the time. You are waiting for a flight and it is well past the scheduled boarding time. No announcement has been made, and no one has told you the reason for the delay or the rescheduled departure time. Or you go to the laundry to collect your clothes and find that your clothes are not ready. You call your advertising agency and they say they are not ready with the presentation scheduled for this afternoon.

You are waiting for a software release that was to happen over the weekend. It is already Monday afternoon; there is no sign of it.

We have all been victims of missed deadlines, but when it comes to us, we all too quickly forget that we may be doing it to others as well.

2. Failing to Be Forthright

Behind the unprofessional conduct of missing deadlines is often poor estimation of the time required to get the job done. Sometimes missing a deadline is due to unavoidable circumstances. The thing to do in either case is to bring the bad news up ahead of time and tell the other person that you are going to miss the deadline—but do it well before the deadline, not after it.

There is always some problem or other that has come in the way of meeting the deadline. But we all know that most problems announce their arrival in advance. But someone, somewhere, pretends that a problem does not exist or, if ignored, it will solve itself. And then it all just blows up.

If communicated proactively to the right person, most problems not only get addressed, but they can be converted into an opportunity.

If it is really bad news, try to deliver it in person. At the very least call personally. No e-mails. No carefully left voice mails at the customer's office past midnight knowing that no one will pick up the phone at that hour. Call the number most likely to be picked up in two rings and tell it like it is. Better still, show up in person.

We do not communicate problems due to fear of reprisal, rejection and sometimes loss of business. The price for such denial can be very heavy in the organizational context.

Noncommunication remains one of the top reasons why things go wrong in delivering products and services to people who depend on us.

3. Withholding Information

A professional is privy to special information by virtue of his or her professional relationship. It is a relationship of trust. A doctor, a lawyer, a counselor may have access to a person's private life. Such people have to notify the recipient of their service of all conflicting interests they may have, and all such disclosure must happen in a proactive manner. Withholding such information is a malafide act.

4. Not Respecting Privacy of Information

Similarly, information received in the course of a professional relationship must be held in confidence, and if it is likely to be remotely confidential or proprietary, it can be parted with only after explicit consent of the other party. Consent should never be assumed. However close and long-standing the relationship may be, consent has to be formally sought and formally recorded each time there is a transaction.

Sometime ago my daughter sent me an e-mail, forwarding a rather excited communication from a friend of hers. This individual, who works for an international bank, had written to her that I had just opened an account with them. MindTree had decided to shift its payroll from one bank to another and that required everyone to open a salary account with the new bank. I was taken aback by the indiscretion of the young man. While the intent may have been innocent, it is unacceptable professional conduct. I do not want my lawyer, my doctor or my banker to go around talking about my business with them to other people.

As businesses become part of the global mainstream, one of the things all professionals will have to understand is the importance of privacy. In some parts of the world, people are not used to privacy. However, that is not how the world works. As a result, every professional, irrespective of the field of work, must understand the inviolability of a client's information privacy.

5. Not Respecting "Need to Know"

There are an increasing number of spouses who work. Sometimes, they may even work in the same organization. When they do so, it is important to avoid what is known as "pillow talk"—sharing official information at home. The first test one must use for this is: But for the marital relationship, would the other person have access to a particular piece of information? And second, does the spouse pass the "need to know" test?

6. Plagiarizing

You cannot pass off as your own what does not belong to you. While taking from someone's ideas and most certainly while quoting from someone's work, a professional must acknowledge the source. Too many people—students, teachers, researchers—

routinely pass off as original work that clearly does not belong to them.

In today's world, there is enough technology to track down piracy and plagiarism, whether in text, in graphics or in any other form. Every professional knows that resources like Google and Wikipedia are universally available. Given the pressures of time to create an impressive presentation, it is easy to succumb to the temptation to "research" the net, download charts and graphs and simply embed them in a slide deck or a report.

A professional must do three things when depending on such sources: state why you have chosen to pull out a specific set of information, acknowledge the source and, most important, state your own reasoned conclusion. A professional must state her point of view, not pass off someone else's as if it were her own.

7. Passing the Blame

When things go wrong, many people fault their subordinate, executive assistant or student researcher. Passing along the blame is an unprofessional thing to do. When a doctor, an engineer, a lawyer or an architect blames his or her people in front of a client, it reflects poorly on the professional's own caliber. The last thing you should be doing is separating yourself from your junior's or assistant's inadequacies. It goes against the grain of inclusion and ownership. Just as we say a poor workman blames his tools, a bad professional blames his associates.

8. Overstating Qualifications and Experience

Unfortunately, job seekers sometimes trivialize the difference between exposure and expertise when it comes to writing résumés. Those who know about something only at a superficial level often write about it in their résumés as if they have mastery over the subject.

No one individual can be great at everything. There is no need to embellish one's capabilities while applying for a job. I have often received résumés of young engineers who have done just an internship of short duration in an organization, but who claimed to have designed an entire enterprise application while being there. In many parts of the world, you are taken at face value. If you say that you know something or have done something, it is not contested. So while it may be easy to bluff one's way into a job or an assignment, it becomes a complete loss of face at the time of execution and leaves a bad taste in everyone's mouth.

9. Frequently Changing Jobs

Close to the malaise of overstated qualifications is that of mindless job changes. In every interview with frequent job-hoppers, the conversation is very predictable. It is either a story of "I was okay, my organization was not," or "I was not looking for a change, I got head-hunted." If someone has had five such changes in a ten-year career, it is unlikely that you are looking at an "organization man"; you are probably looking at a competent mercenary. As one progresses professionally, one must understand that failed job changes have as much to do with the individual as with the organization, and that people who pass off the blame simply demonstrate an opportunistic mind-set.

10. Not Taking Care of Your Appearance

Finally, professional conduct has to do with professional appearance and personal hygiene. It is not just for a doctor, a police officer or a pilot to be properly attired, nor is it only about wearing a uniform. Every profession has its own acceptable dress code that may sometimes be written and sometimes not. When it is not, and just because sometimes your organization does not emphasize it, that does not mean you can take liberties with it.

Your attire must inspire credibility, and whether you are a man or a woman, dressing in a way that distracts is not a great idea, be it flashy ornaments or a dress with a deep neckline.

Chapter 59

Return of the Dead

On March 31, 2009, Bangalore woke up to this headline on the front page of *The Times of India:*

> *4 cops sell unclaimed body, get booted out*

According to the paper, sometime in August 2006, the police station in Madivala, a Bangalore locality, got a call from a hotel owner. The man reported a dead body lying outside his hotel. The policemen picked up the dead body and declared that it was a case of death of a mentally retarded man who had fallen on thorns. They subsequently took the body to a medical college as unclaimed and sold the dead man. Medical colleges buy dead bodies for teaching purposes.

When the dead man's son surfaced and lodged a complaint about his missing father, one thing led to another, and finally the whole story came out.

After the investigations, the police commissioner dismissed an inspector, a subinspector, an assistant subinspector and a constable for their complicity in this shocking case.

These policemen, despite their professional training, their uniform, their power and their authority, can never be considered professionals.

In sharp contrast, Mahadeva, the man whose profession is the burial of unclaimed dead bodies, a man who arrived with no means, possessed no education and did not have organizational mentorship, will always remain a professional's professional.

Because, ultimately, being a professional is a matter of personal choice and the values we opt to live by.

Chapter 60

Platform and Purpose

Your profession is a platform. It is something of a springboard; a place from which you start a journey on the road to someplace else. Your purpose, on the other hand, determines how far you may go on that journey, for whom it is undertaken and how meaningful it is.

When thinking about the relationship between your platform and purpose, imagine a two-by-two matrix in which our existence is defined by the x coordinate that represents the platform and the y coordinate that represents the purpose. Between these two, there are four quadrants. We predominantly live in one of the four.

There are some whose existence can be called "low platform and low purpose." There are those who would fit into "high platform and low purpose." Then there are some who live a life best described as "low platform and high purpose." Only a very few go on to lead a life that can be described as "high platform and high purpose." All four options are available in equal measure as we start life, but as life progresses, as professionals we tend to settle into one or the other. Let us examine each one briefly.

Low Platform and Low Purpose

Imagine an ordinary but good-natured high school student who decides to attend a community college even as his classmates go on to the most competitive schools in the country. He earns an ordinary college degree along the way and upon graduating settles down in the family business. He has a small family, builds a reasonable dwelling and remains in relative anonymity.

In the years ahead, he takes care of his immediate needs; he does not venture beyond the familiar and has no ambition to change the world.

Similarly but somewhat differently, some of us, despite stellar professional qualifications, also settle into a comparable life, content with the on-the-job training, the allocated work, the predictable good salary, the postings and assignments that come along the way. These folks stay close to the median and after a few decades of work, of raising a family, simply fade into oblivion. Theirs is a low-platform and low-purpose life. People who stay here are the vast majority that make the world a predictable, often dependable place in which to live, but they leave no lasting impact.

This life is okay as long as you are comfortable with its consequences and make no false comparison with others. Chances are, you will live a life of contentment, and while you do not leave behind a mark, you also do not make things difficult for people around you.

The redeeming aspect of being low platform and low purpose

is that people in this quadrant do not consume resources disproportionate to the impact they make; they take from life what is minimally needed; they are not overly materialistic and do not amass wealth and fortune at other people's cost. They have a lower carbon footprint in every sense. These are people who are driven by the sheer need to exist. Existence is the driver of their being.

High Platform and Low Purpose

The people who live in this quadrant are usually highly qualified and highly competent, but they lead a self-serving life. They are predominantly concerned with what others have that they do not have. They use their education, experience and contacts to see how best to get ahead in life. Both material acquisition and organizational power are very critical to them, as is social recognition. To them, platform is everything. They are very unlikely to risk it because it largely defines their existence. These are people who are more likely to pursue the predictable.

Consider some of the people you know in your own professional circle who fit into this quadrant. As I said, they are bright and intelligent and they probably planned their entire lives early on in their existence. That life script consisted of first landing a good job, then getting a master's degree or an MBA at a brand-name school after a couple of years, then a global assignment. They married equally covetable, equally qualified spouses and focused on building a high-net-worth life from the very start. These people are often in positions of leadership and they are seldom aware that they are leading a life of high platform and low purpose. They genuinely believe that they are hugely beneficial to life and living. They do not think there could be anything more significant than professional ambition. Aggressive pursuit of their personal goals determines their choice of work and their employer, whose brand value brings them critical social recognition. Such people are well suited to handle a large company assignment, but they are not cut out to start something on their own. They depend on a given organizational path for personal success, but they do not create their own path.

People in this quadrant believe that their biggest contribution to society is their own life. For the most part, they think they are indispensable. But, then, they are also the easiest to replace because for everyone who leaves, there is a long waiting list to get into their shoes. The driver of their existence is consumption.

Low Platform and High Purpose

In the lower extreme quadrant is a handful that decides somewhat early on in life that the rat race of the high-platform, low-purpose life is not what they are cut out for.

These are people who become start-up entrepreneurs. Some are also idealistic individuals who want to make a difference to society in ways other than building a business. They do not believe that institutions like the government and the established large organizations, particularly those in the private sector, have the capacity or the sensitivity to make a difference. They decide, not out of compulsion but out of choice, that they must make a direct personal impact, however small, to a cause dear to them.

There are some real and inherent challenges in being in this quadrant beyond the fact that it entails personal hardship and a minimalist lifestyle. Though the cause may be noble or an entrepreneurial idea may be enticing, working outside an established organizational framework often makes it difficult to succeed.

Professionals in this quadrant need to get their spirit up every morning against all external odds and sustain their souls with none of the rewards or recognition mechanisms that are so readily available to those in the high-platform, low-purpose quadrant. It is often a lonely trek, without any validation that their work is making a tangible difference. That said, those who opt to be in this quadrant do what they do because of who they are. They do not seek instant gratification; they are patient with the change-making process.

But being low platform, their work usually affects some people but does not by itself become scalable. Many things in the world require scale.

The driving force behind people who are low platform and high purpose is path making. Sometimes it is altruistic; sometimes it is driven by a desire to simply own one's destiny.

As we all know, many who begin and sometimes even flounder in this quadrant actually make it to the next one: the one I call high platform and high purpose.

High Platform and High Purpose

Many professionally qualified people start off on a high platform by virtue of their privileged upbringing. For those who did not come from an elevated socioeconomic background, the academic credential from an institution of repute often provides them a high platform. Thus thousands of professionally qualified people actually start off in life from a high platform compared with millions of others in this world. Their academic achievement, at least temporarily, propels them upward in life. That said, whether or not they are able to retain and build on the high platform actually gets determined by many factors and certainly is not a matter of lifelong entitlement.

Many people make wrong career choices; life is a snakes and ladders game. Sometimes people get lazy along the way. And then there are also the unpredictable circumstances that may rob some of their platform or slide them down to a lower one. But assuming that life deals most of us a reasonable hand, the larger question is whether or not we build high purpose on the high platform that has been provided to us.

When one embraces a high purpose staged from a high platform, one makes the greatest difference to the world.

This is the quadrant of professionals like Steve Jobs, Bill Gates, Warren Buffett, Indra Nooyi, Meg Whitman and Arthur Ashe. Sometimes this class of professionals actually starts in the immediate lower quadrant of low platform and high purpose.

Bill Gates started life designing a BASIC language compiler for an obscure machine called the Altair. Steve Jobs was a geeky youngster who wanted to build a personal computer with great calligraphy. Dr. Govindappa Venkataswamy of Aravind Hospital was just another ophthalmologist, and Narayana Murthy started Infosys with a loan of ten thousand rupees—at the time probably equivalent to US $500—from his wife Sudha.

Somewhere along their respective journeys, these professionals received their call, as we all do at some stage or other. The difference is, when the call comes, most of us may not hear it. Among other things, obsessive love for a platform invariably reduces the power of hearing.

So when the call comes, we must make sure we pick up. The trick is to recognize it and to embrace the cause that it conveys. Destiny does not leave a voice message.

Once a professional embraces the cause, she must cease her love for the platform even as she recognizes its power, because from now on she has to externalize its benefits and look at herself as a lifelong custodian, and not the owner, of the platform.

The people with the ability to make the greatest impact see their professional qualifications, experience and accomplishments as a platform to make a difference to the world; it is not something that they own for their personal glory and greatness.

These are people who have the power of vision and they do three things right: They create a vision community to carry people along and make them feel it is a shared vision. Then they give their entire life to the cause it entails and understand that in the top right-hand quadrant, there is the least scope for instant gratification. Finally, they remain steadfast even in the wake of great personal sacrifices and they do not yield to distractions. They do not let adulation get in the way of their work; nor do they fight battles that take them away from their main purpose. For people in the high-platform, high-purpose quadrant, the size of their vision is larger than life; this is where comfort with scale really matters.

Lofty as it may sound, this quadrant often tests the aspirant with highly dangerous situations that leave no margin of error. One mistake and you drop into the left or the bottom quadrants. People around you are so busy showering their admiration and sometimes plain sycophancy on you that they will not be the ones to hold the mirror up to your face. History is full of examples of high achievers whose fall from their pedestal was rapid and irreversible.

The driving force of people in this quadrant is the desire to leave a legacy.

As we grow up as professionals, all the four options are available to most of us. There is nothing right or wrong about being in any of the four quadrants. The only opportunity for error is to not know which one you are in. Education and experience give us the power to make informed choices, and we need to choose where we want to be and at the same time understand the consequences.

Life is about making choices and not simply succumbing to inevitability.

Life constantly tests our resolve and sometimes rotates us from one quadrant to another to see what conviction we carry in fulfillment of our purpose. As we journey into life, we must continue to strengthen our platform. But we must also remind ourselves that it is only meant to help us catch the train; it is not a place to live.

A true professional knows that through his profession he has the power to make a huge difference to the world around him; he has an uplifting purpose that takes him beyond earning a living to making a difference in life.

Acknowledgments

The Professional is the third in a series of books Portfolio / Penguin and I have done together. During the course of the first two, the idea for this book was conceived with my erstwhile editor, Sumitra Srinivasan. Our free-ranging discussions seemed to bring us back to the questions: Am I a true professional? What qualities would make me a better professional? What is professional ethics? What does it mean for me? Faced with a difficult decision and multiple options, how do I know I am making the right choice? We realized that most people ask themselves these same questions, undergoing moments of uncertainty about how they can be the best. Why not write a book, she kept suggesting, for the scores of people who become doctors, engineers, journalists, architects, teachers—in fact anyone who works in a professional capacity—which they can use to help them guide their decisions and actions? I am grateful to Sumitra for the conversation, her subsequent encouragement, critiques and patience.

When the book appeared in print in India, it became an instant bestseller and was subsequently chosen by Portfolio / Penguin as a candidate for international release. This time, I got the opportunity to work with my present editor, Brooke Carey. While Sumitra is an outstanding ideator, in Brooke I met the quintessential editor. She helped make the book relevant for a global audience. In the process, she also suggested valuable additions and some even more valuable deletions. I have been extremely fortunate to have the guidance of Sumitra and Brooke, in the sequence in which they appeared. I am very grateful to both. Thanks also to Michael Burke, who did a superb copyediting job.

Writing, at one level, is a solitary experience. Had it not been for my life partner of three decades, Susmita, I would not have been able to take on the task of being a writer. She feeds my body and my soul and puts up with the emotional highs and lows that a writer must suffer before he is able to produce anything of meaning. I remain eternally grateful to her for many reasons; one among them is helping me bring this book to you.

As I began writing, I was fortunate to meet Prashanth, an unusually gifted young man, from the Spastics Society of Karnataka. He agreed to be my research assistant for this project—verifying sources, seeking information that may be useful, and questioning the prose and sometimes the purpose. Despite being severely affected by his congenital affliction of cerebral palsy, he provided a tremendous amount of help and symbolizes the spirit of a professional. I thank him for his dedication and enthusiasm, without which this work of love would not have been complete.

At MindTree, my colleague Shanti Uday has been running my life with the patience of a quiet, ever-flowing river for a decade now. Her support has been enormous in helping me with versions, backups and the final editing. I remain always grateful to her. Shanti, thank you for everything.

Working for MindTree has many rewards—chief among them

is the space it provides me to be myself. And the resources to write three books within a span of five years!

As I look back on my working career to find examples of professionalism, I realize how greatly I have been influenced by countless people—bosses, peers and subordinates—from whom I have gleaned lessons in professional behavior. To each one of those individuals I remain indebted. Many lessons from some very special people, whom I call "professional's professionals," are included in this book. These are people to whom I reached out to tell me what they think are the attributes of a professional through personal anecdotes, illustrating both professional and unprofessional conduct. A big thank you to Ashok Soota, Anjan Lahiri, Abhi Dhar, Amit Varma, Asha Mathen, Ashutosh Shukla, Bobby Mitra, Dr. Thimmappa Hegde, Kush Desai, K. Ramesh, Lubna Kably, Lu Ellen Schafer, Mahesh Bhat, Manoj Chandran, Maya Chandra, Mukul Pandya, Milind Sathe, Neelesh Prabhu, Olivier Poulard, Pradipt Kapoor, Raj Rawal, Rajesh Nair, Sudhir K. Reddy, Scott Staples, Suresh Gurumurthy, Tridip Saha, Venkatesh Komarla, Rostow Ravanan and Warren Leudecker. I would fail in my duty if I did not make a special mention of two great people: Mukul Pandya, who helped me with the story of Curtis Perrin, and Professor Vijay Govindarajan for his contributions to chapter 44.

And, finally, dear reader, thank you for buying this book.